COMPLEX VARIABLES AND APPLICATIONS

Churchill-Brown Series

Complex Variables and Applications
Fourier Series and Boundary Value Problems
Operational Mathematics

International Series in Pure and Applied Mathematics

Ahlfors: *Complex Analysis*
Bender and Orszag: *Advanced Mathematical Methods for Scientists and Engineers*
Boas: *Invitation to Complex Analysis*
Brown and Churchill: *Complex Variables and Applications*
Brown and Churchill: *Fourier Series and Boundary Value Problems*
Buchanan and Turner: *Numerical Methods and Analysis*
Buck: *Advanced Calculus*
Chartrand and Oellermann: *Applied and Algorithmic Graph Theory*
Colton: *Partial Differential Equations*
Conte and de Boor: *Elementary Numerical Analysis: An Algorithmic Approach*
Edelstein-Keshet: *Mathematical Models in Biology*
Farlow: *An Introduction to Differential Equations and Their Applications*
Goldberg: *Matrix Theory with Applications*
Gulick: *Encounters with Chaos*
Hill: *Experiments in Computational Matrix Algebra*
Kurtz: *Foundations of Abstract Mathematics*
Lewin and Lewin: *An Introduction to Mathematical Analysis*
Morash: *Bridge to Abstract Mathematics: Mathematical Proof and Structures*
Parzynski and Zipse: *Introduction to Mathematical Analysis*
Pinsky: *Partial Differential Equations and Boundary-Value Problems with Applications*
Pinter: *A Book of Abstract Algebra*
Ralston and Rabinowitz: *A First Course in Numerical Analysis*
Ritger and Rose: *Differential Equations with Applications*
Robertson: *Engineering Mathematics with Mathematica*
Rudin: *Functional Analysis*
Rudin: *Principles of Mathematical Analysis*
Rudin: *Real and Complex Analysis*
Simmons: *Differential Equations with Applications and Historical Notes*
Small and Hosack: *Calculus: An Integrated Approach*
Small and Hosack: *Explorations in Calculus with a Computer Algebra System*
Vanden Eynden: *Elementary Number Theory*
Walker: *Introduction to Abstract Algebra*

Also available from McGraw-Hill

Schaum's Outline Series in Mathematics & Statistics

Most outlines include basic theory, definitions and hundreds of example problems solved in step-by-step detail, and supplementary problems with answers.

Related titles on the current list include:

Advanced Calculus
Advanced Mathematics for Engineers and Scientists
Analytic Geometry
Basic Mathematics for Electricity and Electronics
Basic Mathematics with Applications to Science & Technology
Beginning Calculus
Boolean Algebra & Switching Circuits
Calculus
Calculus for Business, Economics, & the Social Sciences
College Algebra
College Mathematics
Combinatorics
Complex Variables
Descriptive Geometry
Differential Equations
Differential Geometry
Discrete Mathematics
Elementary Algebra
Finite Differences & Difference Equations
Finite Element Analysis
Finite Mathematics

Fourier Analysis
General Topology
Geometry
Group Theory
Laplace Transforms
Linear Algebra
Mathematical Handbook of Formulas & Tables
Mathematical Methods for Business & Economics
Mathematics for Nurses
Matrix Operations
Modern Abstract Algebra
Numerical Analysis
Partial Differential Equations
Probability
Probability & Statistics
Real Variables
Review of Elementary Mathematics
Set Theory & Related Topics
Statistics
Technical Mathematics
Tensor Calculus
Trigonometry
Vector Analysis

Schaum's Solved Problems Series

Each title in this series is a complete and expert source of solved problems with solutions worked out in step-by-step detail.

Titles on the current list include:

3000 Solved Problems in Calculus
2500 Solved Problems in College Algebra and Trigonometry
2500 Solved Problems in Differential Equations
2000 Solved Problems in Discrete Mathematics
3000 Solved Problems in Linear Algebra
2000 Solved Problems in Numerical Analysis
3000 Solved Problems in Precalculus

Available at most college bookstores, or for a complete list of titles and prices, write to:
Schaum Division
McGraw-Hill, Inc.
1221 Avenue of the Americas
New York, NY 10020

COMPLEX VARIABLES AND APPLICATIONS

Sixth Edition

James Ward Brown

Professor of Mathematics
The University of Michigan—Dearborn

Ruel V. Churchill

Late Professor of Mathematics
The University of Michigan

McGraw-Hill, Inc.

New York   St. Louis   San Francisco   Auckland   Bogotá
Caracas   Lisbon   London   Madrid   Mexico City   Milan
Montreal   New Delhi   San Juan   Singapore
Sydney   Tokyo   Toronto

This book was set in Times Roman by Publication Services, Inc.
The editors were Jack Shira, Maggie Lanzillo, and James W. Bradley;
the production supervisor was Louise Karam.
The cover was designed by BC Graphics.
R. R. Donnelley & Sons Company was printer and binder.

COMPLEX VARIABLES AND APPLICATIONS

This book is printed on acid-free paper.

9 10 DOC/DOC 0 9 8 7 6 5 4 3 2

P/N 008496-3

Library of Congress Cataloging-in-Publication Data
Brown, James Ward.
   Complex variables and applications / James Ward Brown, Ruel
V. Churchill. —6th ed.
      p.   cm. —(Churchill-Brown series) (International series in
pure and applied mathematics)
   Churchill's name appears first on the previous ed.
   Includes bibliographical refrences (p. –) and index.
   ISBN 0-07-912146-2 (MAC Set)—ISBN 0-07-912147-0 (IBM Set)
   1. Functions of complex variables.   I. Churchill, Ruel Vance,
 (date).   II. Title.   III. Series.   IV. Series: International
series in pure and applied mathematics.
QA331.7.C524   1996
515'.93—dc20   95-9898

INTERNATIONAL EDITION

ABOUT THE AUTHORS

James Ward Brown is Professor of Mathematics at The University of Michigan-Dearborn. He earned his A.B. in Physics at Harvard University and his A.M. and Ph.D. in Mathematics at The University of Michigan in Ann Arbor, where he was an Institute of Science and Technology Predoctoral Fellow. He was coauthor with Dr. Churchill of the fifth edition of *Fourier Series and Boundary Value Problems*. A past director of a research grant from the National Science Foundation, he is the recipient of a Distinguished Teaching Award from his institution, as well as a Distinguished Faculty Award from the Michigan Association of Governing Boards of Colleges and Universities. He is listed in *Who's Who in America.*

Ruel V. Churchill is Late Professor of Mathematics at The University of Michigan, where he began teaching in 1922. He received his B.S. in Physics from the University of Chicago and his M.S. in Physics and Ph.D. in Mathematics from The University of Michigan. He was coauthor with Dr. Brown of the recent fifth edition of *Fourier Series and Boundary Value Problems,* a classic text that he first wrote over fifty years ago. He was also the author of *Operational Mathematics,* now in its third edition. Throughout his long and productive career, Dr. Churchill held various offices in the Mathematical Association of America and in other mathematical societies and councils.

To the memory of my father,
GEORGE H. BROWN,
and of my long-time friend and coauthor,
RUEL V. CHURCHILL.
These distinguished men of science for years influenced
the careers of many people, including myself.

J.W.B.

CONTENTS

PREFACE

This book is a revision of the fifth edition, published in 1990. That edition has served, just as the first four editions did, as a textbook for a one-term introductory course in the theory and applications of functions of a complex variable. This edition preserves the basic content and style of the earlier editions, the first two of which were written by the late Ruel V. Churchill alone.

In this edition, the main changes appear in the first eight chapters. To mention some of the major improvements, the chapter on residues and poles in the last edition is now divided into two chapters, one on the theory of residues and one on applications of residues. The applications chapter contains a substantial amount of new material on the use of residues in finding inverse Laplace transforms, and the material on indented contours has now been brought out of the exercises and given more emphasis. This chapter also contains a completely rewritten section on the argument principle, which was deferred until the final chapter in the earlier editions of the book. In fact, all of the material in the final chapter of the earlier editions now appears in various places throughout the present edition. The proofs of Taylor's and Laurent's theorems have been improved, and the development of properties of power series has been completely revised.

As for certain other improvements, the section on multiplication and division of power series has been enhanced pedagogically, and the discussion of values and Cauchy principal values of improper integrals has been made clearer. Finally, exercises appear more frequently, and there is a substantial number of new figures.

As was the case with the earlier editions, the *first objective* of this edition is to develop those parts of the theory that are prominent in applications of the subject. The *second objective* is to furnish an introduction to applications of residues and conformal mapping. Special emphasis is given to the use of conformal mapping in solving boundary value problems that arise in studies of heat conduction, electrostatic potential, and fluid flow. Hence the book may be considered as a companion volume to the authors' "Fourier Series and Boundary Value Problems" and Ruel V. Churchill's "Operational Mathematics," in which other classical methods for solving boundary value problems in partial differential equations are developed. The latter book also contains applications of residues in connection with Laplace transforms.

The material in the first ten chapters of this book, with various substitutions from the remaining chapters, has for many years formed the content of a three-hour course given each term at The University of Michigan. The classes have consisted mainly of seniors and graduate students majoring in mathematics, engineering, or one of the physical sciences. Before taking the course, the students have completed at least a three-term calculus sequence, a first course in ordinary differential equations, and sometimes a term of advanced calculus. In order to accommodate as wide a range of readers as possible, there are footnotes referring to texts that give proofs and discussions of the more delicate results from calculus that are occasionally needed. Some of the material in the book need not be covered in lectures and can be left for students to read on their own. If mapping by elementary functions and applications of conformal mapping are desired earlier in the course, one can skip to Chapters 8, 9, and 10 immediately after Chapter 3 on elementary functions.

Most of the basic results are stated as theorems or corollaries, followed by examples and exercises illustrating those results. A bibliography of other books, many of which are more advanced, is provided in Appendix 1. A table of conformal transformations useful in applications appears in Appendix 2.

Each copy of this new edition will be packaged with a computer diskette containing an abbreviated version of $f(z)$—The Complex Variable Program, produced and developed by Lascaux Graphics. This software will allow students to generate graphs of complex variables in a four-dimensional space without requiring user programming. These graphs can be easily rotated in real time, zoomed, and scaled to permit close and varied examination. Exercises in the text that can be enhanced by the use of this program are denoted with an asterisk (*). The software is available for both PC and Macintosh platforms.

Preparation of this revision has been influenced by suggestions from a number of people. Specifically, there has been considerable input from the following reviewers: Harry Hochstadt, Polytechnic University; Meyer Jerison, Purdue University; Fred Rispoli, Dowling College; and Calvin Wilcox, University of Utah.

Constant interest and support have also been provided by Jacqueline R. Brown, Margret H. Höft, Michael A. Lachance, Ronald P. Morash, Frank J. Papp, Richard L. Patterson, and Gene G. Rae, as well as Jack Shira, Maggie Lanzillo, and James W. Bradley of the editorial staff at McGraw-Hill.

James Ward Brown

COMPLEX VARIABLES AND APPLICATIONS

CHAPTER

1

COMPLEX NUMBERS

In this chapter, we survey the algebraic and geometric structure of the complex number system. We assume various corresponding properties of real numbers to be known.

1. SUMS AND PRODUCTS

Complex numbers can be defined as ordered pairs (x, y) of real numbers that are to be interpreted as points in the *complex plane*, with rectangular coordinates x and y, just as real numbers x are thought of as points on the real line. When real numbers x are displayed as points $(x, 0)$ on the *real axis*, it is clear that the set of complex numbers includes the real numbers as a subset. Complex numbers of the form $(0, y)$ correspond to points on the y axis and are called *pure imaginary numbers*. The y axis is, then, referred to as the *imaginary axis*.

It is customary to denote a complex number (x, y) by z, so that

$$(1) \qquad z = (x, y).$$

The real numbers x and y are, moreover, known as the *real and imaginary parts* of z, respectively; and we write

$$(2) \qquad \operatorname{Re} z = x, \qquad \operatorname{Im} z = y.$$

Two complex numbers $z_1 = (x_1, y_1)$ and $z_2 = (x_2, y_2)$ are equal whenever they have the same real parts and the same imaginary parts. Thus $z_1 = z_2$ if and only if z_1 and z_2 correspond to the same point in the complex, or z, plane.

The *sum* $z_1 + z_2$ and the *product* $z_1 z_2$ of two complex numbers $z_1 = (x_1, y_1)$ and $z_2 = (x_2, y_2)$ are defined as follows:

$$(3) \qquad (x_1, y_1) + (x_2, y_2) = (x_1 + x_2, y_1 + y_2),$$

$$(4) \qquad (x_1, y_1)(x_2, y_2) = (x_1 x_2 - y_1 y_2, y_1 x_2 + x_1 y_2).$$

Note that the operations defined by equations (3) and (4) become the usual operations of addition and multiplication when restricted to the real numbers:

$$(x_1, 0) + (x_2, 0) = (x_1 + x_2, 0),$$

$$(x_1, 0)(x_2, 0) = (x_1 x_2, 0).$$

The complex number system is, therefore, a natural extension of the real number system.

Any complex number $z = (x, y)$ can be written $z = (x, 0) + (0, y)$, and it is easy to see that $(0, 1)(y, 0) = (0, y)$. Hence

$$z = (x, 0) + (0, 1)(y, 0);$$

and, if we think of a real number as either x or $(x, 0)$ and *let i denote the pure imaginary number* $(0, 1)$, it is clear that*

$$(5) \qquad z = x + iy.$$

Also, with the convention $z^2 = zz$, $z^3 = zz^2$, etc., we find that

$$i^2 = (0, 1)(0, 1) = (-1, 0),$$

or

$$(6) \qquad i^2 = -1.$$

In view of expression (5), definitions (3) and (4) become

$$(7) \qquad (x_1 + iy_1) + (x_2 + iy_2) = (x_1 + x_2) + i(y_1 + y_2),$$

$$(8) \qquad (x_1 + iy_1)(x_2 + iy_2) = (x_1 x_2 - y_1 y_2) + i(y_1 x_2 + x_1 y_2).$$

Observe that the right-hand sides of these equations can be obtained by formally manipulating the terms on the left as if they involved only real numbers and by replacing i^2 by -1 when it occurs.

2. ALGEBRAIC PROPERTIES

Various properties of addition and multiplication of complex numbers are the same as for real numbers. We list here the more basic of these algebraic properties and verify some of them. Most of the others are verified in the exercises.

---

*In electrical engineering, the letter j is used instead of i.

The commutative laws

(1) $$z_1 + z_2 = z_2 + z_1, \qquad z_1 z_2 = z_2 z_1$$

and the associative laws

(2) $$(z_1 + z_2) + z_3 = z_1 + (z_2 + z_3), \qquad (z_1 z_2)z_3 = z_1(z_2 z_3)$$

follow easily from the definitions in Sec. 1 of addition and multiplication of complex numbers and the fact that real numbers obey these laws. For example, if $z_1 = (x_1, y_1)$ and $z_2 = (x_2, y_2)$, then

$$z_1 + z_2 = (x_1 + x_2, y_1 + y_2) = (x_2 + x_1, y_2 + y_1) = z_2 + z_1.$$

Verification of the rest of the above laws, as well as the distributive law

(3) $$z(z_1 + z_2) = zz_1 + zz_2,$$

is similar.

According to the commutative law for multiplication, $iy = yi$. Hence one can write $z = x + yi$ instead of $z = x + iy$. Also, because of the associative laws, a sum $z_1 + z_2 + z_3$ or a product $z_1 z_2 z_3$ is well defined without parentheses, as is the case with real numbers.

The additive identity $0 = (0,0)$ and the multiplicative identity $1 = (1,0)$ for real numbers carry over to the entire complex number system. That is,

(4) $$z + 0 = z \qquad \text{and} \qquad z \cdot 1 = z$$

for every complex number z. Furthermore, 0 and 1 are the only complex numbers with such properties (see Exercise 11).

There is associated with each complex number $z = (x, y)$ an additive inverse

(5) $$-z = (-x, -y),$$

satisfying the equation $z + (-z) = 0$. Moreover, there is only one additive inverse for any given z, since the equation $(x, y) + (u, v) = (0,0)$ implies that $u = -x$ and $v = -y$. Expression (5) can also be written $-z = -x - iy$ without ambiguity since (Exercise 10) $-(iy) = (-i)y = i(-y)$. Additive inverses are used to define subtraction:

(6) $$z_1 - z_2 = z_1 + (-z_2).$$

So if $z_1 = (x_1, y_1)$ and $z_2 = (x_2, y_2)$, then

(7) $$z_1 - z_2 = (x_1 - x_2, y_1 - y_2) = (x_1 - x_2) + i(y_1 - y_2).$$

For any *nonzero* complex number $z = (x, y)$, there is a number z^{-1} such that $zz^{-1} = 1$. This multiplicative inverse is less obvious than the additive one. To find it, we seek real numbers u and v, expressed in terms of x and y, such that

$$(x, y)(u, v) = (1,0).$$

According to equation (4), Sec. 1, which defines the product of two complex numbers, u and v must satisfy the pair

$$xu - yv = 1, \qquad yu + xv = 0$$

of linear simultaneous equations; and simple computation yields the unique solution

$$u = \frac{x}{x^2 + y^2}, \qquad v = \frac{-y}{x^2 + y^2}.$$

So *the* multiplicative inverse of $z = (x, y)$ is

(8)
$$z^{-1} = \left(\frac{x}{x^2 + y^2}, \frac{-y}{x^2 + y^2} \right) \qquad (z \neq 0).$$

The inverse z^{-1} is not defined when $z = 0$. In fact, $z = 0$ means that $x^2 + y^2 = 0$; and this is not permitted in expression (8).

The existence of multiplicative inverses enables us to show that *if a product $z_1 z_2$ is zero, then so is at least one of the factors z_1 and z_2.* For suppose that $z_1 z_2 = 0$ and $z_1 \neq 0$. The inverse z_1^{-1} exists; and, according to the definition of multiplication, any complex number times zero is zero. Hence

(9)
$$z_2 = 1 \cdot z_2 = (z_1^{-1} z_1) z_2 = z_1^{-1} (z_1 z_2) = z_1^{-1} \cdot 0 = 0.$$

That is, if $z_1 z_2 = 0$, either $z_1 = 0$ or $z_2 = 0$; or possibly both z_1 and z_2 equal zero. Another way to state this result is that *if two complex numbers z_1 and z_2 are nonzero, then so is their product $z_1 z_2$.*

Division by a nonzero complex number is defined as follows:

(10)
$$\frac{z_1}{z_2} = z_1 z_2^{-1} \qquad (z_2 \neq 0).$$

If $z_1 = (x_1, y_1)$ and $z_2 = (x_2, y_2)$, equations (10) and (8) tell us that

(11)
$$\frac{z_1}{z_2} = \left(\frac{x_1 x_2 + y_1 y_2}{x_2^2 + y_2^2}, \frac{y_1 x_2 - x_1 y_2}{x_2^2 + y_2^2} \right)$$

$$= \left(\frac{x_1 x_2 + y_1 y_2}{x_2^2 + y_2^2} \right) + i \left(\frac{y_1 x_2 - x_1 y_2}{x_2^2 + y_2^2} \right) \qquad (z_2 \neq 0).$$

Although expression (11) is not easy to remember, it can be obtained by writing [see Exercise 14(*b*)]

(12)
$$\frac{z_1}{z_2} = \frac{(x_1 + i y_1)(x_2 - i y_2)}{(x_2 + i y_2)(x_2 - i y_2)},$$

multiplying out the products in the numerator and denominator on the right, and then using the property

(13)
$$\frac{z_1 + z_2}{z_3} = (z_1 + z_2) z_3^{-1} = z_1 z_3^{-1} + z_2 z_3^{-1} = \frac{z_1}{z_3} + \frac{z_2}{z_3} \qquad (z_3 \neq 0).$$

The motivation for starting with equation (12) appears in the next section.

Finally, we mention some expected identities involving quotients that follow from the relation

(14)
$$\frac{1}{z_2} = z_2^{-1} \qquad (z_2 \neq 0),$$

which is equation (10) when $z_1 = 1$. It enables us, for example, to write that equation in the form

(15)
$$\frac{z_1}{z_2} = z_1\left(\frac{1}{z_2}\right) \qquad (z_2 \neq 0).$$

Also, by observing that (see Exercise 6)

$$(z_1 z_2)(z_1^{-1} z_2^{-1}) = (z_1 z_1^{-1})(z_2 z_2^{-1}) = 1 \qquad (z_1 \neq 0, z_2 \neq 0),$$

and hence that $(z_1 z_2)^{-1} = z_1^{-1} z_2^{-1}$, one can use equation (14) to show that

(16)
$$\frac{1}{z_1 z_2} = (z_1 z_2)^{-1} = z_1^{-1} z_2^{-1} = \left(\frac{1}{z_1}\right)\left(\frac{1}{z_2}\right) \qquad (z_1 \neq 0, z_2 \neq 0).$$

Another useful identity, to be derived in the exercises, is

(17)
$$\frac{z_1 z_2}{z_3 z_4} = \left(\frac{z_1}{z_3}\right)\left(\frac{z_2}{z_4}\right) \qquad (z_3 \neq 0, z_4 \neq 0).$$

EXAMPLE. Computations such as the following are now justified:

$$\left(\frac{1}{2 - 3i}\right)\left(\frac{1}{1 + i}\right) = \frac{1}{(2 - 3i)(1 + i)} = \frac{1}{5 - i} \cdot \frac{5 + i}{5 + i} = \frac{5 + i}{(5 - i)(5 + i)}$$

$$= \frac{5 + i}{26} = \frac{5}{26} + \frac{i}{26} = \frac{5}{26} + \frac{1}{26}i.$$

EXERCISES

1. Verify that

(a) $(\sqrt{2} - i) - i(1 - \sqrt{2}i) = -2i$; (b) $(2, -3)(-2, 1) = (-1, 8)$;

(c) $(3, 1)(3, -1)\left(\frac{1}{5}, \frac{1}{10}\right) = (2, 1)$; (d) $\dfrac{1 + 2i}{3 - 4i} + \dfrac{2 - i}{5i} = -\dfrac{2}{5}$;

(e) $\dfrac{5}{(1 - i)(2 - i)(3 - i)} = \dfrac{i}{2}$; (f) $(1 - i)^4 = -4$.

2. Show that $(1 + z)^2 = 1 + 2z + z^2$.

3. Verify that each of the two numbers $z = 1 \pm i$ satisfies the equation $z^2 - 2z + 2 = 0$.

4. Show that

(a) $\operatorname{Im}(iz) = \operatorname{Re} z$; (b) $\operatorname{Re}(iz) = -\operatorname{Im} z$; (c) $\dfrac{1}{1/z} = z$ $(z \neq 0)$;

(d) $(-1)z = -z$.

5. Prove that multiplication is commutative, as stated in the second of equations (1), Sec. 2.

6. Use the associative and commutative laws for multiplication to show that

$$(z_1 z_2)(z_3 z_4) = (z_1 z_3)(z_2 z_4).$$

7. Prove that if $z_1 z_2 z_3 = 0$, then at least one of the three factors is zero.

   Suggestion: Write $(z_1 z_2)z_3 = 0$ and use a similar result (Sec. 2) involving two factors.

8. Verify
   (*a*) the associative law for addition, stated in the first of equations (2), Sec. 2;
   (*b*) the distributive law (3), Sec. 2.

9. Use the associative law for addition and the distributive law to show that

$$z(z_1 + z_2 + z_3) = zz_1 + zz_2 + zz_3.$$

10. By writing $i = (0, 1)$ and $y = (y, 0)$, show that $-(iy) = (-i)y = i(-y)$.

11. (*a*) Write $(x, y) + (u, v) = (x, y)$ and point out how it follows that the complex number $0 = (0, 0)$ is unique as an additive identity.
   (*b*) Likewise, write $(x, y)(u, v) = (x, y)$ and show that the number $1 = (1, 0)$ is a unique multiplicative identity.

12. Solve the equation $z^2 + z + 1 = 0$ for $z = (x, y)$ by writing

$$(x, y)(x, y) + (x, y) + (1, 0) = (0, 0)$$

   and then solving a pair of simultaneous equations in x and y.

   Suggestion: Use the fact that no real number x satisfies the given equation to show that $y \neq 0$.

$$\text{Ans. } z = \left(-\frac{1}{2}, \pm\frac{\sqrt{3}}{2}\right).$$

13. Derive expression (11), Sec. 2, for the quotient z_1/z_2 by the method described just after it.

14. (*a*) With the aid of relations (15) and (16) in Sec. 2, derive identity (17) there.
   (*b*) Use the identity derived in part (*a*) to establish the cancellation law

$$\frac{z_1 z}{z_2 z} = \frac{z_1}{z_2} \qquad (z_2 \neq 0, z \neq 0).$$

3.   MODULI AND CONJUGATES

It is natural to associate any nonzero complex number $z = x + iy$ with the directed line segment, or vector, from the origin to the point (x, y) that represents z (Sec. 1) in the complex plane. In fact, we often refer to z as the point z or the vector z. In Fig. 1 the numbers $z = x + iy$ and $-2 + i$ are displayed graphically as both points and radius vectors.

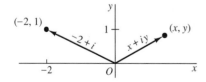

FIGURE 1

According to the definition of the sum of two complex numbers $z_1 = x_1 + iy_1$ and $z_2 = x_2 + iy_2$, the number $z_1 + z_2$ corresponds to the point $(x_1 + x_2, y_1 + y_2)$. It also corresponds to a vector with those coordinates as its components. Hence $z_1 + z_2$ may be obtained vectorially as shown in Fig. 2. The difference $z_1 - z_2 = z_1 + (-z_2)$ corresponds to the sum of the vectors for z_1 and $-z_2$ (Fig. 3). Note that, by translating the radius vector $z_1 - z_2$ in Fig. 3, one can interpret $z_1 - z_2$ as the directed line segment from the point (x_2, y_2) to the point (x_1, y_1).

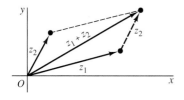

FIGURE 2

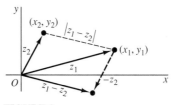

FIGURE 3

Although the product of two complex numbers z_1 and z_2 is itself a complex number represented by a vector, that vector lies in the same plane as the vectors for z_1 and z_2. Evidently, then, this product is neither the scalar nor the vector product used in ordinary vector analysis.

The vector interpretation of complex numbers is especially helpful in extending the concept of absolute values of real numbers to the complex plane. The *modulus,* or absolute value, of a complex number $z = x + iy$ is defined as the nonnegative real number $\sqrt{x^2 + y^2}$ and is denoted by $|z|$; that is,

$$(1) \qquad\qquad |z| = \sqrt{x^2 + y^2}.$$

Geometrically, the number $|z|$ is the distance between the point (x, y) and the origin, or the length of the vector representing z. It reduces to the usual absolute value in the real number system when $y = 0$. Note that, while *the inequality $z_1 < z_2$ is meaningless unless both z_1 and z_2 are real,* the statement $|z_1| < |z_2|$ means that the point z_1 is closer to the origin than the point z_2 is.

EXAMPLE 1.   Since $|-3 + 2i| = \sqrt{13}$ and $|1 + 4i| = \sqrt{17}$, the point $-3 + 2i$ is closer to the origin than $1 + 4i$ is.

The distance between two points $z_1 = x_1 + iy_1$ and $z_2 = x_2 + iy_2$ is $|z_1 - z_2|$. This is clear from Fig. 3, since $|z_1 - z_2|$ is the length of the vector representing $z_1 - z_2$. Alternatively, it follows from definition (1) and the expression

$$z_1 - z_2 = (x_1 - x_2) + i(y_1 - y_2)$$

that

$$|z_1 - z_2| = \sqrt{(x_1 - x_2)^2 + (y_1 - y_2)^2}.$$

The complex numbers z corresponding to the points lying on the circle with center z_0 and radius R thus satisfy the equation $|z - z_0| = R$, and conversely. We refer to this set of points simply as the circle $|z - z_0| = R$.

EXAMPLE 2.   The equation $|z - 1 + 3i| = 2$ represents the circle whose center is $z_0 = (1, -3)$ and whose radius is $R = 2$.

It also follows from definition (1) that the real numbers $|z|$, $\operatorname{Re} z = x$, and $\operatorname{Im} z = y$ are related by the equation

(2) $$|z|^2 = (\operatorname{Re} z)^2 + (\operatorname{Im} z)^2.$$

Thus

(3) $$\operatorname{Re} z \leq |\operatorname{Re} z| \leq |z| \qquad \text{and} \qquad \operatorname{Im} z \leq |\operatorname{Im} z| \leq |z|.$$

The *complex conjugate*, or simply the conjugate, of a complex number $z = x + iy$ is defined as the complex number $x - iy$ and is denoted by $\bar{z}$; that is,

(4) $$\bar{z} = x - iy.$$

The number $\bar{z}$ is represented by the point $(x, -y)$, which is the reflection in the real axis of the point (x, y) representing z (Fig. 4). Note that $\bar{\bar{z}} = z$ and $|\bar{z}| = |z|$ for all z.

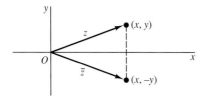

FIGURE 4

If $z_1 = x_1 + iy_1$ and $z_2 = x_2 + iy_2$, then

$$\overline{z_1 + z_2} = (x_1 + x_2) - i(y_1 + y_2) = (x_1 - iy_1) + (x_2 - iy_2).$$

So the conjugate of the sum is the sum of the conjugates:

(5) $$\overline{z_1 + z_2} = \bar{z}_1 + \bar{z}_2.$$

In like manner, it is easy to show that

(6) $$\overline{z_1 - z_2} = \bar{z}_1 - \bar{z}_2,$$

(7) $$\overline{z_1 z_2} = \bar{z}_1 \bar{z}_2,$$

(8) $$\overline{\left(\frac{z_1}{z_2}\right)} = \frac{\bar{z}_1}{\bar{z}_2} \qquad (z_2 \neq 0).$$

The sum $z + \bar{z}$ of a complex number $z = x + iy$ and its conjugate $\bar{z} = x - iy$ is the real number $2x$, and the difference $z - \bar{z}$ is the pure imaginary number $2iy$. Hence

(9) $$\operatorname{Re} z = \frac{z + \bar{z}}{2}, \qquad \operatorname{Im} z = \frac{z - \bar{z}}{2i}.$$

An important identity relating the conjugate of a complex number $z = x + iy$ to its modulus is

(10) $$z\bar{z} = |z|^2,$$

where each side is equal to $x^2 + y^2$. It suggests the method for determining a quotient z_1/z_2 that begins with expression (12), Sec. 2. That method is, of course, based on multiplying both the numerator and the denominator of z_1/z_2 by $\bar{z}_2$, so that the denominator becomes the real number $|z_2|^2$.

EXAMPLE 3. As an illustration,

$$\frac{-1 + 3i}{2 - i} = \frac{(-1 + 3i)(2 + i)}{(2 - i)(2 + i)} = \frac{-5 + 5i}{|2 - i|^2} = \frac{-5 + 5i}{5} = -1 + i.$$

Also, see the example at the end of Sec. 2.

With the aid of identity (10), one can easily obtain various other properties of moduli from properties of conjugates noted above. We mention that

$$(11) \qquad\qquad\qquad |z_1 z_2| = |z_1||z_2|,$$

$$(12) \qquad\qquad\qquad \left|\frac{z_1}{z_2}\right| = \frac{|z_1|}{|z_2|} \qquad (z_2 \neq 0).$$

Property (11) can be established by writing

$$|z_1 z_2|^2 = (z_1 z_2)(\overline{z_1 z_2}) = (z_1 \bar{z}_1)(z_2 \bar{z}_2) = |z_1|^2 |z_2|^2 = (|z_1||z_2|)^2$$

and recalling that a modulus is never negative. Property (12) can be verified in a similar way.

4. TRIANGLE INEQUALITY

Properties of moduli and conjugates in Sec. 3 enable us to give an algebraic derivation of the *triangle inequality,* which provides an upper bound for the modulus of the sum of two complex numbers z_1 and z_2:

$$(1) \qquad\qquad\qquad |z_1 + z_2| \leq |z_1| + |z_2|.$$

This important inequality is geometrically evident in Fig. 2 of Sec. 3. Indeed, it is merely a statement that the length of one side of a triangle is less than or equal to the sum of the lengths of the other two sides. We note from Fig. 2 that inequality (1) is actually an equality when the points z_1, z_2, and 0 are collinear.

We start the algebraic derivation by writing

$$|z_1 + z_2|^2 = (z_1 + z_2)(\overline{z_1 + z_2}) = (z_1 + z_2)(\bar{z}_1 + \bar{z}_2)$$

and multiplying out the far right-hand side. This shows that

$$|z_1 + z_2|^2 = z_1 \bar{z}_1 + (z_1 \bar{z}_2 + \overline{z_1 \bar{z}_2}) + z_2 \bar{z}_2.$$

But

$$z_1 \bar{z}_2 + \overline{z_1 \bar{z}_2} = 2\,\mathrm{Re}(z_1 \bar{z}_2) \leq 2|z_1 \bar{z}_2| = 2|z_1||z_2|;$$

and so

$$|z_1 + z_2|^2 \leq |z_1|^2 + 2|z_1||z_2| + |z_2|^2,$$

or

$$|z_1 + z_2|^2 \leq (|z_1| + |z_2|)^2.$$

Since moduli are nonnegative, inequality (1) now follows.

An immediate consequence of the triangle inequality is the fact that

(2) $$|z_1 + z_2| \geq ||z_1| - |z_2||.$$

To derive inequality (2), we write

$$|z_1| = |(z_1 + z_2) + (-z_2)| \leq |z_1 + z_2| + |-z_2|,$$

which means that

(3) $$|z_1 + z_2| \geq |z_1| - |z_2|.$$

This is inequality (2) when $|z_1| \geq |z_2|$. If $|z_1| < |z_2|$, we need only interchange z_1 and z_2 in inequality (3) to get

$$|z_1 + z_2| \geq -(|z_1| - |z_2|),$$

which is the desired result. Inequality (2) tells us, of course, that the length of one side of a triangle is greater than or equal to the difference of the lengths of the other two sides.

Useful alternative forms of inequalities (1) and (2) are obtained when z_2 is replaced by $-z_2$:

(4) $$|z_1 - z_2| \leq |z_1| + |z_2|,$$

(5) $$|z_1 - z_2| \geq ||z_1| - |z_2||.$$

EXAMPLE 1. If a point z lies on the unit circle $|z| = 1$ about the origin, then

$$|z^3 - 2| \leq |z|^3 + 2 = 3$$

and

$$|z^3 - 2| \geq ||z|^3 - 2| = 1.$$

The triangle inequality can be generalized by means of mathematical induction to sums involving any finite number of terms:

(6) $$|z_1 + z_2 + \cdots + z_n| \leq |z_1| + |z_2| + \cdots + |z_n| \qquad (n = 2, 3, \ldots).$$

To give details of the induction proof here, we note that when $n = 2$, inequality (6) is just inequality (1). Furthermore, if inequality (6) is assumed to be valid when $n = m$, it must also hold when $n = m + 1$ since, by the triangle inequality,

$$|(z_1 + z_2 + \cdots + z_m) + z_{m+1}| \leq |z_1 + z_2 + \cdots + z_m| + |z_{m+1}|$$
$$\leq (|z_1| + |z_2| + \cdots + |z_m|) + |z_{m+1}|.$$

EXAMPLE 2. If z is a point inside the circle centered at the origin and with radius 2, so that $|z| < 2$, then

$$|z^3 + 3z^2 - 2z + 1| \leq |z|^3 + 3|z|^2 + 2|z| + 1 < 25.$$

EXERCISES

1. Locate the numbers $z_1 + z_2$ and $z_1 - z_2$ vectorially when

(a) $z_1 = 2i, z_2 = \dfrac{2}{3} - i$;   (b) $z_1 = (-\sqrt{3}, 1), z_2 = (\sqrt{3}, 0)$;

(c) $z_1 = (-3, 1), z_2 = (1, 4)$;   (d) $z_1 = x_1 + iy_1, z_2 = x_1 - iy_1$.

2. Use properties of conjugates and moduli to show that

(a) $\overline{\bar{z} + 3i} = z - 3i$;   (b) $\overline{iz} = -i\bar{z}$;   (c) $\overline{(2 + i)^2} = 3 - 4i$;

(d) $|(2\bar{z} + 5)(\sqrt{2} - i)| = \sqrt{3}|2z + 5|$.

3. Verify inequalities (3), Sec. 3, involving $\operatorname{Re} z$, $\operatorname{Im} z$, and $|z|$.

4. Prove that $\sqrt{2}|z| \geq |\operatorname{Re} z| + |\operatorname{Im} z|$.

5. Verify properties (6) and (7) of $\bar{z}$ in Sec. 3.

6. Use the property $\overline{z_1 z_2} = \bar{z}_1 \bar{z}_2$ to show that (a) $\overline{z_1 z_2 z_3} = \bar{z}_1 \bar{z}_2 \bar{z}_3$;   (b) $\overline{(z^4)} = (\bar{z})^4$.

7. Verify property (12) of moduli in Sec. 3.

8. Use results in Sec. 3 to show that when z_2 and z_3 are nonzero,

(a) $\overline{\left(\dfrac{z_1}{z_2 z_3}\right)} = \dfrac{\bar{z}_1}{\bar{z}_2 \bar{z}_3}$;   (b) $\left|\dfrac{z_1}{z_2 z_3}\right| = \dfrac{|z_1|}{|z_2||z_3|}$.

9. With the aid of inequalities in Sec. 4, show that when $|z_3| \neq |z_4|$,

$$\left|\frac{z_1 + z_2}{z_3 + z_4}\right| \leq \frac{|z_1| + |z_2|}{||z_3| - |z_4||}.$$

10. In each case, sketch the set of points determined by the given condition:

(a) $|z - 1 + i| = 1$;   (b) $|z + i| \leq 3$;   (c) $\operatorname{Re}(\bar{z} - i) = 2$;   (d) $|2z - i| = 4$.

11. Apply inequalities in Secs. 3 and 4 to show that

$$|\operatorname{Im}(1 - \bar{z} + z^2)| < 3 \qquad \text{when} \qquad |z| < 1.$$

12. By factoring $z^4 - 4z^2 + 3$ into two quadratic factors and then using inequality (5), Sec. 4, show that if z lies on the circle $|z| = 2$, then

$$\left|\frac{1}{z^4 - 4z^2 + 3}\right| \leq \frac{1}{3}.$$

13. It is shown in Sec. 2 that if $z_1 z_2 = 0$, then at least one of the numbers z_1 and z_2 must be zero. Give an alternative proof, based on the corresponding result for real numbers, using identity (11), Sec. 3.

14. Prove that

(a) z is real if and only if $\bar{z} = z$;

(b) z is either real or pure imaginary if and only if $(\bar{z})^2 = z^2$.

15. Use mathematical induction to show that when $n = 2, 3, \ldots,$

(a) $\overline{z_1 + z_2 + \cdots + z_n} = \bar{z}_1 + \bar{z}_2 + \cdots + \bar{z}_n$;   (b) $\overline{z_1 z_2 \cdots z_n} = \bar{z}_1 \bar{z}_2 \cdots \bar{z}_n$.

16. Let $a_0, a_1, a_2, \ldots, a_n (n \geq 1)$ denote *real* numbers, and let z be any complex number. With the aid of the results in Exercise 15, show that

$$\overline{a_0 + a_1 z + a_2 z^2 + \cdots + a_n z^n} = a_0 + a_1 \bar{z} + a_2 \bar{z}^2 + \cdots + a_n \bar{z}^n.$$

17. Show that the equation $|z - z_0| = R$ of a circle, centered at z_0 with radius R, can be written

$$|z|^2 - 2\operatorname{Re}(z\bar{z}_0) + |z_0|^2 = R^2.$$

18. Using expressions (9), Sec. 3, for Re z and Im z, show that the hyperbola $x^2 - y^2 = 1$ can be written

$$z^2 + \bar{z}^2 = 2.$$

19. Using the fact that $|z_1 - z_2|$ is the distance between two points z_1 and z_2, give a geometric argument that

(a) the equation $|z - 4i| + |z + 4i| = 10$ represents an ellipse whose foci are $(0, \pm4)$;

(b) the equation $|z - 1| = |z + i|$ represents the line through the origin whose slope is -1.

5. POLAR COORDINATES AND EULER'S FORMULA

Let r and θ be polar coordinates of the point (x, y) that corresponds to a *nonzero* complex number $z = x + iy$. Since $x = r\cos\theta$ and $y = r\sin\theta$, z can be written in *polar form* as

$$(1) \qquad\qquad z = r(\cos\theta + i\sin\theta).$$

If $z = 0$, the coordinate θ is undefined.

In complex analysis, the real number r is not allowed to be negative and is the length of the radius vector for z; that is, $r = |z|$. The real number θ represents the angle, measured in radians, that z makes with the positive real axis when z is interpreted as a radius vector (Fig. 5). As in calculus, θ has an infinite number of possible values, including negative ones, that differ by integral multiples of 2π. Those values can be determined from the equation $\tan\theta = y/x$, where the quadrant containing the point corresponding to z must be specified. Each value of θ is called an *argument* of z, and the set of all such values is denoted by arg z. The *principal value* of arg z, denoted by Arg z, is that unique value Θ such that $-\pi < \Theta \le \pi$. Note that

$$(2) \qquad\qquad \arg z = \text{Arg } z + 2n\pi \qquad (n = 0, \pm1, \pm2, \ldots).$$

Also, when z is a negative real number, Arg z has value π, not $-\pi$.

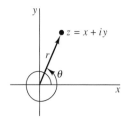

FIGURE 5

EXAMPLE 1. The complex number $-1 - i$, which lies in the third quadrant, has principal argument $-3\pi/4$. That is,

$$\text{Arg}(-1 - i) = -\frac{3\pi}{4};$$

and it follows that

$$\arg(-1-i) = -\frac{3\pi}{4} + 2n\pi \qquad (n = 0, \pm1, \pm2, \ldots).$$

Using the symbol $e^{i\theta}$, or $\exp(i\theta)$, which is defined by *Euler's formula* for any real value of θ as

$$(3) \qquad\qquad e^{i\theta} = \cos\theta + i\sin\theta,$$

we can write the polar form (1) more compactly in *exponential form* as

$$(4) \qquad\qquad z = re^{i\theta}.$$

The choice of the symbol $e^{i\theta}$ will be motivated later on in Sec. 23. Its use in Sec. 6 will, however, suggest that it is a natural choice.

EXAMPLE 2. The number $-1-i$ in Example 1 has exponential form

$$(5) \qquad\qquad -1-i = \sqrt{2}\exp\left[i\left(-\frac{3\pi}{4}\right)\right].$$

With the agreement that $e^{-i\theta} = e^{i(-\theta)}$, this can also be written $-1-i = \sqrt{2}e^{-i3\pi/4}$. Expression (5) is, of course, only one of an infinite number of possibilities for the exponential form of $-1-i$:

$$(6) \qquad -1-i = \sqrt{2}\exp\left[i\left(-\frac{3\pi}{4} + 2n\pi\right)\right] \qquad (n = 0, \pm1, \pm2, \ldots).$$

Consider now a point $z = re^{i\theta}$, lying on a circle centered at the origin and with radius r (Fig. 6). As θ is increased, z moves around the circle in the counterclockwise direction. In particular, when θ is increased by 2π, we arrive at the original point; and the same is true when θ is decreased by 2π. It is, therefore, evident from Fig. 6 that *two nonzero complex numbers*

$$z_1 = r_1 e^{i\theta_1} \qquad \text{and} \qquad z_2 = r_2 e^{i\theta_2}$$

are equal if and only if

$$r_1 = r_2 \qquad \text{and} \qquad \theta_1 = \theta_2 + 2n\pi,$$

where n is some integer $(n = 0, \pm1, \pm2, \ldots).$

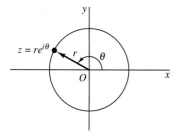

FIGURE 6

Note, too, that values of $e^{i\theta}$ are immediate from Fig. 6, without reference to Euler's formula (3), when $r = 1$ and θ is some integral multiple of $\pi/2$. It is, for instance, geometrically obvious that $e^{i\pi} = -1$, $e^{-i\pi/2} = -i$, and $e^{-i4\pi} = 1$.

Figure 6, with $r = R$, also shows that the equation

$$(7) \qquad\qquad z = Re^{i\theta} \qquad (0 \le \theta \le 2\pi)$$

is a parametric representation of the circle $|z| = R$, centered at the origin with radius R. As the parameter θ in Fig. 6 increases from $\theta = 0$ over the interval $0 \le \theta \le 2\pi$, the point z starts from the positive real axis and traverses the circle once in the counterclockwise direction. More generally, the circle $|z - z_0| = R$, whose center is z_0 and whose radius is R, has the parametric representation

$$(8) \qquad\qquad z = z_0 + Re^{i\theta} \qquad (0 \le \theta \le 2\pi).$$

This can be seen vectorially (Fig. 7) by noting that a point z traversing the circle $|z - z_0| = R$ once in the counterclockwise direction corresponds to the sum of the fixed vector z_0 and a vector of length R whose angle of inclination θ varies from $\theta = 0$ to $\theta = 2\pi$.

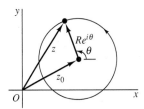

FIGURE 7

6. PRODUCTS AND QUOTIENTS IN EXPONENTIAL FORM

Simple trigonometry tells us that $e^{i\theta}$ has the familiar additive property of the exponential function in calculus:

$$
\begin{aligned}
e^{i\theta_1} e^{i\theta_2} &= (\cos\theta_1 + i\sin\theta_1)(\cos\theta_2 + i\sin\theta_2) \\
&= (\cos\theta_1 \cos\theta_2 - \sin\theta_1 \sin\theta_2) + i(\sin\theta_1 \cos\theta_2 + \cos\theta_1 \sin\theta_2) \\
&= \cos(\theta_1 + \theta_2) + i\sin(\theta_1 + \theta_2) = e^{i(\theta_1 + \theta_2)}.
\end{aligned}
$$

Thus, if $z_1 = r_1 e^{i\theta_1}$ and $z_2 = r_2 e^{i\theta_2}$, the product $z_1 z_2$ has exponential form

$$(1) \qquad\qquad z_1 z_2 = r_1 r_2 e^{i\theta_1} e^{i\theta_2} = r_1 r_2 e^{i(\theta_1 + \theta_2)}.$$

Moreover,

$$(2) \qquad\qquad \frac{z_1}{z_2} = \frac{r_1}{r_2} \cdot \frac{e^{i\theta_1} e^{-i\theta_2}}{e^{i\theta_2} e^{-i\theta_2}} = \frac{r_1}{r_2} \cdot \frac{e^{i(\theta_1 - \theta_2)}}{e^{i0}} = \frac{r_1}{r_2} e^{i(\theta_1 - \theta_2)}.$$

Because $1 = 1e^{i0}$, it follows from expression (2) that the inverse of any nonzero complex number $z = re^{i\theta}$ is

$$(3) \qquad\qquad z^{-1} = \frac{1}{z} = \frac{1}{r}e^{-i\theta}.$$

Expressions (1), (2), and (3) are, of course, easily remembered by applying the usual algebraic rules for real numbers and e^x.

Another important result that can be obtained formally by applying rules for real numbers is

$$(4) \qquad\qquad z^n = r^n e^{in\theta} \qquad (n = 0, \pm 1, \pm 2, \ldots).$$

It is easily verified for positive values of n by mathematical induction. To be specific, we first note that it becomes $z = re^{i\theta}$ when $n = 1$. Next, we assume that it is valid when $n = m$, where m is any positive integer. In view of expression (1) for the product of two nonzero complex numbers in exponential form, it is then valid for $n = m + 1$:

$$z^{m+1} = z z^m = re^{i\theta} r^m e^{im\theta} = r^{m+1} e^{i(m+1)\theta}.$$

Expression (4) is thus verified when n is a positive integer. It also holds when $n = 0$, with the convention that $z^0 = 1$. If $n = -1, -2, \ldots$, on the other hand, we define z^n in terms of the multiplicative inverse of z by writing

$$z^n = (z^{-1})^m, \qquad \text{where} \qquad m = -n = 1, 2, \ldots.$$

Then, since expression (4) is valid for positive integral powers, it follows from the exponential form (3) of z^{-1} that

$$z^n = \left[\frac{1}{r}e^{i(-\theta)}\right]^m = \left(\frac{1}{r}\right)^m e^{im(-\theta)} = \left(\frac{1}{r}\right)^{-n} e^{i(-n)(-\theta)} = r^n e^{in\theta} \quad (n = -1, -2, \ldots).$$

Expression (4) is now established for all integral powers.

Observe that if $r = 1$, expression (4) becomes

$$(5) \qquad\qquad (e^{i\theta})^n = e^{in\theta} \qquad (n = 0, \pm 1, \pm 2, \ldots).$$

When written in the form

$$(6) \qquad (\cos\theta + i\sin\theta)^n = \cos n\theta + i\sin n\theta \qquad (n = 0, \pm 1, \pm 2, \ldots),$$

this is known as *de Moivre's formula*.

Expression (4) can be useful in finding powers of complex numbers even when they are given in rectangular form and the result is desired in that form.

EXAMPLE 1. In order to put $(\sqrt{3} + i)^7$ in rectangular form, one need only write

$$(\sqrt{3} + i)^7 = (2e^{i\pi/6})^7 = 2^7 e^{i7\pi/6} = (2^6 e^{i\pi})(2e^{i\pi/6}) = -64(\sqrt{3} + i).$$

We turn now to an important identity involving arguments (Sec. 5) of products:

$$(7) \qquad\qquad \arg(z_1 z_2) = \arg z_1 + \arg z_2.$$

It is to be interpreted as saying that if values of two of these three (multiple-valued) arguments are specified, then there is a value of the third such that the equation holds.

We start the verification of statement (7) by letting θ_1 and θ_2 denote any values of $\arg z_1$ and $\arg z_2$, respectively. Expression (1) then tells us that $\theta_1 + \theta_2$ is a value of $\arg(z_1 z_2)$. (See Fig. 8.) If, on the other hand, values of $\arg(z_1 z_2)$ and $\arg z_1$ are specified, those values correspond to particular choices of n and n_1 in the expressions

$$\arg(z_1 z_2) = (\theta_1 + \theta_2) + 2n\pi \qquad (n = 0, \pm 1, \pm 2, \ldots)$$

and

$$\arg z_1 = \theta_1 + 2n_1\pi \qquad (n_1 = 0, \pm 1, \pm 2, \ldots).$$

Since

$$(\theta_1 + \theta_2) + 2n\pi = (\theta_1 + 2n_1\pi) + [\theta_2 + 2(n - n_1)\pi],$$

equation (7) is evidently satisfied when the value

$$\arg z_2 = \theta_2 + 2(n - n_1)\pi$$

is chosen. Verification when values of $\arg(z_1 z_2)$ and $\arg z_2$ are specified follows by symmetry.

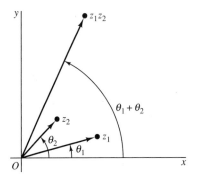

FIGURE 8

Statement (7) is sometimes valid when *arg* is replaced everywhere by *Arg* (see Exercise 6). But, as the following example illustrates, that is *not always* the case.

EXAMPLE 2. When $z_1 = -1$ and $z_2 = i$,

$$\mathrm{Arg}(z_1 z_2) = \mathrm{Arg}(-i) = -\frac{\pi}{2} \qquad \text{but} \qquad \mathrm{Arg}\, z_1 + \mathrm{Arg}\, z_2 = \pi + \frac{\pi}{2} = \frac{3\pi}{2}.$$

If, however, we take the values of $\arg z_1$ and $\arg z_2$ just used and select the value $\arg(z_1 z_2) = 3\pi/2$, we find that equation (7) *is* satisfied.

Another statement, analogous to statement (7), is that

$$(8) \qquad \arg\left(\frac{z_1}{z_2}\right) = \arg z_1 - \arg z_2.$$

It can be verified with the aid of expression (2).

EXERCISES

1. Find the principal argument Arg z when

   (a) $z = \dfrac{-2}{1 + \sqrt{3}i}$;   (b) $z = \dfrac{i}{-2 - 2i}$;   (c) $z = (\sqrt{3} - i)^6$.

   Ans. (a) $2\pi/3$;   (b) $-3\pi/4$;   (c) π.

2. By writing the individual factors on the left in exponential form, performing the needed operations, and finally changing back to rectangular coordinates, show that

   (a) $i(1 - \sqrt{3}i)(\sqrt{3} + i) = 2(1 + \sqrt{3}i)$;   (b) $5i/(2 + i) = 1 + 2i$;

   (c) $(-1 + i)^7 = -8(1 + i)$;   (d) $(1 + \sqrt{3}i)^{-10} = 2^{-11}(-1 + \sqrt{3}i)$.

3. Show that

   (a) $|e^{i\theta}| = 1$;   (b) $\overline{e^{i\theta}} = e^{-i\theta}$;   (c) $e^{i\theta_1}e^{i\theta_2}\cdots e^{i\theta_n} = e^{i(\theta_1 + \theta_2 + \cdots + \theta_n)}$ $(n = 2, 3, \ldots)$.

4. Solve the equation $|e^{i\theta} - 1| = 2$ for $\theta(0 \leq \theta < 2\pi)$ and verify the solution geometrically.

   Ans. π.

5. Use de Moivre's formula (Sec. 6) to derive the following trigonometric identities:

   (a) $\cos 3\theta = \cos^3 \theta - 3\cos \theta \sin^2 \theta$;   (b) $\sin 3\theta = 3\cos^2 \theta \sin \theta - \sin^3 \theta$.

6. Show that if $\operatorname{Re} z_1 > 0$ and $\operatorname{Re} z_2 > 0$, then

   $$\operatorname{Arg}(z_1 z_2) = \operatorname{Arg} z_1 + \operatorname{Arg} z_2,$$

   where $\operatorname{Arg}(z_1 z_2)$ denotes the principal value of $\arg(z_1 z_2)$, etc.

7. Verify the statement (Sec. 6)

   $$\arg\left(\frac{z_1}{z_2}\right) = \arg z_1 - \arg z_2.$$

8. According to Sec. 3, the difference $z - z_0$ of two distinct complex numbers can be interpreted vectorially. (See Fig. 9, where θ denotes the angle of inclination of the vector representing $z - z_0$). By translating the vector for $z - z_0$ so that it is a radius vector, show that the values of $\arg(\overline{z - z_0})$ are the same as the values of $-\arg(z - z_0)$. Use the same method to show that

   $$\operatorname{Arg}(\overline{z - z_0}) = -\operatorname{Arg}(z - z_0)$$

   if and only if $z - z_0$ is not a negative real number.

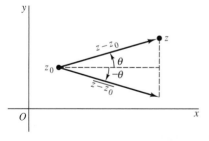

FIGURE 9

9. Given that $z_1 z_2 \neq 0$, use the exponential forms of z_1 and z_2 to prove that

   $$\operatorname{Re}(z_1 \bar{z}_2) = |z_1||z_2|$$

   if and only if $\theta_1 - \theta_2 = 2n\pi$ $(n = 0, \pm 1, \pm 2, \ldots)$, where $\theta_1 = \arg z_1$ and $\theta_2 = \arg z_2$.

10. Given that $z_1 z_2 \neq 0$ and using the result in Exercise 9, modify the algebraic derivation of the triangle inequality in Sec. 4 to show that

$$|z_1 + z_2| = |z_1| + |z_2|$$

if and only if $\theta_1 - \theta_2 = 2n\pi$ $(n = 0, \pm 1, \pm 2, \ldots)$, where $\theta_1 = \arg z_1$ and $\theta_2 = \arg z_2$. Interpret this statement geometrically.

11. Let z be a nonzero complex number and n a negative integer $(n = -1, -2, \ldots)$. Also, write $z = re^{i\theta}$ and $m = -n = 1, 2, \ldots$. Using the expressions $z^m = r^m e^{im\theta}$ and $z^{-1} = (1/r)e^{i(-\theta)}$, verify that $(z^m)^{-1} = (z^{-1})^m$ and hence that the definition $z^n = (z^{-1})^m$ in Sec. 6 could have been written alternatively as $z^n = (z^m)^{-1}$.

12. Prove that two nonzero complex numbers z_1 and z_2 have the same moduli if and only if there are complex numbers c_1 and c_2 such that $z_1 = c_1 c_2$ and $z_2 = c_1 \bar{c}_2$.

Suggestion: Note that

$$\exp\left(i\frac{\theta_1 + \theta_2}{2}\right)\exp\left(i\frac{\theta_1 - \theta_2}{2}\right) = \exp(i\theta_1)$$

and [see Exercise 3(*b*)]

$$\exp\left(i\frac{\theta_1 + \theta_2}{2}\right)\overline{\exp\left(i\frac{\theta_1 - \theta_2}{2}\right)} = \exp(i\theta_2).$$

13. Establish the identity

$$1 + z + z^2 + \cdots + z^n = \frac{1 - z^{n+1}}{1 - z} \qquad (z \neq 1),$$

and then use it to derive *Lagrange's trigonometric identity:*

$$1 + \cos\theta + \cos 2\theta + \cdots + \cos n\theta = \frac{1}{2} + \frac{\sin[(2n + 1)\theta/2]}{2\sin(\theta/2)} \qquad (0 < \theta < 2\pi).$$

Suggestion: As for the first identity, write $S = 1 + z + z^2 + \cdots + z^n$ and consider the difference $S - zS$. To derive the second identity, write $z = e^{i\theta}$ in the first one.

14. Use mathematical induction to establish the binomial formula for complex numbers:

$$(z_1 + z_2)^n = \sum_{k=0}^{n} \binom{n}{k} z_1^{n-k} z_2^k \qquad (n = 1, 2, \ldots),$$

where

$$\binom{n}{k} = \frac{n!}{k!(n - k)!} \qquad (k = 0, 1, 2, \ldots, n)$$

and where it is agreed that $0! = 1$.

15. Use mathematical induction to verify de Moivre's formula (Sec. 6)

$$(\cos\theta + i\sin\theta)^n = \cos n\theta + i\sin n\theta$$

when n is a positive integer $(n = 1, 2, \ldots)$.

16. (*a*) Use the binomial formula (Exercise 14) and de Moivre's formula (see Exercise 15) to write

$$\cos n\theta + i\sin n\theta = \sum_{k=0}^{n} \binom{n}{k} \cos^{n-k}\theta (i\sin\theta)^k \qquad (n = 1, 2, \ldots).$$

Then define the integer m by means of the equations

$$m = \begin{cases} n/2 & \text{if } n \text{ is even,} \\ (n-1)/2 & \text{if } n \text{ is odd,} \end{cases}$$

and use the above sum to obtain the expression [compare Exercise 5(a)]

$$\cos n\theta = \sum_{k=0}^{m} \binom{n}{2k}(-1)^k \cos^{n-2k}\theta \sin^{2k}\theta \qquad (n = 1, 2, \ldots).$$

(b) Write $x = \cos\theta$ and suppose that $0 \le \theta \le \pi$, in which case $-1 \le x \le 1$. Point out how it follows from the final result in part (a) that each of the functions

$$T_n(x) = \cos(n\cos^{-1}x) \qquad (n = 0, 1, 2, \ldots)$$

is a polynomial of degree n in the variable x.*

7. ROOTS OF COMPLEX NUMBERS

The expression $z^n = r^n e^{in\theta}$ in Sec. 6 for integral powers of complex numbers $z = re^{i\theta}$ is useful in finding the nth roots of any nonzero complex number $z_0 = r_0 e^{i\theta_0}$, where n has one of the values $n = 2, 3, \ldots$. The method starts with the observation that an nth root of z_0 is a nonzero number $z = re^{i\theta}$ such that $z^n = z_0$, or

$$r^n e^{in\theta} = r_0 e^{i\theta_0}.$$

Now, according to the statement in italics near the end of Sec. 5,

$$r^n = r_0 \qquad \text{and} \qquad n\theta = \theta_0 + 2k\pi,$$

where k is any integer ($k = 0, \pm 1, \pm 2, \ldots$). So $r = \sqrt[n]{r_0}$, where this radical denotes the unique *positive* nth root of the positive real number r_0, and

$$\theta = \frac{\theta_0 + 2k\pi}{n} = \frac{\theta_0}{n} + \frac{2k\pi}{n} \qquad (k = 0, \pm 1, \pm 2, \ldots).$$

Consequently, the complex numbers

$$z = \sqrt[n]{r_0}\exp\left[i\left(\frac{\theta_0}{n} + \frac{2k\pi}{n}\right)\right] \qquad (k = 0, \pm 1, \pm 2, \ldots)$$

are the nth roots of z_0. We are able to see immediately from this exponential form of the roots that they all lie on the circle $|z| = \sqrt[n]{r_0}$ about the origin and are equally spaced every $2\pi/n$ radians, starting with argument θ_0/n. Evidently, then, all of the *distinct* roots are obtained when $k = 0, 1, 2, \ldots, n-1$, and no further roots arise with other values of k. We let c_k ($k = 0, 1, 2, \ldots, n-1$) denote these distinct roots and write

$$(1) \qquad c_k = \sqrt[n]{r_0}\exp\left[i\left(\frac{\theta_0}{n} + \frac{2k\pi}{n}\right)\right] \qquad (k = 0, 1, 2, \ldots, n-1).$$

---

*These polynomials are called Chebyshev polynomials and are prominent in approximation theory.

The number $\sqrt[n]{r_0}$ is the length of each of the radius vectors representing the n roots. The first root c_0 has argument θ_0/n; and the two roots when $n = 2$ lie at the opposite ends of a diameter of the circle $|z| = \sqrt[n]{r_0}$, the second root being $-c_0$. When $n \geq 3$, the roots lie at the vertices of a regular polygon of n sides inscribed in that circle.

We shall let $z_0^{1/n}$ denote the *set* of nth roots of z_0. If, in particular, z_0 is a positive real number r_0, the symbol $r_0^{1/n}$ denotes the entire set of roots; and the symbol $\sqrt[n]{r_0}$ in expression (1) is reserved for the one positive root. When the value of θ_0 that is used in expression (1) is the principal value of $\arg z_0$ ($-\pi < \theta_0 \leq \pi$), the number c_0 is referred to as the *principal root*. Thus when z_0 is a positive real number r_0, its principal root is $\sqrt[n]{r_0}$.

Finally, a convenient way to remember expression (1) is to write z_0 in its most general exponential form (compare Example 2 in Sec. 5),

$$z_0 = r_0 \exp[i(\theta_0 + 2k\pi)] \qquad (k = 0, \pm1, \pm2, \ldots),$$

and to *formally* apply laws of fractional exponents involving real numbers, keeping in mind that there are precisely n roots:

$$z_0^{1/n} = \{r_0 \exp[i(\theta_0 + 2k\pi)]\}^{1/n}$$

$$= \sqrt[n]{r_0} \exp\left[\frac{i(\theta_0 + 2k\pi)}{n}\right]$$

$$= \sqrt[n]{r_0} \exp\left[i\left(\frac{\theta_0}{n} + \frac{2k\pi}{n}\right)\right] \qquad (k = 0, 1, 2, \ldots, n-1).$$

EXAMPLE 1. In order to determine the nth roots of unity, we write

$$1 = 1 \exp[i(0 + 2k\pi)] \qquad (k = 0, \pm1, \pm2 \ldots)$$

and find that

$$(2) \quad 1^{1/n} = \sqrt[n]{1} \exp\left[i\left(\frac{0}{n} + \frac{2k\pi}{n}\right)\right] = \exp\left(i\frac{2k\pi}{n}\right) \qquad (k = 0, 1, 2, \ldots, n-1).$$

When $n = 2$, these roots are, of course, ±1. When $n \geq 3$, the regular polygon at whose vertices the roots lie is inscribed in the unit circle $|z| = 1$, with one vertex corresponding to the principal root $z = 1$ ($k = 0$).

If we write

$$(3) \qquad \qquad \omega_n = \exp\left(i\frac{2\pi}{n}\right),$$

it follows from property (5), Sec. 6, of $e^{i\theta}$ that

$$\omega_n^k = \exp\left(i\frac{2k\pi}{n}\right) \qquad (k = 0, 1, 2, \ldots, n-1).$$

Hence the distinct nth roots of unity just found are simply

$$1, \omega_n, \omega_n^2, \ldots, \omega_n^{n-1}.$$

See Fig. 10, where the cases $n = 3, 4$, and 6 are illustrated. Note that $\omega_n^n = 1$. Finally, it is worthwhile observing that if c is any particular nth root of a nonzero complex number z_0, the set of nth roots can be put in the form

$$c, c\omega_n, c\omega_n^2, \ldots, c\omega_n^{n-1}.$$

This is because multiplication of any nonzero complex number by ω_n increases the argument of that number by $2\pi/n$, while leaving its modulus unchanged.

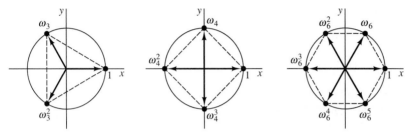

FIGURE 10

EXAMPLE 2.   Let us find all values of $(-8i)^{1/3}$, or the three cube roots of $-8i$. One need only write

$$-8i = 8\exp\left[i\left(-\frac{\pi}{2} + 2k\pi\right)\right] \qquad (k = 0, \pm1, \pm2, \ldots)$$

to see that the desired roots are

(4) $$c_k = 2\exp\left[i\left(-\frac{\pi}{6} + \frac{2k\pi}{3}\right)\right] \qquad (k = 0, 1, 2).$$

They lie at the vertices of an equilateral triangle, inscribed in the circle $|z| = 2$, and are equally spaced around that circle every $2\pi/3$ radians, starting with the principal root (Fig. 11)

$$c_0 = 2\exp\left[i\left(-\frac{\pi}{6}\right)\right] = 2\left(\cos\frac{\pi}{6} - i\sin\frac{\pi}{6}\right) = \sqrt{3} - i.$$

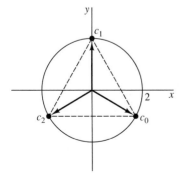

FIGURE 11

Without any further calculations, it is then evident that $c_1 = 2i$; and, since c_2 is symmetric to c_0 with respect to the imaginary axis, we know that $c_2 = -\sqrt{3} - i$.

These roots can, of course, be written

$$c_0, c_0\omega_3, c_0\omega_3^2, \qquad \text{where} \qquad \omega_3 = \exp\left(i\frac{2\pi}{3}\right).$$

(See the remarks at the end of Example 1.)

EXERCISES

1. Find the square roots of (a) $2i$; (b) $1 - \sqrt{3}i$, and express them in rectangular coordinates.

   Ans. (a) $\pm(1 + i)$;   (b) $\pm\dfrac{\sqrt{3} - i}{\sqrt{2}}$.

2. In each case, find all of the roots in rectangular coordinates, exhibit them geometrically, and point out which is the principal root:

   (a) $(-1)^{1/3}$;   (b) $(-16)^{1/4}$;   (c) $8^{1/6}$;   (d) $(-8 - 8\sqrt{3}i)^{1/4}$.

   Ans. (b) $\pm\sqrt{2}(1 + i), \pm\sqrt{2}(1 - i)$;   (c) $\pm\sqrt{2}, \pm\dfrac{1 + \sqrt{3}i}{\sqrt{2}}, \pm\dfrac{1 - \sqrt{3}i}{\sqrt{2}}$;

   (d) $\pm(\sqrt{3} - i), \pm(1 + \sqrt{3}i)$.

3. Let $z = re^{i\theta}$ be any nonzero complex number and n a negative integer ($n = -1, -2, \ldots$). Then define $z^{1/n}$ by means of the equation $z^{1/n} = (z^{-1})^{1/m}$, where $m = -n$. By showing that the m values of $(z^{1/m})^{-1}$ and $(z^{-1})^{1/m}$ are the same, verify that $z^{1/n} = (z^{1/m})^{-1}$. (Compare Exercise 11, Sec. 6.)

4. (a) Let a denote any fixed real number, and show that the two square roots of $a + i$ are

   $$\pm\sqrt{A}\exp\left(i\frac{\alpha}{2}\right),$$

   where $A = \sqrt{a^2 + 1}$ and $\alpha = \operatorname{Arg}(a + i)$.

   (b) With the aid of the trigonometric identities

   $$\cos^2\left(\frac{\alpha}{2}\right) = \frac{1 + \cos\alpha}{2}, \qquad \sin^2\left(\frac{\alpha}{2}\right) = \frac{1 - \cos\alpha}{2},$$

   show that the square roots obtained in part (a) can be written

   $$\pm\frac{1}{\sqrt{2}}\left(\sqrt{A + a} + i\sqrt{A - a}\right).$$

5. According to Sec. 7, the three cube roots of a nonzero complex number z_0 can be written $c_0, c_0\omega_3, c_0\omega_3^2$, where c_0 is the principal cube root of z_0 and

   $$\omega_3 = \exp\left(i\frac{2\pi}{3}\right) = \frac{-1 + \sqrt{3}i}{2}.$$

   Show that if $z_0 = -4\sqrt{2} + 4\sqrt{2}i$, then $c_0 = \sqrt{2}(1 + i)$ and the other two cube roots are, in rectangular form, the numbers

   $$c_0\omega_3 = \frac{-(\sqrt{3} + 1) + (\sqrt{3} - 1)i}{\sqrt{2}}, \qquad c_0\omega_3^2 = \frac{(\sqrt{3} - 1) - (\sqrt{3} + 1)i}{\sqrt{2}}.$$

6. Find the four roots of the equation $z^4 + 4 = 0$ and use them to factor $z^4 + 4$ into quadratic factors with real coefficients.

$Ans.$ $(z^2 + 2z + 2)(z^2 - 2z + 2)$.

7. Show that if c is any nth root of unity other than unity itself, then

$$1 + c + c^2 + \cdots + c^{n-1} = 0.$$

Suggestion: Use the first identity in Exercise 13, Sec. 6.

8. (*a*) Prove that the usual formula solves the quadratic equation

$$az^2 + bz + c = 0 \qquad (a \neq 0)$$

when the coefficients a, b, and c are complex numbers. Specifically, by completing the square on the left-hand side, derive the quadratic formula

$$z = \frac{-b + (b^2 - 4ac)^{1/2}}{2a},$$

where both square roots are to be considered when $b^2 - 4ac \neq 0$.

(*b*) Use the result in part (*a*) to find the roots of the equation $z^2 + 2z + (1 - i) = 0$.

$Ans.$ (*b*) $\left(-1 + \dfrac{1}{\sqrt{2}}\right) + \dfrac{i}{\sqrt{2}}, \quad \left(-1 - \dfrac{1}{\sqrt{2}}\right) - \dfrac{i}{\sqrt{2}}$.

8. REGIONS IN THE COMPLEX PLANE

In this section, we are concerned with sets of complex numbers, or points in the z plane, and their closeness to one another. Our basic tool is the concept of an ε *neighborhood*

(1) $$|z - z_0| < \varepsilon$$

of a given point z_0. It consists of all points z lying inside but not on a circle centered at z_0 and with a specified positive radius ε (Fig. 12). When the value of ε is understood or is immaterial in the discussion, the set (1) is often referred to as just a neighborhood. Occasionally, it is convenient to speak of a *deleted neighborhood*

(2) $$0 < |z - z_0| < \varepsilon,$$

consisting of all points z in an ε neighborhood of z_0 except for the point z_0 itself.

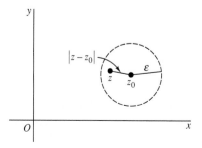

FIGURE 12

A point z_0 is said to be an *interior point* of a set S whenever there is some neighborhood of z_0 that contains only points of S; it is called an *exterior point* of S when there exists a neighborhood of it containing no points of S. If z_0 is neither of these, it is a *boundary point* of S. A boundary point is, therefore, a point all of whose neighborhoods contain points in S and points not in S. The totality of all boundary points is called the *boundary* of S. The circle $|z| = 1$, for instance, is the boundary of each of the sets

(3) $$|z| < 1 \quad \text{and} \quad |z| \le 1.$$

A set is *open* if it contains none of its boundary points. It is left as an exercise to show that a set is open if and only if each of its points is an interior point. A set is *closed* if it contains all of its boundary points; and the *closure* of a set S is the closed set consisting of all points in S together with the boundary of S. Note that the first of the sets (3) is open and that the second is the closure of both of those sets.

Some sets are, of course, neither open nor closed. For a set to be not open, there must be a boundary point that is contained in the set; and if a set is not closed, there exists a boundary point not contained in the set. Observe that the punctured disk $0 < |z| \le 1$ is neither open nor closed. The set of all complex numbers is, on the other hand, both open and closed since it has no boundary points.

An open set S is *connected* if each pair of points z_1 and z_2 in it can be joined by a polygonal line, consisting of a finite number of line segments joined end to end, that lies entirely in S. The open set $|z| < 1$ is connected. The annulus $1 < |z| < 2$ is, of course, open and it is also connected (see Fig. 13). An open set that is connected is called a *domain*. Note that any neighborhood is a domain. A domain together with some, none, or all of its boundary points is referred to as a *region*.

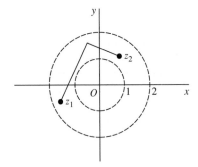

FIGURE 13

A set S is *bounded* if every point of S lies inside some circle $|z| = R$; otherwise, it is *unbounded*. Both of the sets (3) are bounded regions, and the half plane $\operatorname{Re} z \ge 0$ is unbounded.

A point z_0 is said to be an *accumulation point* of a set S if each deleted neighborhood of z_0 contains at least one point of S. It follows that if a set S is closed, then it contains each of its accumulation points. For if an accumulation point z_0 were not in S, it would be a boundary point of S; but this contradicts the fact that a closed set contains all of its boundary points. It is left as an exercise to show that the converse is, in fact, true. Thus, a set is closed if and only if it contains all of its accumulation points.

Evidently, a point z_0 is *not* an accumulation point of a set S whenever there exists some deleted neighborhood of z_0 that does not contain points of S. Note that the origin is the only accumulation point of the set $z_n = i/n \ (n = 1, 2, \ldots)$.

EXERCISES

1. Sketch the following sets and determine which are domains:
   (a) $|z - 2 + i| \le 1$;   (b) $|2z + 3| > 4$;
   (c) $\operatorname{Im} z > 1$;   (d) $\operatorname{Im} z = 1$;
   (e) $0 \le \arg z \le \pi/4 \ (z \ne 0)$;   (f) $|z - 4| \ge |z|$.
   Ans. (b), (c) are domains.

2. Which sets in Exercise 1 are neither open nor closed?
   Ans. (e).

3. Which sets in Exercise 1 are bounded?
   Ans. (a).

4. In each case, sketch the closure of the set:
   (a) $-\pi < \arg z < \pi \ (z \ne 0)$; (b) $|\operatorname{Re} z| < |z|$; (c) $\operatorname{Re}\left(\dfrac{1}{z}\right) \le \dfrac{1}{2}$; (d) $\operatorname{Re}(z^2) > 0$.

5. Let S be the open set consisting of all points z such that $|z| < 1$ or $|z - 2| < 1$. State why S is not connected.

6. Show that a set S is open if and only if each point in S is an interior point.

7. Determine the accumulation points of each of the following sets:
   (a) $z_n = i^n \ (n = 1, 2, \ldots)$;   (b) $z_n = i^n/n \ (n = 1, 2, \ldots)$;
   (c) $0 \le \arg z < \pi/2 \ (z \ne 0)$;   (d) $z_n = (-1)^n(1 + i)(n - 1)/n \ (n = 1, 2, \ldots)$.
   Ans. (a) None; (b) 0; (d) $\pm(1 + i)$.

8. Prove that if a set contains each of its accumulation points, then it must be a closed set.

9. Show that any point z_0 of a domain is an accumulation point of that domain.

10. Prove that a finite set of points $z_1, z_2, \ldots, z_n$ cannot have any accumulation points.

CHAPTER
2

ANALYTIC
FUNCTIONS

We now consider functions of a complex variable and develop a theory of differentiation for them. The main goal of the chapter is to introduce analytic functions, which play a central role in complex analysis.

9. FUNCTIONS OF A COMPLEX VARIABLE

Let S be a set of complex numbers. A *function* f defined on S is a rule that assigns to each z in S a complex number w. The number w is called the *value* of f at z and is denoted by $f(z)$; that is, $w = f(z)$. The set S is called the *domain of definition* of f.*

It is not always convenient to use different notation to distinguish between a given function and its values. For example, if f is defined on the half plane $\operatorname{Re} z > 0$ by means of the equation $w = 1/z$, it may also be referred to as the function $w = 1/z$, or simply the function $1/z$, where $\operatorname{Re} z > 0$.

It must be emphasized that both a domain of definition and a rule are needed in order for a function to be well defined. When the domain of definition is not mentioned, we agree that the largest possible set is to be taken. Thus, if we speak only of the function $1/z$, the domain of definition is understood to be the set of all nonzero points in the plane.

Suppose that $w = u + iv$ is the value of a function f at $z = x + iy$, so that

$$u + iv = f(x + iy).$$

---

*Although the domain of definition is often a domain as defined in Sec. 8, it need not be.

Each of the real numbers u and v depends on the real variables x and y, and it follows that $f(z)$ can be expressed in terms of a pair of real-valued functions of the real variables x and y:

(1) $$f(z) = u(x, y) + iv(x, y).$$

If the polar coordinates r and θ, instead of x and y, are used, then

$$u + iv = f(re^{i\theta}),$$

where $w = u + iv$ and $z = re^{i\theta}$. In that case, we may write

(2) $$f(z) = u(r, \theta) + iv(r, \theta).$$

EXAMPLE. If $f(z) = z^2$, then

$$f(x + iy) = (x + iy)^2 = x^2 - y^2 + i2xy.$$

Hence

$$u(x, y) = x^2 - y^2 \qquad \text{and} \qquad v(x, y) = 2xy.$$

When polar coordinates are used,

$$f(re^{i\theta}) = (re^{i\theta})^2 = r^2 e^{i2\theta} = r^2 \cos 2\theta + ir^2 \sin 2\theta.$$

Consequently,

$$u(r, \theta) = r^2 \cos 2\theta \qquad \text{and} \qquad v(r, \theta) = r^2 \sin 2\theta.$$

If, in either equation (1) or (2), the function v is always zero, then the number $f(z)$ is always real. An example of such a *real-valued function of a complex variable* is

$$f(z) = |z|^2 = x^2 + y^2 + i0.$$

If n is zero or a positive integer and if $a_0, a_1, a_2, \ldots, a_n$ are complex constants, where $a_n \neq 0$, the function

$$P(z) = a_0 + a_1 z + a_2 z^2 + \cdots + a_n z^n$$

is a *polynomial* of degree n. Note that the sum here has a finite number of terms and that the domain of definition is the entire z plane. Quotients $P(z)/Q(z)$ of polynomials are called *rational functions* and are defined at each point z where $Q(z) \neq 0$. Polynomials and rational functions constitute elementary, but important, classes of functions of a complex variable.

A generalization of the concept of function is a rule that assigns more than one value to a point z in the domain of definition. These *multiple-valued functions* occur in the theory of functions of a complex variable, just as they do in the case of real variables. When multiple-valued functions are studied, usually just one of the possible values assigned to each point is taken, in a systematic manner, and a (single-valued) function is constructed from the multiple-valued function. Suppose, for example, that z is any nonzero complex number $z = re^{i\theta}$. We know from Sec. 7

that $z^{1/2}$ has the two values $z^{1/2} = \pm \sqrt{r} e^{i\theta/2}$, where θ is the principal value $(-\pi < \theta \le \pi)$ of arg z. But, if we choose only the positive value of $\pm \sqrt{r}$ and write

$$f(z) = \sqrt{r} e^{i\theta/2} \qquad (r > 0, -\pi < \theta < \pi),$$

we see that this (single-valued) function f is well defined on the indicated domain. Since zero is the only square root of zero, we also write $f(0) = 0$. The function f is, then, well defined on the domain consisting of the entire complex plane except for the ray $\theta = \pi$, which is the negative real axis.

10. MAPPINGS

Properties of a real-valued function of a real variable are often exhibited by the graph of the function. But when $w = f(z)$, where z and w are complex, no such convenient graphical representation of the function f is available because each of the numbers z and w is located in a plane rather than on a line. One can, however, display some information about the function by indicating pairs of corresponding points $z = (x, y)$ and $w = (u, v)$. To do this, it is generally simpler to draw the z and w planes separately.

When a function f is thought of in this way, it is often referred to as a *mapping*, or transformation. The *image* of a point z in the domain of definition S is the point $w = f(z)$, and the set of images of all points in a set T that is contained in S is called the image of T. The image of the entire domain of definition S is called the *range* of f. The *inverse image* of a point w is the set of all points z in the domain of definition of f that have w as their image. The inverse image of a point may contain just one point, many points, or none at all. The last case occurs, of course, when w is not in the range of f.

Terms such as *translation, rotation,* and *reflection* are used to convey dominant geometric characteristics of certain mappings. In such cases, it is sometimes convenient to consider the z and w planes to be the same. For example, the mapping

$$w = z + 1 = (x + 1) + iy,$$

where $z = x + iy$, can be thought of as a translation of each point z one unit to the right. Since $i = e^{i\pi/2}$, the mapping

$$w = iz = r \exp\left[i\left(\theta + \frac{\pi}{2}\right)\right],$$

where $z = re^{i\theta}$, rotates the radius vector for each nonzero point z through a right angle about the origin in the counterclockwise direction; and the mapping

$$w = \bar{z} = x - iy$$

transforms each point $z = x + iy$ into its reflection in the real axis.

More information is usually exhibited by sketching images of curves and regions than by simply indicating images of individual points. In the following examples, we illustrate this with the transformation $w = z^2$. Geometric interpretations of functions as mappings will be developed more extensively in Chap. 8.

EXAMPLE 1. According to the example in Sec. 9, the mapping $w = z^2$ can be thought of as the transformation

(1) $$u = x^2 - y^2, \qquad v = 2xy$$

from the xy plane to the uv plane. This form of the mapping is especially useful in finding the images of certain hyperbolas.

It is easy to show, for instance, that each branch of a hyperbola $x^2 - y^2 = c_1$ ($c_1 > 0$) is mapped in a one to one manner onto the vertical line $u = c_1$. We start by noting from the first of equations (1) that $u = c_1$ when (x, y) is a point lying on either branch. When, in particular, it lies on the right-hand branch, the second of equations (1) tells us that $v = 2y\sqrt{y^2 + c_1}$. Thus the image of the right-hand branch can be expressed parametrically as

$$u = c_1, \qquad v = 2y\sqrt{y^2 + c_1} \qquad (-\infty < y < \infty);$$

and it is evident that the image of a point (x, y) on that branch moves upward along the entire line as (x, y) traces out the branch in the upward direction (Fig. 14). Likewise, since the pair of equations

$$u = c_1 \qquad v = -2y\sqrt{y^2 + c_1} \qquad (-\infty < y < \infty)$$

furnishes a parametric representation for the image of the left-hand branch of the hyperbola, the image of a point going *downward* along the entire left-hand branch is seen to move up the entire line $u = c_1$.

It is left as an exercise to show that each branch of a hyperbola $2xy = c_2$ ($c_2 > 0$) is transformed into the line $v = c_2$, as indicated in Fig. 14. The cases in which c_1 and c_2 are negative are also treated in the exercises.

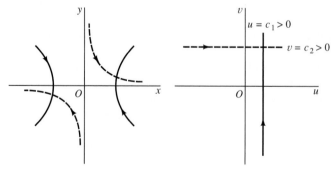

FIGURE 14
$w = z^2$.

EXAMPLE 2. Let us use equations (1) to show that the image of the vertical strip $0 \le x \le 1, y \ge 0$, shown in Fig. 15, is the closed semiparabolic region indicated there.

When $0 < x_1 < 1$, the point (x_1, y) moves up a vertical half line, labeled L_1 in Fig. 15, as y increases from $y = 0$. The image traced out in the uv plane has,

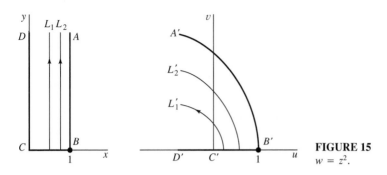

FIGURE 15
$w = z^2$.

according to equations (1), the parametric representation

(2) $$u = x_1^2 - y^2, \qquad v = 2x_1 y \qquad (0 \le y < \infty).$$

Using the second of these equations to substitute for y in the first one, we see that the image points (u, v) must lie on the parabola

(3) $$v^2 = -4x_1^2(u - x_1^2),$$

with vertex at $(x_1^2, 0)$ and focus at the origin. Since v increases with y from $v = 0$, according to the second of equations (2), we also see that as the point (x_1, y) moves up L_1 from the x axis, its image moves up the top half L_1' of the parabola from the u axis. Furthermore, when a number x_2 larger than x_1, but less than 1, is taken, the corresponding half line L_2 has an image L_2' that is a half parabola to the right of L_1', as indicated in Fig. 15. We note, in fact, that the image of the half line BA in that figure is the top half of the parabola $v^2 = -4(u - 1)$, labeled $B'A'$.

   The image of the half line CD is found by observing from equations (1) that a typical point $(0, y)$, where $y \ge 0$, on CD is transformed into the point $(-y^2, 0)$ in the uv plane. So, as a point moves up from the origin along CD, its image moves left from the origin along the u axis. Evidently, then, as the vertical half lines in the xy plane move to the left, the half parabolas that are their images in the uv plane shrink down to become the half line $C'D'$.

   It is now clear that the images of all the half lines between and including CD and BA fill up the closed semiparabolic region bounded by $A'B'C'D'$. Also, each point in that region is the image of only one point in the closed strip bounded by $ABCD$. Hence we may conclude that the semiparabolic region is the image of the strip and that there is a one to one correspondence between points in those closed regions. (Compare Fig. 3 in Appendix 2, where the strip has arbitrary width.)

   EXAMPLE 3.   We saw in the example in Sec. 9 that

$$w = z^2 = r^2 e^{i2\theta}$$

when $z = re^{i\theta}$. Hence if $w = \rho e^{i\phi}$, we have $\rho e^{i\phi} = r^2 e^{i2\theta}$; and the statement in italics near the end of Sec. 5 tells us that

$$\rho = r^2, \qquad \phi = 2\theta + 2n\pi \quad (n = 0, \pm 1, \pm 2, \dots).$$

Evidently, then, the image of any nonzero point z is found by squaring the modulus of z and doubling a value of arg z.

Observe that points $z = r_0 e^{i\theta}$ on a circle $r = r_0$ are transformed into points $w = r_0^2 e^{i2\theta}$ on the circle $\rho = r_0^2$. As a point on the first circle moves counterclockwise from the positive real axis to the positive imaginary axis, its image on the second circle moves counterclockwise from the positive real axis to the negative real axis (see Fig. 16). So, as all possible positive values of r_0 are chosen, the corresponding arcs in the z and w planes fill out the first quadrant and the upper half plane, respectively. The transformation $w = z^2$ is, then, a one to one mapping of the first quadrant $r \geq 0, 0 \leq \theta \leq \pi/2$ in the z plane onto the upper half $\rho \geq 0, 0 \leq \phi \leq \pi$ of the w plane, as indicated in Fig. 16. The point $z = 0$ is, of course, mapped onto the point $w = 0$.

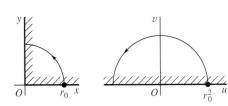

FIGURE 16
$w = z^2$.

The transformation $w = z^2$ also maps the upper half plane $r \geq 0, 0 \leq \theta \leq \pi$ onto the entire w plane. However, in this case, the transformation is not one to one since both the positive and negative real axes in the z plane are mapped onto the positive real axis in the w plane.

When n is a positive integer greater than 2, various mapping properties of the transformation $w = z^n$, or $\rho e^{i\phi} = r^n e^{in\theta}$, are similar to those of $w = z^2$. Such a transformation maps the entire z plane onto the entire w plane, where each nonzero point in the w plane is the image of n distinct points in the z plane. The circle $r = r_0$ is mapped onto the circle $\rho = r_0^n$; and the sector $r \leq r_0, 0 \leq \theta \leq 2\pi/n$ is mapped onto the disk $\rho \leq r_0^n$, but not in a one to one manner.

EXERCISES

1. For each of the functions below, describe the domain of definition that is understood:

(a) $f(z) = \dfrac{1}{z^2 + 1}$;   (b) $f(z) = \text{Arg}\left(\dfrac{1}{z}\right)$;

(c) $f(z) = \dfrac{z}{z + \bar{z}}$;   (d) $f(z) = \dfrac{1}{1 - |z|^2}$.

Ans. (a) $z \neq \pm i$;   (c) $\text{Re}\, z \neq 0$.

2. Write the function $f(z) = z^3 + z + 1$ in the form $f(z) = u(x, y) + iv(x, y)$.
Ans. $(x^3 - 3xy^2 + x + 1) + i(3x^2 y - y^3 + y)$.

3. Suppose that $f(z) = x^2 - y^2 - 2y + i(2x - 2xy)$, where $z = x + iy$. Use the fact (Sec. 3) that

$$x = \frac{z + \bar{z}}{2} \quad \text{and} \quad y = \frac{z - \bar{z}}{2i}$$

to express $f(z)$ in terms of z, and simplify the result.
Ans. $\bar{z}^2 + 2iz$.

4. Write the function

$$f(z) = z + \frac{1}{z} \qquad (z \neq 0)$$

in the form $f(z) = u(r, \theta) + iv(r, \theta)$.

Ans. $\left(r + \dfrac{1}{r}\right)\cos\theta + i\left(r - \dfrac{1}{r}\right)\sin\theta.$

5. Show that each branch of a hyperbola $2xy = c_2 (c_2 > 0)$ is mapped onto the line $v = c_2$ by the transformation $w = z^2$, as indicated in Fig. 14 (Sec. 10).

6. The domain $x > 0, y > 0, xy < 1$ consists of all points lying on the upper branches of hyperbolas from the family $xy = c$, where $0 < c < 1$. Use the result in Exercise 5 to show that the image of this domain under the transformation $w = z^2$ is the horizontal strip $0 < v < 2$.

7. By referring to Example 1, Sec. 10, and Exercise 5, find a domain in the z plane whose image under the transformation $w = z^2$ is the square domain in the w plane bounded by the lines $u = 1, u = 2, v = 1$, and $v = 2$. (See Fig. 2, Appendix 2.)

8. Find and sketch, showing corresponding orientations, the images of the hyperbolas $x^2 - y^2 = c_1 (c_1 < 0)$ and $2xy = c_2 (c_2 < 0)$ under the transformation $w = z^2$.

*9. Show, indicating corresponding orientations, that the mapping $w = z^2$ transforms lines $y = c_2 (c_2 > 0)$ into parabolas $v^2 = 4c_2^2(u + c_2^2)$, all with foci at $w = 0$. (Compare Example 2, Sec. 10.)

10. Use the result in Exercise 9 to show that the transformation $w = z^2$ is a one to one mapping of a strip $a \leq y \leq b$ above the x axis onto the closed region between the two parabolas $v^2 = 4a^2(u + a^2), v^2 = 4b^2(u + b^2)$.

*11. Point out how it follows from the discussion in Example 2, Sec. 10, that the transformation $w = z^2$ maps a vertical strip $0 \leq x \leq c, y \geq 0$ of arbitrary width onto a closed semiparabolic region, as shown in Fig. 3, Appendix 2.

12. Modify the discussion in Example 2, Sec. 10, to show that when $w = z^2$, the image of the closed triangular region formed by the lines $y = \pm x$ and $x = 1$ is the closed parabolic region bounded on the left by the segment $-2 \leq v \leq 2$ of the v axis and on the right by a portion of the parabola $v^2 = -4(u - 1)$. Verify the corresponding points on the two boundaries shown in Fig. 17.

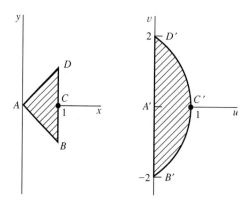

FIGURE 17
$w = z^2.$

13. Sketch the region onto which the sector $r \leq 1, 0 \leq \theta \leq \pi/4$ is mapped by the transformation (*a*) $w = z^2$; (*b*) $w = z^3$; (*c*) $w = z^4$.

*14. Another interpretation of a function $w = f(z) = u(x, y) + iv(x, y)$ is that of a *vector field* in the domain of definition of f. The function assigns a vector w, with components $u(x, y)$ and $v(x, y)$, to each point z at which it is defined. Indicate graphically the vector fields represented by (*a*) $w = iz$; (*b*) $w = z/|z|$.

11. LIMITS

Let a function f be defined at all points z in some deleted neighborhood (Sec. 8) of z_0. The statement that the *limit* of $f(z)$ as z approaches z_0 is a number w_0, or

$$(1) \qquad\qquad \lim_{z \to z_0} f(z) = w_0,$$

means that the point $w = f(z)$ can be made arbitrarily close to w_0 if we choose the point z close enough to z_0 but distinct from it. We now express the definition of limit in a precise and usable form.

  Statement (1) means that, for each positive number ε, there is a positive number δ such that

$$(2) \qquad\qquad |f(z) - w_0| < \varepsilon \qquad \text{whenever} \qquad 0 < |z - z_0| < \delta.$$

Geometrically, this definition says that, for each ε neighborhood $|w - w_0| < \varepsilon$ of w_0, there is a deleted δ neighborhood $0 < |z - z_0| < \delta$ of z_0 such that every point z in it has an image w lying in the ε neighborhood (Fig. 18). Note that even though all points in the deleted neighborhood $0 < |z - z_0| < \delta$ are to be considered, their images need not constitute the entire neighborhood $|w - w_0| < \varepsilon$. If f has the constant value w_0, for instance, the image of z is always the center of that neighborhood. Note, too, that once a δ has been found, it can be replaced by any smaller positive number, such as $\delta/2$.

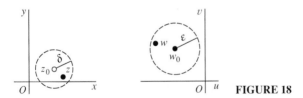

FIGURE 18

  Definition (2) requires that f be defined at all points in some deleted neighborhood of z_0. Such a deleted neighborhood, of course, always exists when z_0 is an interior point of a region on which f is defined. We can extend the definition of limit to the case in which z_0 is a boundary point of the region by agreeing that the first of inequalities (2) need be satisfied by only those points z that lie in *both* the region and the domain $0 < |z - z_0| < \delta$.

  EXAMPLE 1. Let us show that if $f(z) = iz/2$ in the open disk $|z| < 1$, then

$$(3) \qquad\qquad \lim_{z \to 1} f(z) = \frac{i}{2},$$

the point $z = 1$ being on the boundary of the domain of definition of f. Observe that when z is in the region $|z| < 1$,

$$\left| f(z) - \frac{i}{2} \right| = \left| \frac{iz}{2} - \frac{i}{2} \right| = \frac{|z - 1|}{2}.$$

Hence, for any such z and any positive number ε,

$$\left| f(z) - \frac{i}{2} \right| < \varepsilon \qquad \text{whenever} \qquad 0 < |z - 1| < 2\varepsilon.$$

Thus condition (2) is satisfied by points in the region $|z| < 1$ when δ is equal to 2ε (Fig. 19) or any smaller positive number.

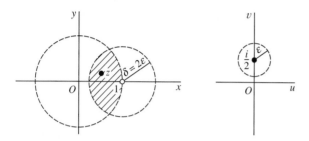

FIGURE 19

In introducing the concept of limit in the first paragraph of this section, we tacitly assumed that *when a limit of a function $f(z)$ exists at a point z_0, it is unique.* This is, indeed, the case. For suppose that

$$\lim_{z \to z_0} f(z) = w_0 \qquad \text{and} \qquad \lim_{z \to z_0} f(z) = w_1.$$

Then, for any positive number ε, there are positive numbers δ_0 and δ_1 such that

$$|f(z) - w_0| < \varepsilon \qquad \text{whenever} \qquad 0 < |z - z_0| < \delta_0$$

and

$$|f(z) - w_1| < \varepsilon \qquad \text{whenever} \qquad 0 < |z - z_0| < \delta_1.$$

So if $0 < |z - z_0| < \delta$, where δ denotes the smaller of the two numbers δ_0 and δ_1, then

$$|[f(z) - w_0] - [f(z) - w_1]| \le |f(z) - w_0| + |f(z) - w_1| < 2\varepsilon;$$

that is, $|w_1 - w_0| < 2\varepsilon$. But $|w_1 - w_0|$ is a nonnegative constant, and ε can be chosen arbitrarily small. Hence $w_1 - w_0 = 0$, or $w_1 = w_0$.

If z_0 is an interior point of the domain of definition of f, and limit (1) is to exist, the first of inequalities (2) must hold for *all* points in the deleted neighborhood $0 < |z - z_0| < \delta$. Thus the symbol $z \to z_0$ implies that z is allowed to approach z_0 in an arbitrary manner, not just from some particular direction. The next example emphasizes this.

EXAMPLE 2.   If $f(z) = z/\bar{z}$, then

(4) $$\lim_{z\to 0} f(z)$$

does not exist. For, if it did exist, it could be found by letting the point $z = (x, y)$ approach the origin in any manner. But when $z = (x, 0)$ is a nonzero point on the real axis,

$$f(z) = \frac{x + i0}{x - i0} = 1;$$

and when $z = (0, y)$ is a nonzero point on the imaginary axis,

$$f(z) = \frac{0 + iy}{0 - iy} = -1.$$

Thus, by letting z approach the origin along the real axis, we would find that the desired limit is 1. An approach along the imaginary axis would, on the other hand, yield the limit -1. Since a limit is unique, we must conclude that limit (4) does not exist.

While definition (2) provides a means of testing whether a given point w_0 is a limit, it does not directly provide a method for determining that limit. Theorems on limits, presented in the next section, will enable us to actually find many limits.

12.   THEOREMS ON LIMITS

We can expedite our treatment of limits by establishing a connection between limits of functions of a complex variable and limits of real-valued functions of two real variables. Since limits of the latter type are studied in calculus, we use their definition and properties freely.

Theorem 1. *Suppose that*

$$f(z) = u(x, y) + iv(x, y), \qquad z_0 = x_0 + iy_0, \qquad and \qquad w_0 = u_0 + iv_0.$$

Then

(1) $$\lim_{z\to z_0} f(z) = w_0$$

if and only if

(2) $$\lim_{(x,y)\to(x_0,y_0)} u(x, y) = u_0 \qquad and \qquad \lim_{(x,y)\to(x_0,y_0)} v(x, y) = v_0.$$

To prove the theorem, we first assume that limits (2) hold and obtain limit (1). Limits (2) tell us that, for each positive number ε, there exist positive numbers δ_1 and δ_2 such that

(3) $$|u - u_0| < \frac{\varepsilon}{2} \qquad \text{whenever} \qquad 0 < \sqrt{(x - x_0)^2 + (y - y_0)^2} < \delta_1$$

and

$$(4) \qquad |v - v_0| < \frac{\varepsilon}{2} \qquad \text{whenever} \qquad 0 < \sqrt{(x - x_0)^2 + (y - y_0)^2} < \delta_2.$$

Let δ denote the smaller of the two numbers δ_1 and δ_2. Since

$$|(u + iv) - (u_0 + iv_0)| = |(u - u_0) + i(v - v_0)| \leq |u - u_0| + |v - v_0|$$

and

$$\sqrt{(x - x_0)^2 + (y - y_0)^2} = |(x - x_0) + i(y - y_0)| = |(x + iy) - (x_0 + iy_0)|,$$

it follows from statements (3) and (4) that

$$|(u + iv) - (u_0 + iv_0)| < \frac{\varepsilon}{2} + \frac{\varepsilon}{2} = \varepsilon$$

whenever

$$0 < |(x + iy) - (x_0 + iy_0)| < \delta.$$

That is, limit (1) holds.

Let us now start with the assumption that limit (1) holds. With that assumption, we know that, for each positive number ε, there is a positive number δ such that

$$(5) \qquad\qquad\qquad\qquad |(u + iv) - (u_0 + iv_0)| < \varepsilon$$

whenever

$$(6) \qquad\qquad\qquad\qquad 0 < |(x + iy) - (x_0 + iy_0)| < \delta.$$

But

$$|u - u_0| \leq |(u - u_0) + i(v - v_0)| = |(u + iv) - (u_0 + iv_0)|,$$
$$|v - v_0| \leq |(u - u_0) + i(v - v_0)| = |(u + iv) - (u_0 + iv_0)|,$$

and

$$|(x + iy) - (x_0 + iy_0)| = |(x - x_0) + i(y - y_0)| = \sqrt{(x - x_0)^2 + (y - y_0)^2}.$$

Hence it follows from inequalities (5) and (6) that

$$|u - u_0| < \varepsilon \qquad \text{and} \qquad |v - v_0| < \varepsilon$$

whenever

$$0 < \sqrt{(x - x_0)^2 + (y - y_0)^2} < \delta.$$

This establishes limits (2), and the proof of the theorem is complete.

Theorem 2. *Suppose that*

$$(7) \qquad\qquad \lim_{z \to z_0} f(z) = w_0 \qquad and \qquad \lim_{z \to z_0} F(z) = W_0.$$

Then

(8)
$$\lim_{z \to z_0} [f(z) + F(z)] = w_0 + W_0,$$

(9)
$$\lim_{z \to z_0} [f(z)F(z)] = w_0 W_0;$$

and, if $W_0 \neq 0$,

(10)
$$\lim_{z \to z_0} \frac{f(z)}{F(z)} = \frac{w_0}{W_0}.$$

This important theorem can be proved directly by using the definition of the limit of a function of a complex variable. But, with the aid of Theorem 1, it follows almost immediately from theorems on limits of real-valued functions of two real variables.

To verify property (9), for example, we write

$$f(z) = u(x, y) + iv(x, y), \qquad F(z) = U(x, y) + iV(x, y),$$

$$z_0 = x_0 + iy_0, \qquad w_0 = u_0 + iv_0, \qquad W_0 = U_0 + iV_0.$$

Then, according to hypotheses (7) and Theorem 1, the limits as (x, y) approaches (x_0, y_0) of the functions u, v, U, and V exist and have the values u_0, v_0, U_0, and V_0, respectively. So the real and imaginary components of the product

$$f(z)F(z) = (uU - vV) + i(vU + uV)$$

have the limits $u_0 U_0 - v_0 V_0$ and $v_0 U_0 + u_0 V_0$, respectively, as (x, y) approaches (x_0, y_0). Hence, by Theorem 1 again, $f(z)F(z)$ has the limit

$$(u_0 U_0 - v_0 V_0) + i(v_0 U_0 + u_0 V_0)$$

as z approaches z_0; and this is equal to $w_0 W_0$. Property (9) is thus established. Corresponding verifications of properties (8) and (10) can be given.

An immediate consequence of Theorem 1 is the fact that

$$\lim_{z \to z_0} c = c$$

for any complex constant $c = a + bi$ and any z_0. Also,

$$\lim_{z \to z_0} z = z_0;$$

and, by property (9) and mathematical induction, it follows that

$$\lim_{z \to z_0} z^n = z_0^n \qquad (n = 1, 2, \ldots).$$

So, in view of properties (8) and (9), the limit of a polynomial

$$P(z) = a_0 + a_1 z + a_2 z^2 + \cdots + a_n z^n$$

as z approaches a point z_0 is the value of the polynomial at that point:

(11)
$$\lim_{z \to z_0} P(z) = P(z_0).$$

Another useful property of limits is that

(12)   if $\lim_{z \to z_0} f(z) = w_0$,   then $\lim_{z \to z_0} |f(z)| = |w_0|$.

This is easily proved by using the definition of limit and the fact that (see Sec. 4)

$$\big||f(z)| - |w_0|\big| \le |f(z) - w_0|.$$

13. LIMITS INVOLVING THE POINT AT INFINITY

It is sometimes convenient to include with the complex plane the *point at infinity*, denoted by ∞, and to use limits involving it. The complex plane together with this point is called the *extended complex plane*. To visualize the point at infinity, one can think of the complex plane as passing through the equator of a unit sphere centered at the point $z = 0$ (Fig. 20). To each point z in the plane there corresponds exactly one point P on the surface of the sphere. The point P is determined by the intersection of the line through the point z and the north pole N of the sphere with that surface. In like manner, to each point P on the surface of the sphere, other than the north pole N, there corresponds exactly one point z in the plane. By letting the point N of the sphere correspond to the point at infinity, we obtain a one to one correspondence between the points of the sphere and the points of the extended complex plane. The sphere is known as the *Riemann sphere*, and the correspondence is called a *stereographic projection*.

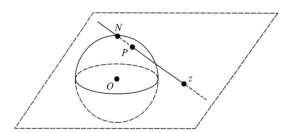

FIGURE 20

Observe that the exterior of the unit circle centered at the origin in the complex plane corresponds to the upper hemisphere with the equator and the point N deleted. Moreover, for each small positive number ε, those points in the complex plane exterior to the circle $|z| = 1/\varepsilon$ correspond to points on the sphere close to N. We thus call the set $|z| > 1/\varepsilon$ an ε *neighborhood*, or neighborhood, of ∞.

Let us agree that, in referring to a point z, we mean a point in the *finite* plane. Hereafter, when the point at infinity is to be considered, it will be specifically mentioned.

A meaning is now readily given to the statement

$$\lim_{z \to z_0} f(z) = w_0$$

when either z_0 or w_0, or possibly each of these numbers, is replaced by the point at infinity. In the definition of limit in Sec. 11, we simply replace the appropriate neighborhoods of z_0 and w_0 by neighborhoods of ∞.

The statement

$$\lim_{z \to z_0} f(z) = \infty,$$

for instance, means that, for each positive number ε, there is a positive number δ such that

$$(1) \qquad\qquad |f(z)| > \frac{1}{\varepsilon} \qquad \text{whenever} \qquad 0 < |z - z_0| < \delta.$$

That is, the point $w = f(z)$ lies in the ε neighborhood $|w| > 1/\varepsilon$ of ∞ whenever z lies in the deleted neighborhood $0 < |z - z_0| < \delta$ of z_0. Since statement (1) can be written

$$\left| \frac{1}{f(z)} - 0 \right| < \varepsilon \qquad \text{whenever} \qquad 0 < |z - z_0| < \delta,$$

we see that

$$(2) \qquad\qquad \lim_{z \to z_0} f(z) = \infty \qquad \text{if and only if} \qquad \lim_{z \to z_0} \frac{1}{f(z)} = 0.$$

EXAMPLE 1. Observe that

$$\lim_{z \to -1} \frac{iz + 3}{z + 1} = \infty \qquad \text{since} \qquad \lim_{z \to -1} \frac{z + 1}{iz + 3} = 0.$$

To say that

$$\lim_{z \to \infty} f(z) = w_0$$

means that, for each positive number ε, a positive number δ exists such that

$$(3) \qquad\qquad |f(z) - w_0| < \varepsilon \qquad \text{whenever} \qquad |z| > \frac{1}{\delta}.$$

Replacing z by $1/z$ in statement (3) and then writing the result as

$$\left| f\left(\frac{1}{z}\right) - w_0 \right| < \varepsilon \qquad \text{whenever} \qquad 0 < |z - 0| < \delta,$$

we find that

$$(4) \qquad\qquad \lim_{z \to \infty} f(z) = w_0 \qquad \text{if and only if} \qquad \lim_{z \to 0} f\left(\frac{1}{z}\right) = w_0.$$

EXAMPLE 2. According to statement (4),

$$\lim_{z \to \infty} \frac{2z + i}{z + 1} = 2 \qquad \text{since} \qquad \lim_{z \to 0} \frac{(2/z) + i}{(1/z) + 1} = \lim_{z \to 0} \frac{2 + iz}{1 + z} = 2.$$

Finally, the statement

$$\lim_{z \to \infty} f(z) = \infty$$

is to be interpreted as saying that, for each positive number ε, there is a positive number δ such that

(5) $$|f(z)| > \frac{1}{\varepsilon} \quad \text{whenever} \quad |z| > \frac{1}{\delta}.$$

When z is replaced by $1/z$, this statement can be put in the form

$$\left| \frac{1}{f(1/z)} - 0 \right| < \varepsilon \quad \text{whenever} \quad 0 < |z - 0| < \delta;$$

and so

(6) $$\lim_{z \to \infty} f(z) = \infty \quad \text{if and only if} \quad \lim_{z \to 0} \frac{1}{f(1/z)} = 0.$$

EXAMPLE 3. Evidently,

$$\lim_{z \to \infty} \frac{2z^3 - 1}{z^2 + 1} = \infty \quad \text{since} \quad \lim_{z \to 0} \frac{(1/z^2) + 1}{(2/z^3) - 1} = \lim_{z \to 0} \frac{z + z^3}{2 - z^3} = 0.$$

14. CONTINUITY

A function f is *continuous* at a point z_0 if all three of the following conditions are satisfied:

(1) $$\lim_{z \to z_0} f(z) \text{ exists,}$$

(2) $$f(z_0) \text{ exists,}$$

(3) $$\lim_{z \to z_0} f(z) = f(z_0).$$

Observe that statement (3) actually contains statements (1) and (2), since the existence of the quantity on each side of the equation there is implicit. Statement (3) says that, for each positive number ε, there is a positive number δ such that

(4) $$|f(z) - f(z_0)| < \varepsilon \quad \text{whenever} \quad |z - z_0| < \delta.$$

A function of a complex variable is said to be continuous in a region R if it is continuous at each point in R.

If two functions are continuous at a point, their sum and product are also continuous at that point; their quotient is continuous at any such point where the denominator is not zero. These observations are immediate consequences of Theorem 2, Sec. 12. Note, too, that a polynomial is continuous in the entire plane because of equation (11), Sec. 12.

It follows directly from definition (4) that *a composition of continuous functions is continuous*. To show this, we let $w = f(z)$ be a function defined for all z in a neighborhood of a point z_0; and we let $g(w)$ be a function whose domain of definition contains the image (Sec. 10) of that neighborhood. The composition $g[f(z)]$ is, then, defined for all z in the neighborhood of z_0. Suppose now that f is continuous at z_0 and that g is continuous at the point $w_0 = f(z_0)$. In view of the continuity of g at w_0, we know that, for each positive number ε, there is a positive number γ such that

$$|g[f(z)] - g[f(z_0)]| < \varepsilon \quad \text{whenever} \quad |f(z) - f(z_0)| < \gamma.$$

But, corresponding to γ, there exists a positive number δ such that the second of these inequalities is satisfied whenever $|z - z_0| < \delta$. The continuity of the composition $g[f(z)]$ at z_0 is thus established.

It is also easy to see from definition (4) that *if a function $f(z)$ is continuous and nonzero at a point z_0, then $f(z) \neq 0$ throughout some neighborhood of that point*. For, when $f(z_0) \neq 0$ and the positive number ε in the first of inequalities (4) is assigned the value $\varepsilon = |f(z_0)|/2$, we know that there is a positive number δ such that

$$|f(z) - f(z_0)| < \frac{|f(z_0)|}{2} \quad \text{whenever} \quad |z - z_0| < \delta.$$

So if there is a point z in the neighborhood $|z - z_0| < \delta$ at which $f(z) = 0$, we have the contradiction $|f(z_0)| < |f(z_0)|/2$.

From Theorem 1, Sec. 12, it follows that a function f of a complex variable is continuous at a point $z_0 = (x_0, y_0)$ if and only if its component functions u and v are continuous there.

EXAMPLE. The function

$$f(z) = \cos(x^2 - y^2)\cosh 2xy - i\sin(x^2 - y^2)\sinh 2xy$$

is continuous everywhere in the complex plane since its real and imaginary components are continuous at each point (x, y). The continuity of the component functions is a consequence of the continuity of polynomials in x and y as well as the continuity of the trigonometric and hyperbolic functions appearing.

Various properties of continuous functions of a complex variable can be deduced from corresponding properties of continuous real-valued functions of two real variables.* Suppose, for example, that a function $f(z) = u(x, y) + iv(x, y)$ is continuous in a region R that is both closed and bounded. The function

$$\sqrt{[u(x, y)]^2 + [v(x, y)]^2}$$

is then continuous in R and thus reaches a maximum value somewhere in that region. That is, f is *bounded* on R and $|f(z)|$ reaches a maximum value somewhere in R. More

---

*For such properties quoted here, see, for instance, A. E. Taylor and W. R. Mann, "Advanced Calculus," 3d ed., pp. 125–126 and 529–532, 1983.

precisely, there exists a nonnegative real number M such that

(5)
$$|f(z)| \leq M \qquad \text{for all } z \text{ in } R,$$

where equality holds for at least one such z.

   Another result which follows from the corresponding one for real-valued functions of two real variables is that a function f which is continuous in a closed bounded region R is *uniformly continuous* there. That is, a single value of δ, independent of z_0, may be chosen such that condition (4) is satisfied at each point z_0 in R.

EXERCISES

1. Let a, b, c, and z_0 denote complex constants. Use definition (2), Sec. 11, of limit to prove that

   (a) $\lim\limits_{z \to z_0} c = c$;

   (b) $\lim\limits_{z \to z_0} (az + b) = az_0 + b$ $(a \neq 0)$;

   (c) $\lim\limits_{z \to z_0} (z^2 + c) = z_0^2 + c$;

   (d) $\lim\limits_{z \to z_0} \text{Re} z = \text{Re} z_0$;

   (e) $\lim\limits_{z \to z_0} \bar{z} = \bar{z}_0$;

   (f) $\lim\limits_{z \to 1-i} [x + i(2x + y)] = 1 + i$ $(z = x + iy)$;

   (g) $\lim\limits_{z \to 0} (\bar{z}^2/z) = 0$.

2. Let n be a positive integer and let $P(z)$ and $Q(z)$ be polynomials, where $Q(z_0) \neq 0$. Use Theorem 2, Sec. 12, and limits appearing in that section to find

   (a) $\lim\limits_{z \to z_0} \dfrac{1}{z^n}$ $(z_0 \neq 0)$;   (b) $\lim\limits_{z \to i} \dfrac{iz^3 - 1}{z + i}$;   (c) $\lim\limits_{z \to z_0} \dfrac{P(z)}{Q(z)}$.

   Ans. (a) $1/z_0^n$;   (b) 0;   (c) $P(z_0)/Q(z_0)$.

3. Use property (9), Sec. 12, of limits and mathematical induction to show that

$$\lim\limits_{z \to z_0} z^n = z_0^n$$

   when n is a positive integer.

4. Show that the limit of the function $f(z) = (z/\bar{z})^2$ as z tends to 0 does not exist. Do this by letting nonzero points $z = (x, 0)$ and $z = (x, x)$ approach the origin. [Note that it is not sufficient to simply consider points $z = (x, 0)$ and $z = (0, y)$, as it was in Example 2, Sec. 11.]

5. Prove statement (8) in Theorem 2, Sec. 12, using (a) Theorem 1, Sec. 12, and properties of limits of real-valued functions of two real variables; (b) definition (2), Sec. 11, of limit.

6. Write $\Delta z = z - z_0$ and show that

$$\lim\limits_{z \to z_0} f(z) = w_0 \qquad \text{if and only if} \qquad \lim\limits_{\Delta z \to 0} f(z_0 + \Delta z) = w_0.$$

7. Show that

$$\lim\limits_{z \to z_0} f(z)g(z) = 0 \qquad \text{if} \qquad \lim\limits_{z \to z_0} f(z) = 0$$

   and if there exists a positive number M such that $|g(z)| \leq M$ for all z in some neighborhood of z_0.

8. Prove property (12), Sec. 12, of limits.

9. By using properties (2), (4), and (6) of limits in Sec. 13, show that

(*a*) $\lim\limits_{z \to \infty} \dfrac{4z^2}{(z - 1)^2} = 4;$ (*b*) $\lim\limits_{z \to 1} \dfrac{1}{(z - 1)^3} = \infty;$ (*c*) $\lim\limits_{z \to \infty} \dfrac{z^2 + 1}{z - 1} = \infty.$

10. Use properties (2), (4), and (6) of limits in Sec. 13 to show that when

$$T(z) = \frac{az + b}{cz + d} \qquad (ad - bc \neq 0),$$

(*a*) $\lim\limits_{z \to \infty} T(z) = \infty$ if $c = 0$;

(*b*) $\lim\limits_{z \to \infty} T(z) = \dfrac{a}{c}$ and $\lim\limits_{z \to -d/c} T(z) = \infty$ if $c \neq 0$.

11. Use definitions (1) and (3), Sec. 13, of limits involving the point at infinity to show that

$$\lim_{z \to 0} \frac{1}{z} = \infty \qquad \text{and} \qquad \lim_{z \to \infty} \frac{1}{z} = 0.$$

12. Consider the function f defined on the extended plane by means of the equations

$$f(z) = \begin{cases} 1/z & \text{when} \quad z \neq 0, \\ \infty & \text{when} \quad z = 0, \\ 0 & \text{when} \quad z = \infty. \end{cases}$$

By allowing the numbers z_0 and w_0 in definition (3), Sec. 14, of continuity to be the point at infinity and by referring to the limits in Exercise 11, state why f is continuous everywhere in the extended plane.

13. State why limits involving the point at infinity are unique.

14. Show that a set S is unbounded (Sec. 8) if and only if every neighborhood of the point at infinity contains at least one point in S.

15.   DERIVATIVES

Let f be a function whose domain of definition contains a neighborhood of a point z_0. The *derivative* of f at z_0, written $f'(z_0)$, is defined by the equation

(1) $$f'(z_0) = \lim_{z \to z_0} \frac{f(z) - f(z_0)}{z - z_0},$$

provided this limit exists. The function f is said to be *differentiable* at z_0 when its derivative at z_0 exists.

By expressing the variable z in definition (1) in terms of the new complex variable

$$\Delta z = z - z_0,$$

we can write that definition as

(2) $$f'(z_0) = \lim_{\Delta z \to 0} \frac{f(z_0 + \Delta z) - f(z_0)}{\Delta z}.$$

Note that, because f is defined throughout a neighborhood of z_0, the number

$$f(z_0 + \Delta z)$$

is always defined for $|\Delta z|$ sufficiently small (Fig. 21).

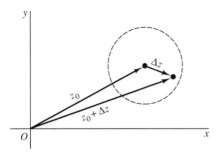

FIGURE 21

When taking form (2) of the definition of derivative, we often drop the subscript on z_0 and introduce the number

$$\Delta w = f(z + \Delta z) - f(z),$$

which denotes the change in the value of f corresponding to a change Δz in the point at which f is evaluated. Then, if we write dw/dz for $f'(z)$, equation (2) becomes

(3)
$$\frac{dw}{dz} = \lim_{\Delta z \to 0} \frac{\Delta w}{\Delta z}.$$

EXAMPLE 1. Suppose that $f(z) = z^2$. At any point z,

$$\lim_{\Delta z \to 0} \frac{\Delta w}{\Delta z} = \lim_{\Delta z \to 0} \frac{(z + \Delta z)^2 - z^2}{\Delta z} = \lim_{\Delta z \to 0} (2z + \Delta z) = 2z,$$

since $2z + \Delta z$ is a polynomial in Δz. Hence $dw/dz = 2z$, or $f'(z) = 2z$.

EXAMPLE 2. Consider now the function $f(z) = |z|^2$. Here

$$\frac{\Delta w}{\Delta z} = \frac{|z + \Delta z|^2 - |z|^2}{\Delta z} = \frac{(z + \Delta z)(\bar{z} + \overline{\Delta z}) - z\bar{z}}{\Delta z} = \bar{z} + \overline{\Delta z} + z\frac{\overline{\Delta z}}{\Delta z}.$$

If the limit of $\Delta w/\Delta z$ exists, it may be found by letting the point $\Delta z = (\Delta x, \Delta y)$ approach the origin in the Δz plane in any manner. In particular, when Δz approaches the origin horizontally through the points $(\Delta x, 0)$ on the real axis (Fig. 22), we may write $\overline{\Delta z} = \Delta z$. Hence if the limit of $\Delta w/\Delta z$ exists, its value must be $\bar{z} + z$. However, when Δz approaches the origin vertically through the points $(0, \Delta y)$ on the imaginary axis, so that $\overline{\Delta z} = -\Delta z$, we find that the limit must be $\bar{z} - z$ if it exists. Since limits are unique, it follows that $\bar{z} + z = \bar{z} - z$, or $z = 0$, if dw/dz is to exist.

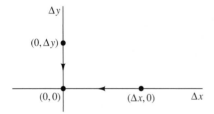

FIGURE 22

To show that dw/dz does, in fact, exist at $z = 0$, we need only observe that our expression for $\Delta w/\Delta z$ reduces to $\overline{\Delta z}$ when $z = 0$. We conclude, therefore, that dw/dz exists *only* at that point $z = 0$, its value there being 0.

Example 2 shows that a function can be differentiable at a certain point but nowhere else in any neighborhood of that point. Since the real and imaginary parts of $f(z) = |z|^2$ are

(4) $$u(x, y) = x^2 + y^2 \quad \text{and} \quad v(x, y) = 0,$$

respectively, it shows that the real and imaginary components of a function of a complex variable can have continuous partial derivatives of all orders at a point and yet the function may not be differentiable there.

The function $f(z) = |z|^2$ is continuous at each point in the plane since its components (4) are continuous at each point. So the continuity of a function at a point does not imply the existence of a derivative there. It is, however, true that *the existence of the derivative of a function at a point implies the continuity of the function at that point*. To see this, we assume that $f'(z_0)$ exists and write

$$\lim_{z \to z_0}[f(z) - f(z_0)] = \lim_{z \to z_0}\frac{f(z) - f(z_0)}{z - z_0} \lim_{z \to z_0}(z - z_0) = f'(z_0) \cdot 0 = 0,$$

from which it follows that

$$\lim_{z \to z_0} f(z) = f(z_0).$$

This is the statement of continuity of f at z_0 (Sec. 14).

Geometric interpretations of derivatives of functions of a complex variable are not as immediate as they are for derivatives of functions of a real variable. We defer the development of such interpretations until Chap. 9.

16. DIFFERENTIATION FORMULAS

The definition of derivative in Sec. 15 is identical in form to that of the derivative of a real-valued function of a real variable. In fact, the basic differentiation formulas given below can be derived from that definition, together with various theorems on limits, by essentially the same steps as the ones used in calculus. In these formulas, the derivative of a function f at a point z is denoted by either

$$\frac{d}{dz}f(z) \quad \text{or} \quad f'(z),$$

depending on which notation is more convenient.

Let c be a complex constant, and let f be a function whose derivative exists at a point z. It is easy to show that

(1) $$\frac{d}{dz}c = 0, \quad \frac{d}{dz}z = 1, \quad \frac{d}{dz}[cf(z)] = cf'(z).$$

Also, if n is a positive integer,

$$(2) \qquad \frac{d}{dz}z^n = nz^{n-1}.$$

This formula remains valid when n is a negative integer, provided $z \neq 0$.

If the derivatives of two functions f and F exist at a point z, then

$$(3) \qquad \frac{d}{dz}[f(z) + F(z)] = f'(z) + F'(z),$$

$$(4) \qquad \frac{d}{dz}[f(z)F(z)] = f(z)F'(z) + f'(z)F(z);$$

and, when $F(z) \neq 0$,

$$(5) \qquad \frac{d}{dz}\left[\frac{f(z)}{F(z)}\right] = \frac{F(z)f'(z) - f(z)F'(z)}{[F(z)]^2}.$$

Let us derive formula (4). To do this, we write the following expression for the change in the product $f(z)F(z)$:

$$f(z + \Delta z)F(z + \Delta z) - f(z)F(z) = f(z)[F(z + \Delta z) - F(z)]$$
$$+ [f(z + \Delta z) - f(z)]F(z + \Delta z).$$

If we divide both sides of this equation by Δz and then let Δz tend to zero, we arrive at the desired formula for the derivative of $f(z)F(z)$. Here we have used the fact that F is continuous at the point z, since $F'(z)$ exists; thus $F(z + \Delta z)$ tends to $F(z)$ as Δz tends to zero (see Exercise 6, Sec. 14).

There is also a chain rule for differentiating composite functions. Suppose that f has a derivative at z_0 and that g has a derivative at the point $f(z_0)$. Then the function $F(z) = g[f(z)]$ has a derivative at z_0, and

$$(6) \qquad F'(z_0) = g'[f(z_0)]f'(z_0).$$

If we write $w = f(z)$ and $W = g(w)$, so that $W = F(z)$, the chain rule becomes

$$\frac{dW}{dz} = \frac{dW}{dw}\frac{dw}{dz}.$$

EXAMPLE.   To find the derivative of $(2z^2 + i)^5$, write $w = 2z^2 + i$ and $W = w^5$. Then

$$\frac{d}{dz}(2z^2 + i)^5 = 5w^4 4z = 20z(2z^2 + i)^4.$$

To start the proof of formula (6), choose a specific point z_0 at which $f'(z_0)$ exists. Write $w_0 = f(z_0)$ and also assume that $g'(w_0)$ exists. There is, then, some ε neighborhood $|w - w_0| < \varepsilon$ of w_0 such that, for all points w in that neighborhood, we can define a function Φ which has the values $\Phi(w_0) = 0$ and

$$(7) \qquad \Phi(w) = \frac{g(w) - g(w_0)}{w - w_0} - g'(w_0) \qquad \text{when} \qquad w \neq w_0.$$

Note that, in view of the definition of derivative,

(8)
$$\lim_{w \to w_0} \Phi(w) = 0.$$

Hence Φ is continuous at w_0.

Now expression (7) can be put in the form

(9) $g(w) - g(w_0) = [g'(w_0) + \Phi(w)](w - w_0)$ $(|w - w_0| < \varepsilon),$

which is valid even when $w = w_0$; and, since $f'(z_0)$ exists and f is, therefore, continuous at z_0, we can choose a positive number δ such that the point $f(z)$ lies in the ε neighborhood $|w - w_0| < \varepsilon$ of w_0 if z lies in the δ neighborhood $|z - z_0| < \delta$ of z_0. Thus it is legitimate to replace the variable w in equation (9) by $f(z)$ when z is any point in the neighborhood $|z - z_0| < \delta$. With that substitution, as well as the one $w_0 = f(z_0)$, equation (9) becomes

(10)
$$\frac{g[f(z)] - g[f(z_0)]}{z - z_0} = \{g'[f(z_0)] + \Phi[f(z)]\}\frac{f(z) - f(z_0)}{z - z_0}$$
$$(0 < |z - z_0| < \delta),$$

where we must stipulate that $z \neq z_0$ so that we are not dividing by zero. As already noted, f is continuous at z_0 and Φ is continuous at the point $w_0 = f(z_0)$. Thus the composition $\Phi[f(z)]$ is continuous at z_0; and, since $\Phi(w_0) = 0$,

$$\lim_{z \to z_0} \Phi[f(z)] = 0.$$

So equation (10) becomes

$$F'(z_0) = g'[f(z_0)]f'(z_0)$$

in the limit as z approaches z_0.

EXERCISES

1. Use results in Sec. 16 to find $f'(z)$ when
  (a) $f(z) = 3z^2 - 2z + 4;$ (b) $f(z) = (1 - 4z^2)^3;$
  (c) $f(z) = \dfrac{z - 1}{2z + 1} \left(z \neq -\dfrac{1}{2}\right);$ (d) $f(z) = \dfrac{(1 + z^2)^4}{z^2}$ $(z \neq 0).$

2. Using results in Sec. 16, show that
  (a) a polynomial

$$P(z) = a_0 + a_1 z + a_2 z^2 + \cdots + a_n z^n   (a_n \neq 0)$$

  of degree n $(n \geq 1)$ is differentiable everywhere, with derivative

$$P'(z) = a_1 + 2a_2 z + \cdots + n a_n z^{n-1};$$

  (b) the coefficients in the polynomial $P(z)$ in part (a) can be written

$$a_0 = P(0), \qquad a_1 = \frac{P'(0)}{1!}, \qquad a_2 = \frac{P''(0)}{2!}, \qquad \ldots, \qquad a_n = \frac{P^{(n)}(0)}{n!}.$$

3. Apply definition (3), Sec. 15, of derivative to give a direct proof that $f'(z) = -1/z^2$ when $f(z) = 1/z \ (z \neq 0)$.

4. Suppose that $f(z_0) = g(z_0) = 0$ and that $f'(z_0)$ and $g'(z_0)$ exist, where $g'(z_0) \neq 0$. Use definition (1), Sec. 15, of derivative to show that

$$\lim_{z \to z_0} \frac{f(z)}{g(z)} = \frac{f'(z_0)}{g'(z_0)}.$$

5. Derive formula (3), Sec. 16, for the derivative of the sum of two functions.

6. Derive expression (2), Sec. 16, for the derivative of z^n when n is a positive integer by using
   (a) mathematical induction and formula (4), Sec. 16, for the derivative of the product of two functions;
   (b) definition (3), Sec. 15, of derivative and the binomial formula (Exercise 14, Sec. 6).

7. Prove that expression (2), Sec. 16, for the derivative of z^n remains valid when n is a negative integer ($n = -1, -2, \ldots$), provided $z \neq 0$.
   Suggestion: Write $m = -n$ and use the formula for the derivative of a quotient of two functions.

8. Use the method in Example 2, Sec. 15, to show that $f'(z)$ does not exist at any point z when
   (a) $f(z) = \bar{z}$;   (b) $f(z) = \operatorname{Re} z$;   (c) $f(z) = \operatorname{Im} z$.

9. Let f denote the function whose values are

$$f(z) = \begin{cases} \dfrac{(\bar{z})^2}{z} & \text{when} \quad z \neq 0, \\ 0 & \text{when} \quad z = 0. \end{cases}$$

Show that if $z = 0$, then $\Delta w/\Delta z = 1$ at each nonzero point on the real and imaginary axes in the Δz, or $(\Delta x, \Delta y)$, plane. Then show that $\Delta w/\Delta z = -1$ at each nonzero point $(\Delta x, \Delta x)$ on the line $\Delta y = \Delta x$ in that plane. Conclude from these observations that $f'(0)$ does not exist. (Note that, to obtain this result, it is not sufficient to consider only horizontal and vertical approaches to the origin in the Δz plane.)

17.   CAUCHY-RIEMANN EQUATIONS

In this section, we obtain a pair of equations that the first-order partial derivatives of the component functions u and v of a function

$$(1) \qquad\qquad f(z) = u(x, y) + iv(x, y)$$

must satisfy at a point $z_0 = (x_0, y_0)$ when the derivative of f exists there. We also show how to write $f'(z_0)$ in terms of those partial derivatives.

Suppose that the derivative

$$(2) \qquad\qquad f'(z_0) = \lim_{\Delta z \to 0} \frac{f(z_0 + \Delta z) - f(z_0)}{\Delta z}$$

exists. Writing $z_0 = x_0 + iy_0$ and $\Delta z = \Delta x + i\Delta y$, we then have, by Theorem 1 in Sec. 12, the expressions

$$(3) \qquad \mathrm{Re}[f'(z_0)] = \lim_{(\Delta x, \Delta y) \to (0,0)} \mathrm{Re}\left[\frac{f(z_0 + \Delta z) - f(z_0)}{\Delta z}\right],$$

$$(4) \qquad \mathrm{Im}[f'(z_0)] = \lim_{(\Delta x, \Delta y) \to (0,0)} \mathrm{Im}\left[\frac{f(z_0 + \Delta z) - f(z_0)}{\Delta z}\right],$$

where

$$(5) \qquad \frac{f(z_0 + \Delta z) - f(z_0)}{\Delta z} = \frac{u(x_0 + \Delta x, y_0 + \Delta y) - u(x_0, y_0)}{\Delta x + i\Delta y}$$
$$+ \frac{i[v(x_0 + \Delta x, y_0 + \Delta y) - v(x_0, y_0)]}{\Delta x + i\Delta y}.$$

It is important now to keep in mind that expressions (3) and (4) are valid as $(\Delta x, \Delta y)$ tends to $(0,0)$ in any manner that we may choose.

In particular, let $(\Delta x, \Delta y)$ tend to $(0,0)$ horizontally through the points $(\Delta x, 0)$, as indicated in Fig. 22 (Sec. 15). This means that $\Delta y = 0$ in equation (5), and we find that

$$\mathrm{Re}[f'(z_0)] = \lim_{\Delta x \to 0} \frac{u(x_0 + \Delta x, y_0) - u(x_0, y_0)}{\Delta x},$$

$$\mathrm{Im}[f'(z_0)] = \lim_{\Delta x \to 0} \frac{v(x_0 + \Delta x, y_0) - v(x_0, y_0)}{\Delta x}.$$

That is,

$$(6) \qquad f'(z_0) = u_x(x_0, y_0) + iv_x(x_0, y_0),$$

where $u_x(x_0, y_0)$ and $v_x(x_0, y_0)$ denote the first-order partial derivatives with respect to x of the functions u and v at (x_0, y_0).

We might have let $(\Delta x, \Delta y)$ tend to zero vertically through the points $(0, \Delta y)$. In that case, $\Delta x = 0$ in equation (5); and we obtain the expression

$$(7) \qquad f'(z_0) = v_y(x_0, y_0) - iu_y(x_0, y_0)$$

for $f'(z_0)$, this time in terms of the first-order partial derivatives of u and v with respect to y. Note that equation (7) can also be written

$$f'(z_0) = -i[u_y(x_0, y_0) + iv_y(x_0, y_0)].$$

Equations (6) and (7) not only give $f'(z_0)$ in terms of partial derivatives of the component functions u and v, but they also provide necessary conditions for the existence of $f'(z_0)$. For, on equating the real and imaginary parts on the right-hand sides of these equations, we see that the existence of $f'(z_0)$ requires that

$$(8) \qquad u_x(x_0, y_0) = v_y(x_0, y_0) \qquad \text{and} \qquad u_y(x_0, y_0) = -v_x(x_0, y_0).$$

Equations (8) are the *Cauchy-Riemann equations*, so named in honor of the French mathematician A. L. Cauchy (1789–1857), who discovered and used them, and in honor of the German mathematician G. F. B. Riemann (1826–1866), who made them fundamental in his development of the theory of functions of a complex variable.

We summarize the above results as follows.

Theorem. *Suppose that*

$$f(z) = u(x, y) + iv(x, y)$$

and that $f'(z)$ exists at a point $z_0 = x_0 + iy_0$. Then the first-order partial derivatives of u and v must exist at (x_0, y_0), and they must satisfy the Cauchy-Riemann equations

$$(9) \qquad\qquad u_x = v_y, \qquad u_y = -v_x$$

there. Also, $f'(z_0)$ can be written

$$(10) \qquad\qquad f'(z_0) = u_x + iv_x,$$

where these partial derivatives are to be evaluated at (x_0, y_0).

EXAMPLE 1.   In Example 1, Sec. 15, we showed that the function

$$f(z) = z^2 = x^2 - y^2 + i2xy$$

is differentiable everywhere and that $f'(z) = 2z$. To verify that the Cauchy-Riemann equations are satisfied everywhere, we note that $u(x, y) = x^2 - y^2$ and $v(x, y) = 2xy$. Thus

$$u_x = 2x = v_y, \qquad u_y = -2y = -v_x.$$

Moreover, according to equation (10),

$$f'(z) = 2x + i2y = 2(x + iy) = 2z.$$

Since the Cauchy-Riemann equations are necessary conditions for the existence of the derivative of a function f at a point z_0, they can often be used to locate points at which f does *not* have a derivative.

EXAMPLE 2.   When $f(z) = |z|^2$, we have $u(x, y) = x^2 + y^2$ and $v(x, y) = 0$. If the Cauchy-Riemann equations are to hold at a point (x, y), it follows that $2x = 0$ and $2y = 0$, or that $x = y = 0$. Consequently, $f'(z)$ does not exist at any nonzero point, as we already know from Example 2 in Sec. 15. Note that the above theorem does not ensure the existence of $f'(0)$. The theorem in the next section will, however, do this.

18.   SUFFICIENT CONDITIONS FOR DIFFERENTIABILITY

Satisfaction of the Cauchy-Riemann equations at a point $z_0 = (x_0, y_0)$ is not sufficient to ensure the existence of the derivative of a function $f(z)$ at that point. (See Exercise 6, Sec. 19.) But, with certain continuity conditions, we have the following useful theorem.

Theorem. *Let the function*

$$f(z) = u(x, y) + iv(x, y)$$

be defined throughout some ε neighborhood of a point $z_0 = x_0 + iy_0$. Suppose that the first-order partial derivatives of the functions u and v with respect to x and y exist everywhere in that neighborhood and that they are continuous at (x_0, y_0). Then, if those partial derivatives satisfy the Cauchy-Riemann equations

$$u_x = v_y, \qquad u_y = -v_x$$

at (x_0, y_0), the derivative $f'(z_0)$ exists.

To start the proof, we write $\Delta z = \Delta x + i\Delta y$, where $0 < |\Delta z| < \varepsilon$, and

$$\Delta w = f(z_0 + \Delta z) - f(z_0).$$

Thus

$$\Delta w = \Delta u + i\Delta v,$$

where

$$\begin{aligned}
(1) \qquad \Delta u &= u(x_0 + \Delta x, y_0 + \Delta y) - u(x_0, y_0), \\
\Delta v &= v(x_0 + \Delta x, y_0 + \Delta y) - v(x_0, y_0).
\end{aligned}$$

Now, in view of the continuity of the first-order partial derivatives of u and v at the point (x_0, y_0),

$$\begin{aligned}
(2) \qquad \Delta u &= u_x(x_0, y_0)\Delta x + u_y(x_0, y_0)\Delta y + \varepsilon_1 \sqrt{(\Delta x)^2 + (\Delta y)^2}, \\
\Delta v &= v_x(x_0, y_0)\Delta x + v_y(x_0, y_0)\Delta y + \varepsilon_2 \sqrt{(\Delta x)^2 + (\Delta y)^2},
\end{aligned}$$

where ε_1 and ε_2 tend to 0 as $(\Delta x, \Delta y)$ approaches $(0, 0)$ in the Δz plane. Hence

$$\begin{aligned}
(3) \qquad \Delta w &= u_x(x_0, y_0)\Delta x + u_y(x_0, y_0)\Delta y + \varepsilon_1 \sqrt{(\Delta x)^2 + (\Delta y)^2} \\
&\quad + i[v_x(x_0, y_0)\Delta x + v_y(x_0, y_0)\Delta y + \varepsilon_2 \sqrt{(\Delta x)^2 + (\Delta y)^2}].
\end{aligned}$$

The existence of expressions of type (2) for functions of two real variables with continuous first-order partial derivatives is established in advanced calculus in connection with differentials.*

Assuming that the Cauchy-Riemann equations are satisfied at (x_0, y_0), we can replace $u_y(x_0, y_0)$ by $-v_x(x_0, y_0)$ and $v_y(x_0, y_0)$ by $u_x(x_0, y_0)$ in equation (3) and then divide through by Δz to get

$$(4) \qquad \frac{\Delta w}{\Delta z} = u_x(x_0, y_0) + iv_x(x_0, y_0) + (\varepsilon_1 + i\varepsilon_2)\frac{\sqrt{(\Delta x)^2 + (\Delta y)^2}}{\Delta z}.$$

---

*See, for instance, A. E. Taylor and W. R. Mann, "Advanced Calculus," 3d ed., pp. 150–151 and 197–198, 1983.

But $\sqrt{(\Delta x)^2 + (\Delta y)^2} = |\Delta z|$, and so

$$\left| \frac{\sqrt{(\Delta x)^2 + (\Delta y)^2}}{\Delta z} \right| = 1.$$

Also, $\varepsilon_1 + i\varepsilon_2$ tends to 0 as $(\Delta x, \Delta y)$ approaches $(0,0)$. So the last term on the right in equation (4) tends to 0 as the variable $\Delta z = \Delta x + i\Delta y$ tends to 0. This means that the limit of the left-hand side of equation (4) exists and that

(5) $$f'(z_0) = u_x(x_0, y_0) + iv_x(x_0, y_0).$$

EXAMPLE 1. Suppose that

$$f(z) = e^x(\cos y + i \sin y),$$

where y is to be taken in radians when $\cos y$ and $\sin y$ are evaluated. Then

$$u(x, y) = e^x \cos y \quad \text{and} \quad v(x, y) = e^x \sin y.$$

Since $u_x = v_y$ and $u_y = -v_x$ everywhere and since those derivatives are every-where continuous, the conditions in the theorem are satisfied at all points in the complex plane. Thus $f'(z)$ exists everywhere, and

$$f'(z) = u_x + iv_x = e^x(\cos y + i \sin y).$$

Note that $f'(z) = f(z)$.

EXAMPLE 2. It also follows from the theorem in this section that the function $f(z) = |z|^2$, whose components are

$$u(x, y) = x^2 + y^2 \quad \text{and} \quad v(x, y) = 0,$$

has a derivative at $z = 0$. In fact, $f'(0) = 0 + i0$ (compare Example 2, Sec. 15). We saw in Example 2, Sec. 17, that this function cannot have a derivative at any nonzero point since the Cauchy-Riemann equations are not satisfied at such points.

19. POLAR COORDINATES

When $z_0 \neq 0$, the theorem in Sec. 18 is given in polar coordinates by means of the coordinate transformation (Sec. 5)

(1) $$x = r \cos \theta, \quad y = r \sin \theta.$$

Depending on whether we write

$$z = x + iy \quad \text{or} \quad z = re^{i\theta} \quad (z \neq 0)$$

when $w = f(z)$, the real and imaginary parts of $w = u+iv$ are expressed in terms of either the variables x and y or r and θ. Suppose that the first-order partial derivatives of u and v with respect to x and y exist everywhere in some neighborhood of a given nonzero point z_0 and are continuous at that point. The first-order partial derivatives with respect to r and θ also have these properties, and the chain rule for differentiating real-valued functions of two real variables can be used to write them in terms of the ones with respect to x and y. More precisely, since

$$\frac{\partial u}{\partial r} = \frac{\partial u}{\partial x}\frac{\partial x}{\partial r} + \frac{\partial u}{\partial y}\frac{\partial y}{\partial r}, \qquad \frac{\partial u}{\partial \theta} = \frac{\partial u}{\partial x}\frac{\partial x}{\partial \theta} + \frac{\partial u}{\partial y}\frac{\partial y}{\partial \theta},$$

one can write

(2) $u_r = u_x \cos \theta + u_y \sin \theta, \qquad u_\theta = -u_x r \sin \theta + u_y r \cos \theta.$

Likewise,

(3) $v_r = v_x \cos \theta + v_y \sin \theta, \qquad v_\theta = -v_x r \sin \theta + v_y r \cos \theta.$

If the partial derivatives with respect to x and y also satisfy the Cauchy-Riemann equations

(4) $u_x = v_y, \qquad u_y = -v_x$

at z_0, equations (3) become

(5) $v_r = -u_y \cos \theta + u_x \sin \theta, \qquad v_\theta = u_y r \sin \theta + u_x r \cos \theta$

at that point. It is then clear from equations (2) and (5) that

(6) $u_r = \dfrac{1}{r} v_\theta, \qquad \dfrac{1}{r} u_\theta = -v_r$

at the point z_0.

If, on the other hand, equations (6) are known to hold at z_0, it is straightforward to show (Exercise 7) that equations (4) must hold there. Equations (6) are, therefore, an alternative form of the Cauchy-Riemann equations (4).

We can now restate the theorem in Sec. 18 using polar coordinates.

Theorem. *Let the function*

$$f(z) = u(r, \theta) + iv(r, \theta)$$

be defined throughout some ε neighborhood of a nonzero point $z_0 = r_0 \exp(i\theta_0)$. Suppose that the first-order partial derivatives of the functions u and v with respect to r and θ exist everywhere in that neighborhood and that they are continuous at (r_0, θ_0). Then, if those partial derivatives satisfy the polar form (6) of the Cauchy-Riemann equations at (r_0, θ_0), the derivative $f'(z_0)$ exists.

The derivative $f'(z_0)$ here can be written (see Exercise 8)

(7) $f'(z_0) = e^{-i\theta}(u_r + iv_r),$

where the right-hand side is evaluated at (r_0, θ_0).

EXAMPLE.   Consider the function

$$f(z) = \frac{1}{z} = \frac{1}{re^{i\theta}}.$$

Since

$$u(r,\theta) = \frac{\cos\theta}{r} \quad \text{and} \quad v(r,\theta) = -\frac{\sin\theta}{r},$$

the conditions in the theorem are satisfied at any nonzero point $z = re^{i\theta}$ in the plane. Hence the derivative of f exists there; and, according to expression (7),

$$f'(z) = e^{-i\theta}\left(-\frac{\cos\theta}{r^2} + i\frac{\sin\theta}{r^2}\right) = -\frac{1}{(re^{i\theta})^2} = -\frac{1}{z^2}.$$

EXERCISES

1. Use the theorem in Sec. 17 to show that $f'(z)$ does not exist at any point if
   (a) $f(z) = \bar{z}$;   (b) $f(z) = z - \bar{z}$;   (c) $f(z) = 2x + ixy^2$;   (d) $e^x e^{-iy}$.
2. Use the theorem in Sec. 18 to show that $f'(z)$ and its derivative $f''(z)$ exist everywhere, and find $f''(z)$ when
   (a) $f(z) = iz + 2$;   (b) $f(z) = e^{-x}e^{-iy}$;
   (c) $f(z) = z^3$;   (d) $f(z) = \cos x \cosh y - i \sin x \sinh y$.
      Ans. (b) $f''(z) = f(z)$;   (d) $f''(z) = -f(z)$.
3. From results obtained in Secs. 17 and 18, determine where $f'(z)$ exists and find its value when
   (a) $f(z) = 1/z$;   (b) $f(z) = x^2 + iy^2$;   (c) $f(z) = z\,\mathrm{Im}\,z$.
      Ans. (a) $f'(z) = -1/z^2 \ (z \neq 0)$;   (b) $f'(x+ix) = 2x$;   (c) $f'(0) = 0$.
4. Use the theorem in Sec. 19 to show that each of these functions is differentiable in the indicated domain of definition, and then use expression (7) in that section to find $f'(z)$:
   (a) $f(z) = 1/z^4 \ (z \neq 0)$;   (b) $f(z) = \sqrt{r}e^{i\theta/2} \ (r > 0, -\pi < \theta < \pi)$;
   (c) $f(z) = e^{-\theta}\cos(\ln r) + ie^{-\theta}\sin(\ln r) \ (r > 0, 0 < \theta < 2\pi)$.
      Ans. (b) $f'(z) = 1/[2f(z)]$;   (c) $f'(z) = if(z)/z$.
5. Show that when $f(z) = x^3 + i(1-y)^3$, it is legitimate to write
   $$f'(z) = u_x + iv_x = 3x^2$$
   only when $z = i$.
6. Let u and v denote the real and imaginary components of the function f defined by the equations
   $$f(z) = \begin{cases} \dfrac{(\bar{z})^2}{z} & \text{when} \quad z \neq 0, \\ 0 & \text{when} \quad z = 0. \end{cases}$$
   Verify that the Cauchy-Riemann equations $u_x = v_y$ and $u_y = -v_x$ are satisfied at the origin $z = (0,0)$. [Compare Exercise 9, Sec. 16, where it is shown that $f'(0)$ nevertheless fails to exist.]
7. Solve equations (2), Sec. 19, for u_x and u_y to show that
   $$u_x = u_r \cos\theta - u_\theta \frac{\sin\theta}{r}, \qquad u_y = u_r \sin\theta + u_\theta \frac{\cos\theta}{r}.$$
   Then use these equations and similar ones for v_x and v_y to show that, in Sec. 19, equations (4) are satisfied at a point z_0 if equations (6) are satisfied there. Thus complete the verification that equations (6), Sec. 19, are the Cauchy-Riemann equations in polar form.

8. Suppose that a function $f(z) = u + iv$ is differentiable at a nonzero point $z_0 = r_0 \exp(i\theta_0)$. Use the expressions for u_x and v_x found in Exercise 7, together with the polar form (6), Sec. 19, of the Cauchy-Riemann equations, to show that $f'(z_0)$ can be written

$$f'(z_0) = e^{-i\theta}(u_r + iv_r),$$

where u_r and v_r are evaluated at (r_0, θ_0).

9. (a) With the aid of the polar form (6), Sec. 19, of the Cauchy-Riemann equations, derive the alternative form

$$f'(z_0) = \frac{-i}{z_0}(u_\theta + iv_\theta)$$

of the expression for $f'(z_0)$ in Exercise 8.

(b) Use the expression for $f'(z_0)$ found in part (a) to show that the derivative of the function $f(z) = 1/z$ $(z \neq 0)$ in the example of Sec. 19 is $f'(z) = -1/z^2$.

10. (a) Recall (Sec. 3) that if $z = x + iy$, then

$$x = \frac{z + \bar{z}}{2} \qquad \text{and} \qquad y = \frac{z - \bar{z}}{2i}.$$

By *formally* applying the chain rule in calculus to a function $F(x, y)$ of two variables, derive the expression

$$\frac{\partial F}{\partial \bar{z}} = \frac{\partial F}{\partial x}\frac{\partial x}{\partial \bar{z}} + \frac{\partial F}{\partial y}\frac{\partial y}{\partial \bar{z}} = \frac{1}{2}\left(\frac{\partial F}{\partial x} + i\frac{\partial F}{\partial y}\right).$$

(b) Define the operator

$$\frac{\partial}{\partial \bar{z}} = \frac{1}{2}\left(\frac{\partial}{\partial x} + i\frac{\partial}{\partial y}\right),$$

suggested by part (a), to show that if the first-order partial derivatives of the real and imaginary parts of a function $f(z) = u(x, y) + iv(x, y)$ satisfy the Cauchy-Riemann equations, then

$$\frac{\partial f}{\partial \bar{z}} = \frac{1}{2}[(u_x - v_y) + i(v_x + u_y)] = 0.$$

Thus derive the *complex form* $\partial f/\partial \bar{z} = 0$ of the Cauchy-Riemann equations.

20. ANALYTIC FUNCTIONS

We are now ready to introduce the concept of an analytic function. A function f of the complex variable z is *analytic* in an open set if it has a derivative at each point in that set.* If we should speak of a function f that is analytic in a set S which is not open, it is to be understood that f is analytic in an open set containing S. In particular, f is *analytic at a point* z_0 if it is analytic in a neighborhood of z_0.

---

*The terms *regular* and *holomorphic* are also used in the literature to denote analyticity.

We note, for instance, that the function $f(z) = 1/z$ is analytic at each nonzero point in the finite plane. But the function $f(z) = |z|^2$ is not analytic at any point since its derivative exists only at $z = 0$ and not throughout any neighborhood. (See Example 2, Sec. 15.)

An *entire* function is a function that is analytic at each point in the entire finite plane. Since the derivative of a polynomial exists everywhere, it follows that *every polynomial is an entire function.*

If a function f fails to be analytic at a point z_0 but is analytic at some point in every neighborhood of z_0, then z_0 is called a *singular point*, or singularity, of f. The point $z = 0$ is evidently a singular point of the function $f(z) = 1/z$. The function $f(z) = |z|^2$, on the other hand, has no singular points since it is nowhere analytic.

A necessary, but by no means sufficient, condition for a function f to be analytic in a domain D is clearly the continuity of f throughout D. Satisfaction of the Cauchy-Riemann equations is also necessary, but not sufficient. Sufficient conditions for analyticity in D are provided by the theorems in Secs. 18 and 19.

Other useful sufficient conditions are obtained from the differentiation formulas in Sec. 16. The derivatives of the sum and product of two functions exist wherever the functions themselves have derivatives. Thus, *if two functions are analytic in a domain D, their sum and their product are both analytic in D*. Similarly, *their quotient is analytic in D provided the function in the denominator does not vanish at any point in D*. In particular, the quotient $P(z)/Q(z)$ of two polynomials is analytic in any domain throughout which $Q(z) \neq 0$.

From the chain rule for the derivative of a composite function, we find that *a composition of two analytic functions is analytic*. More precisely, suppose that a function $f(z)$ is analytic in a domain D and that the image (Sec. 10) of D under the transformation $w = f(z)$ is contained in the domain of definition of a function $g(w)$. Then the composition $g[f(z)]$ is analytic in D, with derivative

$$\frac{d}{dz} g[f(z)] = g'[f(z)]f'(z).$$

We conclude this section with an expected and especially useful property of analytic functions.

Theorem. *If* $f'(z) = 0$ *everywhere in a domain D, then* $f(z)$ *must be constant throughout D.*

To prove this, we write $f(z) = u(x, y) + iv(x, y)$. Then, assuming that $f'(z) = 0$ in D, we note that $u_x + iv_x = 0$; and, in view of the Cauchy-Riemann equations, $v_y - iu_y = 0$. Consequently,

$$u_x = u_y = v_x = v_y = 0$$

at each point in D.

Now u_x and u_y are the x and y components of the vector grad u, and the component of grad u at a given point in a specific direction is the directional derivative of u there in that direction. So the fact that u_x and u_y are always zero means that grad u is always the zero vector and hence that any directional derivative of u is zero. Consequently, u is constant along any line segment lying entirely in D; and, since there is always a finite number of such line segments, joined end to end, connecting any two points in D (Sec. 8), the values of u at those points must be the same. We may conclude, then, that there is a real constant a such that $u(x, y) = a$ throughout D. Similarly, $v(x, y) = b$; and it follows that $f(z) = a + bi$ at each point in D.

21. REFLECTION PRINCIPLE

In the remaining two sections of this chapter, we develop some important properties of analytic functions that, in addition to being of considerable theoretical interest, will be valuable later on in applications.

The theorem in this section concerns the fact that some analytic functions possess the property that $\overline{f(z)} = f(\bar{z})$ for all points z in certain domains while others do not. We note, for example that $z + 1$ and z^2 have that property when D is the entire finite plane; but the same is not true of $z + i$ and iz^2. The theorem, which is known as the *reflection principle*, provides a way of predicting when the reflection of $f(z)$ in the real axis corresponds to the reflection of z.

Theorem. *Suppose that a function f is analytic in some domain D which contains a segment of the x axis and is symmetric to that axis. Then*

$$(1) \qquad \overline{f(z)} = f(\bar{z})$$

for each point z in the domain if and only if $f(x)$ is real for each point x on the segment.

We start the proof by assuming that $f(x)$ is real at each point x on the segment. Once we show that the function

$$(2) \qquad F(z) = \overline{f(\bar{z})}$$

is analytic in D, we shall use that assumption to obtain equation (1). To establish the analyticity of $F(z)$, we write

$$f(z) = u(x, y) + iv(x, y), \qquad F(z) = U(x, y) + iV(x, y)$$

and observe how it follows from equation (2) that, since

$$(3) \qquad \overline{f(\bar{z})} = u(x, -y) - iv(x, -y),$$

the components of $F(z)$ and $f(z)$ are related by the equations

$$(4) \qquad U(x, y) = u(x, t) \qquad \text{and} \qquad V(x, y) = -v(x, t),$$

where $t = -y$. Now, because $f(x + it)$ is an analytic function of $x + it$, the first-order partial derivatives of the functions $u(x, t)$ and $v(x, t)$ are continuous throughout

D and satisfy the Cauchy-Riemann equations

(5) $$u_x = v_t, \qquad u_t = -v_x.$$

Furthermore, in view of equations (4),

$$U_x = u_x, \qquad V_y = -v_t \frac{dt}{dy} = v_t;$$

and it follows from these and the first of equations (5) that $U_x = V_y$. Similarly,

$$U_y = u_t \frac{dt}{dy} = -u_t, \qquad V_x = -v_x;$$

and the second of equations (5) tells us that $U_y = -V_x$. Inasmuch as the first-order partial derivatives of $U(x, y)$ and $V(x, y)$ are now shown to satisfy the Cauchy-Riemann equations and since those derivatives are continuous, we find that the function $F(z)$ is analytic in D. Moreover, since $f(x)$ is real on the segment of the real axis lying in D, $v(x, 0) = 0$ on that segment; and, in view of equations (4), this means that

$$F(x) = U(x, 0) + iV(x, 0) = u(x, 0) - iv(x, 0) = u(x, 0).$$

That is,

(6) $$F(z) = f(z)$$

at each point $z = x$ on the segment. We now refer to a result that will be obtained in Chap. 6 (Sec. 58). Namely, a function that is analytic in a domain D is uniquely determined by its values along any line segment lying in D. Thus equation (6) actually holds throughout D. Because of definition (2) of the function $F(z)$, then,

(7) $$\overline{f(\bar{z})} = f(z);$$

and this is the same as equation (1).

To prove the converse of the theorem, we assume that equation (1) holds and note that, in view of expression (3), the form (7) of equation (1) can be written

$$u(x, -y) - iv(x, -y) = u(x, y) + iv(x, y).$$

In particular, if $(x, 0)$ is a point on the segment of the real axis that lies in D,

$$u(x, 0) - iv(x, 0) = u(x, 0) + iv(x, 0);$$

and, by equating imaginary parts here, we see that $v(x, 0) = 0$. Hence $f(x)$ is real on the segment of the real axis lying in D.

EXAMPLES.   Just prior to the statement of the theorem, we noted that

$$\overline{z + 1} = \bar{z} + 1 \qquad \text{and} \qquad \overline{z^2} = \bar{z}^2$$

for all z in the finite plane. The theorem tells us, of course, that this is true, since $x + 1$ and x^2 are real when x is real. We also noted that $z + i$ and iz^2 do not have the reflection property throughout the plane, and we now know that this is because $x + i$ and ix^2 are *not* real when x is real.

22. HARMONIC FUNCTIONS

A real-valued function H of two real variables x and y is said to be *harmonic* in a given domain of the xy plane if, throughout that domain, it has continuous partial derivatives of the first and second order and satisfies the partial differential equation

$$(1) \qquad\qquad H_{xx}(x, y) + H_{yy}(x, y) = 0,$$

known as *Laplace's equation.*

Harmonic functions play an important role in applied mathematics. For example, the temperatures $T(x, y)$ in thin plates lying in the xy plane are often harmonic. A function $V(x, y)$ is harmonic when it denotes an electrostatic potential that varies only with x and y in the interior of a region of three-dimensional space that is free of charges.

EXAMPLE 1.   It is easy to verify that the function $T(x, y) = e^{-y} \sin x$ is harmonic in any domain of the xy plane and, in particular, in the semi-infinite vertical strip $0 < x < \pi, y > 0$. It also assumes the values on the edges of the strip that are indicated in Fig. 23. More precisely, it satisfies all of the conditions

$$T_{xx}(x, y) + T_{yy}(x, y) = 0,$$

$$T(0, y) = 0, \qquad T(\pi, y) = 0,$$

$$T(x, 0) = \sin x, \qquad \lim_{y \to \infty} T(x, y) = 0,$$

which describe the temperatures $T(x, y)$ in a thin homogeneous plate in the xy plane that has no heat sources or sinks and is insulated except for the stated conditions along the edges.

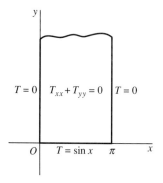

FIGURE 23

The use of the theory of functions of a complex variable in discovering solutions, such as the one in Example 1, of temperature and other problems is described

in considerable detail later on in Chap. 10 and in parts of chapters following it.*
That theory is based on the theorem below, which provides a source of harmonic
functions.

Theorem 1. *If a function $f(z) = u(x, y) + iv(x, y)$ is analytic in a domain D,
then its component functions u and v are harmonic in D.*

To show this, we need a result that is proved in Chap. 4 (Sec. 40). Namely, if
a function of a complex variable is analytic at a point, then its real and imaginary
components have continuous partial derivatives of all orders at that point.

Assuming that f is analytic in D, we start with the observation that the
first-order partial derivatives of its component functions must satisfy the Cauchy-
Riemann equations throughout D:

$$(2) \qquad\qquad u_x = v_y, \qquad u_y = -v_x.$$

Differentiating both sides of these equations with respect to x, we have

$$(3) \qquad\qquad u_{xx} = v_{yx}, \qquad u_{yx} = -v_{xx}.$$

Likewise, differentiation with respect to y yields

$$(4) \qquad\qquad u_{xy} = v_{yy}, \qquad u_{yy} = -v_{xy}.$$

Now, by a theorem in advanced calculus,[†] the continuity of the partial derivatives
of u and v ensures that $u_{yx} = u_{xy}$ and $v_{yx} = v_{xy}$. It then follows from equations (3)
and (4) that

$$u_{xx} + u_{yy} = 0 \qquad \text{and} \qquad v_{xx} + v_{yy} = 0.$$

That is, u and v are harmonic in D.

EXAMPLE 2.   The function

$$f(z) = e^{-y}\sin x - ie^{-y}\cos x$$

is entire, as is shown in Exercise 1(c). Hence the temperature function $T(x, y) = e^{-y}\sin x$ in Example 1 must be harmonic in every domain of the xy plane.

EXAMPLE 3.   Since the function

$$g(z) = z^2 = (x + iy)^2 = x^2 - y^2 + i2xy$$

is entire, so is the product $f(z)g(z)$, where $f(z)$ is the same as in Example 2. The
function

$$\text{Re}[f(z)g(z)] = e^{-y}[(x^2 - y^2)\sin x + 2xy\cos x]$$

is, therefore, also harmonic throughout the xy plane.

---

*Another important method is developed in the authors' "Fourier Series and Boundary Value Problems,"
5th ed., 1993.

†See, for instance, A. E. Taylor and W. R. Mann, "Advanced Calculus," 3d ed., pp. 199–201, 1983.

If two given functions u and v are harmonic in a domain D and their first-order partial derivatives satisfy the Cauchy-Riemann equations (2) throughout D, v is said to be a *harmonic conjugate* of u. The meaning of the word conjugate here is, of course, different from that in Sec. 3, where $\bar{z}$ is defined.

Theorem 2. *A function $f(z) = u(x, y) + iv(x, y)$ is analytic in a domain D if and only if v is a harmonic conjugate of u.*

The proof is easy. If v is a harmonic conjugate of u in D, the theorem in Sec. 18 tells us that f is analytic in D. Conversely, if f is analytic in D, we know from Theorem 1 above that u and v are harmonic in D; and, in view of the theorem in Sec. 17, the Cauchy-Riemann equations are satisfied.

The following example shows that if v is a harmonic conjugate of u in some domain, it is *not*, in general, true that u is a harmonic conjugate of v there.

EXAMPLE 4.   Suppose that

$$u(x, y) = x^2 - y^2 \qquad \text{and} \qquad v(x, y) = 2xy.$$

Since these are the real and imaginary components, respectively, of the entire function $f(z) = z^2$, we know that v is a harmonic conjugate of u throughout the plane. But u cannot be a harmonic conjugate of v since, as verified in Exercise 2(b), the function $2xy + i(x^2 - y^2)$ is not analytic anywhere.

It can be shown [Exercise 11(b)] that if two functions u and v are to be harmonic conjugates of each other, then both u and v must be constant functions. It is, however, true that if v is a harmonic conjugate of u in a domain D, then $-u$ is a harmonic conjugate of v in D, and conversely. This is seen by writing

$$f(z) = u(x, y) + iv(x, y), \qquad -if(z) = v(x, y) - iu(x, y)$$

and noting that $f(z)$ is analytic in D if and only if $-if(z)$ is analytic there.

In Chap. 9 (Sec. 75) we shall show that a function u which is harmonic in a domain of a certain type always has a harmonic conjugate. Thus, in such domains, every harmonic function is the real part of an analytic function. It is also true that a harmonic conjugate, when it exists, is unique except for an additive constant.

EXAMPLE 5.   We now illustrate one method of obtaining a harmonic conjugate of a given harmonic function. The function

$$(5) \qquad\qquad u(x, y) = y^3 - 3x^2 y$$

is readily seen to be harmonic throughout the entire xy plane. To find a harmonic conjugate $v(x, y)$, we note that

$$u_x(x, y) = -6xy.$$

So, in view of the condition $u_x = v_y$, we may write

$$v_y(x, y) = -6xy.$$

Holding x fixed and integrating both sides of this equation with respect to y, we find that

$$(6) \qquad v(x, y) = -3xy^2 + \phi(x),$$

where ϕ is, at present, an arbitrary function of x. Since the condition $u_y = -v_x$ must hold, it follows from equations (5) and (6) that

$$3y^2 - 3x^2 = 3y^2 - \phi'(x).$$

So $\phi'(x) = 3x^2$; and this means that $\phi(x) = x^3 + c$, where c is an arbitrary real number. Hence the function

$$v(x, y) = x^3 - 3xy^2 + c$$

is a harmonic conjugate of $u(x, y)$.

The corresponding analytic function is

$$(7) \qquad f(z) = (y^3 - 3x^2 y) + i(x^3 - 3xy^2 + c).$$

It is easily verified that

$$f(z) = i(z^3 + c).$$

This form is suggested by noting that when $y = 0$, equation (7) becomes

$$f(x) = i(x^3 + c).$$

EXERCISES

1. Apply the theorem in Sec. 18 to verify that each of these functions is entire:
   (a) $f(z) = 3x + y + i(3y - x)$; (b) $f(z) = \sin x \cosh y + i \cos x \sinh y$;
   (c) $f(z) = e^{-y} \sin x - i e^{-y} \cos x$; (d) $f(z) = (z^2 - 2)e^{-x}e^{-iy}$.

2. With the aid of the theorem in Sec. 17, show that each of these functions is nowhere analytic:
   (a) $f(z) = xy + iy$; (b) $f(z) = 2xy + i(x^2 - y^2)$; (c) $f(z) = e^y e^{ix}$.

3. State why a composition of two entire functions is entire. Also, state why any *linear combination* $c_1 f_1(z) + c_2 f_2(z)$ of two entire functions, where c_1 and c_2 are complex constants, is entire.

4. In each case, determine the singular points of the function and state why the function is analytic everywhere except at those points:
   (a) $f(z) = \dfrac{2z + 1}{z(z^2 + 1)}$; (b) $f(z) = \dfrac{z^3 + i}{z^2 - 3z + 2}$; (c) $f(z) = \dfrac{z^2 + 1}{(z + 2)(z^2 + 2z + 2)}$.

   Ans. (a) $z = 0, \pm i$; (b) $z = 1, 2$; (c) $z = -2, -1 \pm i$.

5. According to Exercise 4(b), Sec. 19, the function

$$g(z) = \sqrt{r} e^{i\theta/2} \qquad (r > 0, -\pi < \theta < \pi)$$

is analytic in its domain of definition, with derivative $g'(z) = 1/[2g(z)]$. Show that the composite function $g(2z - 2 + i)$ is analytic in the half plane $x > 1$, with derivative $1/g(2z - 2 + i)$.

   Suggestion: Observe that $\text{Re}(2z - 2 + i) > 0$ when $x > 1$.

6. Use results in Sec. 19 to verify that the function

$$g(z) = \ln r + i\theta \qquad (r > 0, 0 < \theta < 2\pi)$$

is analytic in the indicated domain of definition, with derivative $g'(z) = 1/z$. Then show that the composite function $g(z^2 + 1)$ is an analytic function of z in the quadrant $x > 0, y > 0$, with derivative $2z/(z^2 + 1)$.

Suggestion: Observe that $\text{Im}(z^2 + 1) > 0$ when $x > 0, y > 0$.

7. Let a function $f(z)$ be analytic in a domain D. Prove that $f(z)$ must be constant in D if

(a) $f(z)$ is real-valued for all z in D;

(b) $\overline{f(z)}$ is analytic in D;

(c) $|f(z)|$ is constant in D.

Suggestion: Use the Cauchy-Riemann equations and the theorem in Sec. 20 to prove parts (a) and (b). To prove part (c), observe that $\overline{f(z)} = c^2/f(z)$ if $|f(z)| = c$, where $c \neq 0$; then use part (b).

8. We know from Example 1, Sec. 18, that the function $f(z) = e^x(\cos y + i \sin y)$ has a derivative everywhere in the finite plane. Point out how it follows from the reflection principle (Sec. 21) that, for each z, $\overline{f(z)} = f(\bar{z})$. Then verify this fact directly.

9. Show that if the condition that $f(x)$ be real in the reflection principle (Sec. 21) is replaced by the condition that $f(x)$ be pure imaginary, then equation (1) in the statement of the principle is changed to $\overline{f(z)} = -f(\bar{z})$.

10. Show that $u(x, y)$ is harmonic in some domain and find a harmonic conjugate $v(x, y)$ when

(a) $u(x, y) = 2x(1 - y)$;   (b) $u(x, y) = 2x - x^3 + 3xy^2$;

(c) $u(x, y) = \sinh x \sin y$;   (d) $u(x, y) = y/(x^2 + y^2)$.

       Ans. (a) $v(x, y) = x^2 - y^2 + 2y$;   (b) $v(x, y) = 2y - 3x^2y + y^3$;

             (c) $v(x, y) = -\cosh x \cos y$;   (d) $v(x, y) = x/(x^2 + y^2)$.

11. Show that

(a) if v and V are harmonic conjugates of u in a domain D, then $v(x, y)$ and $V(x, y)$ can differ at most by an additive constant;

(b) if v is a harmonic conjugate of u in a domain D and also u is a harmonic conjugate of v, then $u(x, y)$ and $v(x, y)$ must be constant throughout D.

12. Let the function $f(z) = u(r, \theta) + iv(r, \theta)$ be analytic in a domain D that does not include the origin. Using the Cauchy-Riemann equations in polar coordinates (Sec. 19) and assuming continuity of partial derivatives, show that, throughout D, the function $u(r, \theta)$ satisfies the partial differential equation

$$r^2 u_{rr}(r, \theta) + r u_r(r, \theta) + u_{\theta\theta}(r, \theta) = 0,$$

which is the *polar form of Laplace's equation*. Show that the same is true of the function $v(r, \theta)$.

13. Verify that the function $u(r, \theta) = \ln r$ is harmonic in the domain $r > 0, 0 < \theta < 2\pi$ by showing that it satisfies the polar form of Laplace's equation, obtained in Exercise 12. Then use the technique in Example 5, Sec. 22, but involving the Cauchy-Riemann equations in polar form (Sec. 19), to derive the harmonic conjugate $v(r, \theta) = \theta$. (Compare Exercise 6.)

14. Let the function $f(z) = u(x, y) + iv(x, y)$ be analytic in a domain D, and consider the families of *level curves* $u(x, y) = c_1$ and $v(x, y) = c_2$, where c_1 and c_2 are arbitrary real constants. Prove that these families are orthogonal. More precisely, show that if $z_0 = (x_0, y_0)$ is a point in D which is common to two particular curves $u(x, y) = c_1$ and $v(x, y) = c_2$ and if $f'(z_0) \neq 0$, then the lines tangent to those curves at (x_0, y_0) are perpendicular.

   Suggestion: Note how it follows from the equations $u(x, y) = c_1$ and $v(x, y) = c_2$ that

$$\frac{\partial u}{\partial x} + \frac{\partial u}{\partial y}\frac{dy}{dx} = 0 \quad \text{and} \quad \frac{\partial v}{\partial x} + \frac{\partial v}{\partial y}\frac{dy}{dx} = 0.$$

*15. Show that when $f(z) = z^2$, the level curves $u(x, y) = c_1$ and $v(x, y) = c_2$ of the component functions are the hyperbolas indicated in Fig. 24. Note the orthogonality of the two families, described in Exercise 14. Observe that the curves $u(x, y) = 0$ and $v(x, y) = 0$ intersect at the origin but are not, however, orthogonal to each other. Why is this fact in agreement with the result in Exercise 14?

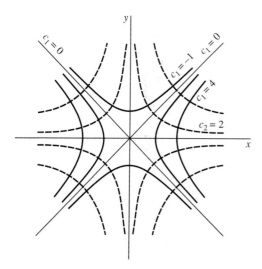

FIGURE 24

*16. Sketch the families of level curves of the component functions u and v when $f(z) = 1/z$, and note the orthogonality described in Exercise 14.

  17. Do Exercise 16 using polar coordinates.

*18. Sketch the families of level curves of the component functions u and v when

$$f(z) = \frac{z - 1}{z + 1},$$

and note how the result in Exercise 14 is illustrated here.

3

ELEMENTARY FUNCTIONS

We consider here various elementary functions studied in calculus and define corresponding functions of a complex variable. To be specific, we define analytic functions of a complex variable z that reduce to the elementary functions in calculus when $z = x + i0$. We start by defining the complex exponential function and then use it to develop the others.

23. THE EXPONENTIAL FUNCTION

If a function f of the complex variable $z = x + iy$ is to reduce to the familiar exponential function in calculus when z is real, we must require that

(1) $$f(x + i0) = e^x$$

for all real numbers x. Inasmuch as $(e^x)' = e^x$ for all real x, it is also natural to impose the following conditions:

(2) $\qquad\qquad f$ is entire $\qquad$ and $\qquad f'(z) = f(z)$ for all z.

As already pointed out in Example 1 of Sec. 18, the function

$$f(z) = e^x(\cos y + i \sin y),$$

where y is to be taken in radians, is differentiable everywhere in the complex plane and $f'(z) = f(z)$. So conditions (1) and (2) are clearly satisfied by this function. It can be shown, moreover, that this is the only function satisfying conditions (1) and (2) (see Exercise 15); and we write $f(z) = e^z$. Sometimes, for convenience, we use the notation $\exp z$ instead of e^z.

The exponential function of complex analysis is thus defined for all z by means of the equation

(3) $$e^z = e^x(\cos y + i \sin y),$$

where $z = x + iy$. As we have just seen, it reduces to the usual exponential function in calculus when $y = 0$, it is entire, and

(4) $$\frac{d}{dz}e^z = e^z$$

everywhere in the z plane.

In calculus, the value $\sqrt[n]{e}$, which is the *positive* nth root of e, is assigned to e^x when $x = 1/n$ $(n = 2, 3, \ldots)$. Thus, according to definition (3), the value of the complex exponential function e^z is also $\sqrt[n]{e}$ when $z = 1/n$ $(n = 2, 3, \ldots)$. This is an exception to the convention (Sec. 7) that would ordinarily require us to interpret $e^{1/n}$ as the set of all nth roots of e.

When z is the pure imaginary number $i\theta$, expression (3) becomes

$$e^{i\theta} = \cos \theta + i \sin \theta.$$

This is Euler's formula, which was introduced in Sec. 5. The definition of the symbol $e^{i\theta}$ given there thus agrees with definition (3) and enables us to express e^z in the more compact form

(5) $$e^z = e^x e^{iy}.$$

This form can, in turn, be written as

(6) $$e^z = \rho e^{i\phi}, \qquad \text{where} \qquad \rho = e^x \text{ and } \phi = y.$$

The number $\rho = e^x$ is, of course, positive for each value of x; and, according to statement (6), the modulus of e^z is e^x and an argument of e^z is y. That is,

(7) $$|e^z| = e^x \qquad \text{and} \qquad \arg(e^z) = y + 2n\pi \quad (n = 0, \pm 1, \pm 2, \ldots).$$

Note that, since $|e^z|$ is always positive,

(8) $$e^z \neq 0 \qquad \text{for any complex number } z.$$

Expression (5) for e^z makes it especially easy to verify the additive property

(9) $$(\exp z_1)(\exp z_2) = \exp(z_1 + z_2).$$

To do this, we write $z_1 = x_1 + iy_1$ and $z_2 = x_2 + iy_2$. Then

$$(\exp z_1)(\exp z_2) = (e^{x_1}e^{iy_1})(e^{x_2}e^{iy_2}) = (e^{x_1}e^{x_2})(e^{iy_1}e^{iy_2}).$$

But x_1 and x_2 are both real, and we also know from Sec. 6 that

$$e^{iy_1}e^{iy_2} = e^{i(y_1+y_2)}.$$

Hence

$$(\exp z_1)(\exp z_2) = e^{(x_1+x_2)}e^{i(y_1+y_2)};$$

and, since

$$(x_1 + x_2) + i(y_1 + y_2) = (x_1 + iy_1) + (x_2 + iy_2) = z_1 + z_2,$$

the right-hand side of this last equation becomes $\exp(z_1 + z_2)$, in accordance with expression (5). Property (9) is now established.

Property (9) enables us to write $\exp(z_1 - z_2) \exp z_2 = \exp z_1$, or

$$(10) \qquad\qquad \frac{\exp z_1}{\exp z_2} = \exp(z_1 - z_2).$$

From this and the fact that $e^0 = 1$, it follows that $1/e^z = e^{-z}$. Another expected property, whose verification is left as an exercise, is

$$(11) \qquad (\exp z)^n = \exp(nz) \qquad (n = 0, \pm 1, \pm 2, \ldots).$$

The function e^z has, in addition, a number of properties that are not shared by e^x. For example, since $e^{z+2\pi i} = e^z e^{2\pi i}$ and $e^{2\pi i} = 1$, we can see that e^z *is periodic with a pure imaginary period of* $2\pi i$:

$$(12) \qquad\qquad \exp(z + 2\pi i) = \exp z$$

for all z. The following example illustrates another property that e^x does not have. Namely, while e^x is never negative, there are values of e^z that are.

EXAMPLE.   There are values of z, for instance, such that

$$(13) \qquad\qquad\qquad e^z = -1.$$

To find them, we write equation (13) as $e^x e^{iy} = 1e^{i\pi}$. Then, in view of the statement in italics in Sec. 5 regarding the equality of two nonzero complex numbers expressed in exponential form,

$$e^x = 1 \qquad \text{and} \qquad y = \pi + 2n\pi \quad (n = 0, \pm 1, \pm 2, \ldots).$$

Thus $x = 0$, and we find that

$$(14) \qquad\qquad z = (2n + 1)\pi i \qquad (n = 0, \pm 1, \pm 2, \ldots).$$

EXERCISES

1. Show that

    (a) $\exp(2 \pm 3\pi i) = -e^2$;   (b) $\exp\left(\dfrac{2 + \pi i}{4}\right) = \sqrt{\dfrac{e}{2}}(1 + i)$;

    (c) $\exp(z + \pi i) = -\exp z$.

2. State why the function $2z^2 - 3 - ze^z + e^{-z}$ is entire.

3. Prove that the function $\exp \bar{z}$ is not analytic anywhere.

4. Show in two ways that the function $\exp(z^2)$ is entire. What is its derivative?
      Ans. $2z \exp(z^2)$.

5. Write $|\exp(2z + i)|$ and $|\exp(iz^2)|$ in terms of x and y. Then show that

$$|\exp(2z + i) + \exp(iz^2)| \le e^{2x} + e^{-2xy}.$$

6. Show that $|\exp(z^2)| \le \exp(|z|^2)$.

7. Prove that $|\exp(-2z)| < 1$ if and only if $\operatorname{Re} z > 0$.

8. Find all values of z such that

(a) $e^z = -2$; (b) $e^z = 1 + \sqrt{3}i$; (c) $\exp(2z - 1) = 1$.

Ans. (a) $z = \ln 2 + (2n + 1)\pi i$ $(n = 0, \pm 1, \pm 2, \ldots)$;

(b) $z = \ln 2 + \left(2n + \dfrac{1}{3}\right)\pi i$ $(n = 0, \pm 1, \pm 2, \ldots)$;

(c) $z = \dfrac{1}{2} + n\pi i$ $(n = 0, \pm 1, \pm 2, \ldots)$.

9. Show that $\overline{\exp(iz)} = \exp(i\bar{z})$ if and only if $z = n\pi$ $(n = 0, \pm 1, \pm 2, \ldots)$. (Compare Exercise 8, Sec. 22.)

10. (a) Show that if e^z is real, then $\operatorname{Im} z = n\pi$ $(n = 0, \pm 1, \pm 2, \ldots)$.

(b) If e^z is pure imaginary, what restriction is placed on z?

*11. Describe the behavior of $\exp(x + iy)$ as (a) x tends to $-\infty$; (b) y tends to ∞.

12. Write $\operatorname{Re}(e^{1/z})$ in terms of x and y. Why is this function harmonic in every domain that does not contain the origin?

13. Let the function $f(z) = u(x, y) + iv(x, y)$ be analytic in some domain D. State why the functions

$$U(x, y) = e^{u(x,y)} \cos v(x, y), \qquad V(x, y) = e^{u(x,y)} \sin v(x, y)$$

are harmonic in D and why $V(x, y)$ is, in fact, a harmonic conjugate of $U(x, y)$.

14. Establish the identity (Sec. 23)

$$(\exp z)^n = \exp(nz) \qquad (n = 0, \pm 1, \pm 2, \ldots)$$

in the following way:

(a) Use mathematical induction to show that it is valid when $n = 0, 1, 2, \ldots$.

(b) Verify it for negative integers by first recalling from Sec. 6 that $z^n = (z^{-1})^{-n}$ $(n = -1, -2, \ldots)$ when $z \ne 0$ and writing $(\exp z)^n = (1/\exp z)^m$, where $m = -n = 1, 2, \ldots$. Then use the result in part (a), together with the property $1/e^z = e^{-z}$ (Sec. 23) of the exponential function.

15. Suppose that a function $f(z) = u(x, y) + iv(x, y)$ satisfies conditions (1) and (2) in Sec. 23. Follow the steps described below to show that $f(z)$ must be the function

$$f(z) = e^x(\cos y + i \sin y).$$

(a) Obtain the equations $u_x = u$, $v_x = v$ and then use them to show that there exist real-valued functions ϕ and ψ of the real variable y such that

$$u(x, y) = e^x \phi(y) \qquad \text{and} \qquad v(x, y) = e^x \psi(y).$$

(b) Use the fact that u is harmonic (Sec. 22) to obtain the differential equation

$$\phi''(y) + \phi(y) = 0$$

and thus show that $\phi(y) = A \cos y + B \sin y$, where A and B are real constants.

(c) After pointing out why $\psi(y) = A \sin y - B \cos y$ and noting that $\phi(0) + i\psi(0) = 1$, find A and B. Conclude that

$$u(x, y) = e^x \cos y \qquad \text{and} \qquad v(x, y) = e^x \sin y.$$

24.   TRIGONOMETRIC FUNCTIONS

Euler's formula (Sec. 5) tells us that

$$e^{ix} = \cos x + i \sin x, \qquad e^{-ix} = \cos x - i \sin x$$

for every real number x; and it follows from these equations that

$$e^{ix} - e^{-ix} = 2i \sin x, \qquad e^{ix} + e^{-ix} = 2 \cos x.$$

Hence it is natural to *define* the sine and cosine functions of a complex variable z as follows:

$$(1) \qquad \sin z = \frac{e^{iz} - e^{-iz}}{2i}, \qquad \cos z = \frac{e^{iz} + e^{-iz}}{2}.$$

These functions are entire since they are linear combinations (Exercise 3, Sec. 22) of the entire functions e^{iz} and e^{-iz}. Knowing the derivatives of those exponential functions, we find from equations (1) that

$$(2) \qquad \frac{d}{dz} \sin z = \cos z, \qquad \frac{d}{dz} \cos z = - \sin z.$$

It is easy to see from definitions (1) that

$$(3) \qquad \sin(-z) = - \sin z \qquad \text{and} \qquad \cos(-z) = \cos z;$$

and a variety of other identities from trigonometry are valid with complex variables.

EXAMPLE.   In order to show that

$$(4) \qquad 2 \sin z_1 \cos z_2 = \sin(z_1 + z_2) + \sin(z_1 - z_2),$$

using definitions (1) and properties of the exponential function, we first write

$$2 \sin z_1 \cos z_2 = 2 \left(\frac{e^{iz_1} - e^{-iz_1}}{2i} \right) \left(\frac{e^{iz_2} + e^{-iz_2}}{2} \right).$$

Multiplication then reduces the right-hand side here to

$$\frac{e^{i(z_1 + z_2)} - e^{-i(z_1 + z_2)}}{2i} + \frac{e^{i(z_1 - z_2)} - e^{-i(z_1 - z_2)}}{2i},$$

or

$$\sin(z_1 + z_2) + \sin(z_1 - z_2);$$

and identity (4) is established.

Identity (4) leads to the following ones, whose verifications are left to the exercises:

$$(5) \qquad \sin(z_1 + z_2) = \sin z_1 \cos z_2 + \cos z_1 \sin z_2,$$

$$(6) \qquad \cos(z_1 + z_2) = \cos z_1 \cos z_2 - \sin z_1 \sin z_2,$$

(7) $$\sin^2 z + \cos^2 z = 1,$$

(8) $$\sin 2z = 2 \sin z \cos z, \qquad \cos 2z = \cos^2 z - \sin^2 z,$$

(9) $$\sin\left(z + \frac{\pi}{2}\right) = \cos z, \qquad \sin\left(z - \frac{\pi}{2}\right) = -\cos z.$$

When y is any real number, one can use definitions (1) and the hyperbolic functions

$$\sinh y = \frac{e^y - e^{-y}}{2} \qquad \text{and} \qquad \cosh y = \frac{e^y + e^{-y}}{2}$$

from calculus to write

(10) $$\sin(iy) = i \sinh y \qquad \text{and} \qquad \cos(iy) = \cosh y.$$

The real and imaginary parts of $\sin z$ and $\cos z$ are then readily displayed by writing $z_1 = x$ and $z_2 = iy$ in identities (5) and (6):

(11) $$\sin z = \sin x \cosh y + i \cos x \sinh y,$$

(12) $$\cos z = \cos x \cosh y - i \sin x \sinh y,$$

where $z = x + iy$.

A number of important properties of $\sin z$ and $\cos z$ follow immediately from expressions (11) and (12). The periodic character of these functions, for example, is evident:

(13) $$\sin(z + 2\pi) = \sin z, \qquad \sin(z + \pi) = -\sin z,$$

(14) $$\cos(z + 2\pi) = \cos z, \qquad \cos(z + \pi) = -\cos z.$$

Also (see Exercise 7)

(15) $$|\sin z|^2 = \sin^2 x + \sinh^2 y,$$

(16) $$|\cos z|^2 = \cos^2 x + \sinh^2 y.$$

It is clear from these two equations that $\sin z$ and $\cos z$ are not bounded in absolute value, whereas the absolute values of $\sin x$ and $\cos x$ are less than or equal to unity for all real values of x.

A *zero* of a given function $f(z)$ is a number z_0 such that $f(z_0) = 0$. Since $\sin z$ becomes the usual sine function in calculus when z is real, we know that the real numbers $z = n\pi$ $(n = 0, \pm 1, \pm 2, \ldots)$ are all zeros of $\sin z$. To show that *there are no other zeros*, we assume that $\sin z = 0$ and note how it follows from equation (15) that

$$\sin^2 x + \sinh^2 y = 0.$$

Thus

$$\sin x = 0 \qquad \text{and} \qquad \sinh y = 0.$$

Evidently, then, $x = n\pi$ $(n = 0, \pm 1, \pm 2, \ldots)$ and $y = 0$; that is,

(17) $$\sin z = 0 \qquad \text{if and only if} \qquad z = n\pi \quad (n = 0, \pm 1, \pm 2, \ldots).$$

Since

$$\cos z = -\sin\left(z - \frac{\pi}{2}\right),$$

according to the second of identities (9),

(18) $\cos z = 0$ if and only if $z = \dfrac{\pi}{2} + n\pi$ $(n = 0, \pm 1, \pm 2, \ldots)$.

So, as was the case with $\sin z$, the zeros of $\cos z$ are all real.

The other four trigonometric functions are defined in terms of the sine and cosine functions by the usual relations:

(19)
$$\tan z = \frac{\sin z}{\cos z}, \qquad \cot z = \frac{\cos z}{\sin z},$$

$$\sec z = \frac{1}{\cos z}, \qquad \csc z = \frac{1}{\sin z}.$$

Observe that $\tan z$ and $\sec z$ are analytic everywhere except at the singularities (Sec. 20) $z = (\pi/2) + n\pi$ $(n = 0, \pm 1, \pm 2, \ldots)$, which are the zeros of $\cos z$. Likewise, $\cot z$ and $\csc z$ have singularities at the zeros of $\sin z$, namely $z = n\pi$ $(n = 0, \pm 1, \pm 2, \ldots)$. By differentiating the right-hand sides of equations (19), we obtain the expected differentiation formulas

(20)
$$\frac{d}{dz}\tan z = \sec^2 z, \qquad \frac{d}{dz}\cot z = -\csc^2 z,$$

$$\frac{d}{dz}\sec z = \sec z \tan z, \qquad \frac{d}{dz}\csc z = -\csc z \cot z.$$

The periodicity of each of the trigonometric functions defined by equations (19) follows readily from equations (13) and (14). For example,

(21) $\tan(z + \pi) = \tan z.$

Mapping properties of the transformation $w = \sin z$ are especially important in the applications later on. A reader who wishes at this time to learn some of those properties is sufficiently prepared to read Sec. 74 (Chap. 8), where they are discussed.

EXERCISES

1. (a) Give details verifying expressions (2) in Sec. 24 for the derivatives of $\sin z$ and $\cos z$.

   (b) Let the function $f(z)$ be analytic in a domain D. State why the functions $\sin f(z)$ and $\cos f(z)$ are analytic there. Also, write $w = f(z)$ and state why

$$\frac{d}{dz}\sin w = \cos w \frac{dw}{dz}, \qquad \frac{d}{dz}\cos w = -\sin w \frac{dw}{dz}.$$

2. Show that $e^{iz} = \cos z + i \sin z$ for every complex number z.

3. (*a*) In Sec. 24, interchange z_1 and z_2 in equation (4) and then add corresponding sides of the resulting equation and equation (4) to derive expression (5) for $\sin(z_1 + z_2)$.

(*b*) By differentiating each side of equation (5) in Sec. 24 with respect to z_1, verify expression (6) in that section for $\cos(z_1 + z_2)$.

4. Show how each of the trigonometric identities (7), (8), and (9) in Sec. 24 follows from one of the identities (5) and (6) in that section.

5. Use identity (7) in Sec. 24 to show that

(*a*) $1 + \tan^2 z = \sec^2 z$; (*b*) $1 + \cot^2 z = \csc^2 z$.

6. Establish differentiation formulas (20), Sec. 24.

7. In Sec. 24, use expressions (11) and (12) to derive expressions (15) and (16) for $|\sin z|^2$ and $|\cos z|^2$.

Suggestion: Recall the identities $\sin^2 x + \cos^2 x = 1$ and $\cosh^2 y - \sinh^2 y = 1$.

8. Point out how it follows from expressions (15) and (16) in Sec. 24 for $|\sin z|^2$ and $|\cos z|^2$ that

(*a*) $|\sin z| \geq |\sin x|$; (*b*) $|\cos z| \geq |\cos x|$.

9. With the aid of expressions (15) and (16) in Sec. 24 for $|\sin z|^2$ and $|\cos z|^2$, show that

(*a*) $|\sinh y| \leq |\sin z| \leq \cosh y$; (*b*) $|\sinh y| \leq |\cos z| \leq \cosh y$.

10. (*a*) Use definitions (1), Sec. 24, of $\sin z$ and $\cos z$ to show that

$$2\sin(z_1 + z_2)\sin(z_1 - z_2) = \cos 2z_2 - \cos 2z_1.$$

(*b*) With the aid of the identity obtained in part (*a*), show that if $\cos z_1 = \cos z_2$, then at least one of the numbers $z_1 + z_2$ and $z_1 - z_2$ is an integral multiple of 2π.

11. Show that neither $\sin \bar{z}$ nor $\cos \bar{z}$ is an analytic function of z anywhere.

12. Use the reflection principle (Sec. 21) to show that, for all z,

(*a*) $\overline{\sin z} = \sin \bar{z}$; (*b*) $\overline{\cos z} = \cos \bar{z}$.

13. With the aid of expressions (11) and (12) in Sec. 24, give direct verifications of the relations obtained in Exercise 12.

14. Show that

(*a*) $\overline{\cos(iz)} = \cos(i\bar{z})$ for all z;

(*b*) $\overline{\sin(iz)} = \sin(i\bar{z})$ if and only if $z = n\pi i$ $(n = 0, \pm 1, \pm 2, \ldots)$.

15. Show in two ways that each of the following functions is everywhere harmonic:

(*a*) $\sin x \sinh y$; (*b*) $\cos 2x \sinh 2y$.

16. Find all roots of the equation $\sin z = \cosh 4$ by equating the real parts and the imaginary parts of $\sin z$ and $\cosh 4$.

$$Ans. \left(\frac{\pi}{2} + 2n\pi\right) \pm 4i \quad (n = 0, \pm 1, \pm 2, \ldots).$$

17. Find all roots of the equation $\cos z = 2$.

$Ans.$ $2n\pi + i \cosh^{-1} 2$, or $2n\pi \pm i \ln(2 + \sqrt{3})$ $(n = 0, \pm 1, \pm 2, \ldots)$.

25. HYPERBOLIC FUNCTIONS

The hyperbolic sine and the hyperbolic cosine of a complex variable are defined as they are with a real variable; that is,

(1) $$\sinh z = \frac{e^z - e^{-z}}{2}, \qquad \cosh z = \frac{e^z + e^{-z}}{2}.$$

Since e^z and e^{-z} are entire, it follows from definitions (1) that $\sinh z$ and $\cosh z$ are entire. Furthermore,

$$\text{(2)} \qquad \frac{d}{dz}\sinh z = \cosh z, \qquad \frac{d}{dz}\cosh z = \sinh z.$$

Because of the way in which the exponential function appears in definitions (1) and in the definitions (Sec. 24)

$$\sin z = \frac{e^{iz} - e^{-iz}}{2i}, \qquad \cos z = \frac{e^{iz} + e^{-iz}}{2}$$

of $\sin z$ and $\cos z$, the hyperbolic sine and cosine functions are closely related to those trigonometric functions:

$$\text{(3)} \qquad -i\sinh(iz) = \sin z, \qquad \cosh(iz) = \cos z,$$

$$\text{(4)} \qquad -i\sin(iz) = \sinh z, \qquad \cos(iz) = \cosh z.$$

Some of the most frequently used identities involving hyperbolic sine and cosine functions are

$$\text{(5)} \qquad \sinh(-z) = -\sinh z, \qquad \cosh(-z) = \cosh z,$$

$$\text{(6)} \qquad \cosh^2 z - \sinh^2 z = 1,$$

$$\text{(7)} \qquad \sinh(z_1 + z_2) = \sinh z_1 \cosh z_2 + \cosh z_1 \sinh z_2,$$

$$\text{(8)} \qquad \cosh(z_1 + z_2) = \cosh z_1 \cosh z_2 + \sinh z_1 \sinh z_2$$

and

$$\text{(9)} \qquad \sinh z = \sinh x \cos y + i\cosh x \sin y,$$

$$\text{(10)} \qquad \cosh z = \cosh x \cos y + i\sinh x \sin y,$$

$$\text{(11)} \qquad |\sinh z|^2 = \sinh^2 x + \sin^2 y,$$

$$\text{(12)} \qquad |\cosh z|^2 = \sinh^2 x + \cos^2 y,$$

where $z = x + iy$. While these identities follow directly from definitions (1), they are often more easily obtained from related trigonometric identities, with the aid of relations (3) and (4).

EXAMPLE. To illustrate the method of proof just suggested, let us verify identity (11). According to the first of relations (4), $|\sinh z|^2 = |\sin(iz)|^2$. That is,

$$\text{(13)} \qquad |\sinh z|^2 = |\sin(-y + ix)|^2,$$

where $z = x + iy$. But from equation (15), Sec. 24, we know that

$$|\sin(x + iy)|^2 = \sin^2 x + \sinh^2 y;$$

and this enables us to write equation (13) in the desired form (11).

In view of the periodicity of $\sin z$ and $\cos z$, it follows immediately from relations (4) that $\sinh z$ and $\cosh z$ are periodic with period $2\pi i$. Relations (4) also reveal that

(14) $\sinh z = 0$ if and only if $z = n\pi i$ $(n = 0, \pm1, \pm2, \ldots)$

and

(15) $\cosh z = 0$ if and only if $z = \left(\dfrac{\pi}{2} + n\pi\right)i$ $(n = 0, \pm1, \pm2, \ldots)$.

The hyperbolic tangent of z is defined by the equation

(16)
$$\tanh z = \frac{\sinh z}{\cosh z}$$

and is analytic in every domain in which $\cosh z \neq 0$. The functions $\coth z$, $\operatorname{sech} z$, and $\operatorname{csch} z$ are the reciprocals of $\tanh z$, $\cosh z$, and $\sinh z$, respectively. It is straightforward to verify the following differentiation formulas, which are the same as those established in calculus for the corresponding functions of a real variable:

(17) $\dfrac{d}{dz}\tanh z = \operatorname{sech}^2 z$, $\dfrac{d}{dz}\coth z = -\operatorname{csch}^2 z$,

(18) $\dfrac{d}{dz}\operatorname{sech} z = -\operatorname{sech} z \tanh z$, $\dfrac{d}{dz}\operatorname{csch} z = -\operatorname{csch} z \coth z$.

EXERCISES

1. Verify that the derivatives of $\sinh z$ and $\cosh z$ are as stated in equations (2), Sec. 25.
2. Prove that $\sinh 2z = 2\sinh z \cosh z$ by starting with
   (a) definitions (1), Sec. 25, of $\sinh z$ and $\cosh z$;
   (b) the identity $\sin 2z = 2\sin z \cos z$ (Sec. 24) and using relations (3) in Sec. 25.
3. Show how identities (6) and (8) in Sec. 25 follow from identities (7) and (6), respectively, in Sec. 24.
4. Write $\sinh z = \sinh(x + iy)$ and $\cosh z = \cosh(x + iy)$, and show how expressions (9) and (10) in Sec. 25 follow from identities (7) and (8), respectively, in that section.
5. Verify expression (12), Sec. 25, for $|\cosh z|^2$.
6. Show that $|\sinh x| \leq |\cosh z| \leq \cosh x$ by using (a) identity (12), Sec. 25; (b) the inequalities obtained in Exercise 9(b), Sec. 24.
7. Show that
   (a) $\sinh(z + \pi i) = -\sinh z$;   (b) $\cosh(z + \pi i) = -\cosh z$;
   (c) $\tanh(z + \pi i) = \tanh z$.
8. Give details showing that the zeros of $\sinh z$ and $\cosh z$ are as in statements (14) and (15) in Sec. 25.
9. Using the results proved in Exercise 8, locate all zeros and singularities of the hyperbolic tangent function.
10. Derive differentiation formulas (17), Sec. 25.
11. Use the reflection principle (Sec. 21) to show that, for all z,
    (a) $\overline{\sinh z} = \sinh \bar{z}$;   (b) $\overline{\cosh z} = \cosh \bar{z}$.
12. Use the results in Exercise 11 to show that $\overline{\tanh z} = \tanh \bar{z}$ at points where $\cosh z \neq 0$.

13. Why is the function $\sinh(e^z)$ entire? Write its real part as a function of x and y, and state why that function must be harmonic everywhere.

14. Find all roots of the equation

(a) $\cosh z = \dfrac{1}{2}$;   (b) $\sinh z = i$;   (c) $\cosh z = -2$.

   Suggestion: Compare part (c) with Exercise 17, Sec. 24.

   Ans. (a) $\left(2n \pm \dfrac{1}{3}\right)\pi i$ $(n = 0, \pm 1, \pm 2, \ldots)$;

   (b) $\left(2n + \dfrac{1}{2}\right)\pi i$ $(n = 0, \pm 1, \pm 2, \ldots)$;

   (c) $\pm \ln(2 + \sqrt{3}) + (2n + 1)\pi i$ $(n = 0, \pm 1, \pm 2, \ldots)$.

26. THE LOGARITHMIC FUNCTION AND ITS BRANCHES

Our motivation for the definition of the logarithmic function is based on solving the equation

$$\text{(1)} \qquad\qquad\qquad e^w = z$$

for w, where z is any *nonzero* complex number. To do this, we note that when z and w are written $z = re^{i\Theta}(-\pi < \Theta \leq \pi)$ and $w = u + iv$, equation (1) becomes

$$e^u e^{iv} = re^{i\Theta}.$$

Then, in view of the statement in italics in Sec. 5 regarding the equality of two complex numbers expressed in exponential form, $e^u = r$ and $v = \Theta + 2n\pi$, where n is any integer. Since the equation $e^u = r$ is the same as $u = \ln r$, it follows that equation (1) is satisfied if and only if w has one of the values

$$w = \ln r + i(\Theta + 2n\pi) \qquad (n = 0, \pm 1, \pm 2, \ldots).$$

Thus, if we write

$$\text{(2)} \qquad \log z = \ln r + i(\Theta + 2n\pi) \qquad (n = 0, \pm 1, \pm 2, \ldots),$$

we have the simple relation

$$\text{(3)} \qquad\qquad\qquad e^{\log z} = z,$$

which serves to motivate expression (2) as the *definition* of the (multiple-valued) logarithmic function of a nonzero complex variable $z = re^{i\Theta}$.

   If z is a nonzero complex number, with exponential form $z = re^{i\theta}$, then θ has any one of the values $\theta = \Theta + 2n\pi$ $(n = 0, \pm 1, \pm 2, \ldots)$, where $\Theta = \text{Arg } z$. Hence equation (2) can be put in the form

$$\text{(4)} \qquad\qquad\qquad \log z = \ln r + i\theta.$$

That is,

$$\text{(5)} \qquad\qquad \log z = \ln |z| + i \arg z \, , \qquad (z \neq 0).$$

It should be emphasized that it is not always true that the left-hand side of equation (3) with the order of the exponential and logarithmic functions reversed is equal to z. This is evident from the fact that $\log(e^z)$ has an infinite number of values for any given z. More precisely, since (Sec. 23)

$$|e^z| = e^x \quad \text{and} \quad \arg(e^z) = y + 2n\pi \quad (n = 0, \pm, 1\pm, 2, \ldots)$$

when $z = x + iy$, expression (5) tells us that

$$\log(e^z) = \ln|e^z| + i\arg(e^z) = x + i(y + 2n\pi),$$

or

(6) $$\log(e^z) = z + 2n\pi i \quad (n = 0, \pm 1, \pm 2, \ldots).$$

The *principal value* of $\log z$ is the value obtained from equation (2) when $n = 0$ and is denoted by $\operatorname{Log} z$. Thus

(7) $$\operatorname{Log} z = \ln r + i\Theta,$$

or

(8) $$\operatorname{Log} z = \ln|z| + i\operatorname{Arg} z \quad (z \neq 0).$$

Note that

$$\log z = \operatorname{Log} z + 2n\pi i \quad (n = 0, \pm 1, \pm 2, \ldots).$$

The function $\operatorname{Log} z$ is evidently well defined and single-valued when $z \neq 0$. It reduces to the usual natural logarithm in calculus when z is a positive real number $z = r$. To see this, one need only write $z = re^{i0}$, in which case equation (7) becomes $\operatorname{Log} z = \ln r$; that is, $\operatorname{Log} r = \ln r$.

EXAMPLES.   From expression (2), we find that

$$\log 1 = 2n\pi i \quad (n = 0, \pm 1, \pm 2, \ldots)$$

and

$$\log(-1) = (2n + 1)\pi i \quad (n = 0, \pm 1, \pm 2, \ldots).$$

In particular, $\operatorname{Log} 1 = 0$ and $\operatorname{Log}(-1) = \pi i$.

If we let α denote any real number and restrict the values of θ in expression (4) to the interval $\alpha < \theta < \alpha + 2\pi$, the function

(9) $$\log z = \ln r + i\theta \quad (r > 0, \alpha < \theta < \alpha + 2\pi),$$

with components

(10) $$u(r, \theta) = \ln r \quad \text{and} \quad v(r, \theta) = \theta,$$

is *single-valued* and continuous in the stated domain (Fig. 25). Note that if the function (9) were to be defined on the ray $\theta = \alpha$, it would not be continuous there. For,

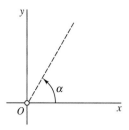

FIGURE 25

if z is a point on that ray, there are points arbitrarily close to z at which the values of v are near α and also points such that the values of v are near $\alpha + 2\pi$.

The function (9) is not only continuous but also analytic in the domain $r > 0$, $\alpha < \theta < \alpha + 2\pi$ since the first-order partial derivatives of u and v are continuous there and satisfy the polar form (Sec. 19)

$$u_r = \frac{1}{r}v_\theta, \qquad \frac{1}{r}u_\theta = -v_r$$

of the Cauchy-Riemann equations. Furthermore, according to Sec. 19,

$$\frac{d}{dz}\log z = e^{-i\theta}(u_r + iv_r) = e^{-i\theta}\left(\frac{1}{r} + i0\right) = \frac{1}{re^{i\theta}};$$

that is,

(11) $$\frac{d}{dz}\log z = \frac{1}{z} \qquad (|z| > 0, \alpha < \arg z < \alpha + 2\pi).$$

In particular,

(12) $$\frac{d}{dz}\operatorname{Log} z = \frac{1}{z} \qquad (|z| > 0, -\pi < \operatorname{Arg} z < \pi).$$

A *branch* of a multiple-valued function f is any single-valued function F that is analytic in some domain at each point z of which the value $F(z)$ is one of the values $f(z)$. The requirement of analyticity, of course, prevents F from taking on a random selection of the values of f. Observe that, for each fixed α, the single-valued function (9) is a branch of the multiple-valued function (4). The function

(13) $$\operatorname{Log} z = \ln r + i\Theta \qquad (r > 0, -\pi < \Theta < \pi)$$

is called the *principal branch*.

A *branch cut* is a portion of a line or curve that is introduced in order to define a branch F of a multiple-valued function f. Points on the branch cut for F are singular points (Sec. 20) of F, and any point that is common to all branch cuts of f is called a *branch point*. The origin and the ray $\theta = \alpha$ make up the branch cut for the branch (9) of the logarithmic function. The branch cut for the principal branch (13) consists of the origin and the ray $\Theta = \pi$. The origin is evidently a branch point for branches of the multiple-valued logarithmic function.

27. SOME IDENTITIES INVOLVING LOGARITHMS

As suggested by relations (3) and (6) in Sec. 26, various identities involving loga-rithms of positive real numbers carry over to complex analysis, with certain modifi-cations. In this section we derive a few of them.

If z_1 and z_2 denote any two nonzero complex numbers, it is straightforward to show that

$$(1) \qquad\qquad \log(z_1 z_2) = \log z_1 + \log z_2.$$

This statement, involving a multiple-valued function, is to be interpreted in the same way that the statement

$$(2) \qquad\qquad \arg(z_1 z_2) = \arg z_1 + \arg z_2$$

was in Sec. 6. That is, if values of two of the three logarithms are specified, then there is a value of the third logarithm such that equation (1) holds.

The proof of statement (1) can be based on statement (2) in the following way. Since $|z_1 z_2| = |z_1||z_2|$ and since these moduli are all positive real numbers, we know from experience with logarithms of such numbers in calculus that

$$\ln|z_1 z_2| = \ln|z_1| + \ln|z_2|.$$

So it follows from this and equation (2) that

$$(3) \qquad \ln|z_1 z_2| + i\arg(z_1 z_2) = (\ln|z_1| + i\arg z_1) + (\ln|z_2| + i\arg z_2).$$

Finally, because of the way in which equations (1) and (2) are to be interpreted, equation (3) is the same as equation (1).

In like manner, it can be shown that

$$(4) \qquad\qquad \log\left(\frac{z_1}{z_2}\right) = \log z_1 - \log z_2.$$

EXAMPLE.   To illustrate statement (1), write $z_1 = z_2 = -1$ and note that $z_1 z_2 = 1$. If the values $\log z_1 = \pi i$ and $\log z_2 = -\pi i$ are specified, equation (1) is evidently satisfied when the value $\log(z_1 z_2) = 0$ is chosen.

Observe that, for the same numbers z_1 and z_2,

$$\mathrm{Log}(z_1 z_2) = 0 \qquad \text{and} \qquad \mathrm{Log}\, z_1 + \mathrm{Log}\, z_2 = 2\pi i.$$

Thus statement (1) is not, in general, valid when *log* is replaced everywhere by *Log*. A similar remark applies to statement (4).

We include here two other properties of $\log z$ that will be of special interest in Sec. 28. If z is a nonzero complex number, then

$$(5) \qquad\qquad z^n = e^{n\log z} \qquad (n = 0, \pm 1, \pm 2, \ldots)$$

for any value of $\log z$ that is taken. When $n = 1$, this reduces, of course, to relation (3), Sec. 26. Equation (5) is readily verified by writing $z = re^{i\theta}$ and noting that each side becomes $r^n e^{in\theta}$.

It is also true that when $z \neq 0$,

(6)
$$z^{1/n} = \exp\left(\frac{1}{n} \log z\right) \qquad (n = 1, 2, \ldots).$$

That is, the term on the right here has n distinct values, and those values are the nth roots of z. To prove this, we write $z = r \exp(i\Theta)$, where Θ is the principal value of $\arg z$. Then, in view of expression (2), Sec. 26, for $\log z$,

$$\exp\left(\frac{1}{n} \log z\right) = \exp\left[\frac{1}{n} \ln r + \frac{i(\Theta + 2k\pi)}{n}\right],$$

where $k = 0, \pm 1, \pm 2, \ldots$. Thus

(7)
$$\exp\left(\frac{1}{n} \log z\right) = \sqrt[n]{r} \exp\left[i\left(\frac{\Theta}{n} + \frac{2k\pi}{n}\right)\right] \qquad (k = 0, \pm 1, \pm 2, \ldots).$$

Because $\exp(i2k\pi/n)$ has distinct values only when $k = 0, 1, \ldots, n - 1$, the right-hand side of equation (7) has only n values. That right-hand side is, in fact, an expression for the nth roots of z (Sec. 7), and so it can be written $z^{1/n}$. This establishes property (6), which is actually valid when n is a negative integer too (see Exercise 16).

Mappings by means of logarithms play a significant role in the applications later on. A reader who is interested in learning some mapping properties of the logarithmic function, as well as of the exponential function, now has adequate background to read all but the last example in Sec. 73 (Chap. 8).

EXERCISES

1. Show that

    (a) $\operatorname{Log}(-ei) = 1 - \dfrac{\pi}{2}i$;   (b) $\operatorname{Log}(1 - i) = \dfrac{1}{2} \ln 2 - \dfrac{\pi}{4}i$.

2. Verify that when $n = 0, \pm 1, \pm 2, \ldots$,

    (a) $\log e = 1 + 2n\pi i$;   (b) $\log i = \left(2n + \dfrac{1}{2}\right)\pi i$;

    (c) $\log(-1 + \sqrt{3}i) = \ln 2 + 2\left(n + \dfrac{1}{3}\right)\pi i$.

3. Show that

    (a) $\operatorname{Log}(1 + i)^2 = 2\operatorname{Log}(1 + i)$;   (b) $\operatorname{Log}(-1 + i)^2 \neq 2\operatorname{Log}(-1 + i)$.

4. Show that

    (a) $\log(i^2) = 2\log i$ when $\log z = \ln r + i\theta$ $\left(r > 0, \dfrac{\pi}{4} < \theta < \dfrac{9\pi}{4}\right)$;

    (b) $\log(i^2) \neq 2\log i$ when $\log z = \ln r + i\theta$ $\left(r > 0, \dfrac{3\pi}{4} < \theta < \dfrac{11\pi}{4}\right)$.

5. Show that

    (a) the set of values of $\log(i^{1/2})$ is $(n + \frac{1}{4})\pi i$ $(n = 0, \pm 1, \pm 2, \ldots)$ and that the same is true of $\frac{1}{2}\log i$;

    (b) the set of values of $\log(i^2)$ is *not* the same as the set of values of $2\log i$.

6. Given that the branch $\log z = \ln r + i\theta$ $(r > 0, \alpha < \theta < \alpha + 2\pi)$ of the logarithmic function is analytic at each point z in the stated domain, obtain its derivative by differentiating each side of the identity $\exp(\log z) = z$ (Sec. 26) and using the chain rule.

7. Find all roots of the equation $\log z = (\pi/2)i$.

     Ans. $z = i$.

8. Suppose that the point $z = x + iy$ lies in the horizontal strip $\alpha < y < \alpha + 2\pi$. Show that when the branch $\log z = \ln r + i\theta$ $(r > 0, \alpha < \theta < \alpha + 2\pi)$ of the logarithmic function is used, $\log(e^z) = z$.

9. Show that if $\operatorname{Re} z_1 > 0$ and $\operatorname{Re} z_2 > 0$, then

$$\operatorname{Log}(z_1 z_2) = \operatorname{Log} z_1 + \operatorname{Log} z_2.$$

10. Show that, for any two nonzero complex numbers z_1 and z_2,

$$\operatorname{Log}(z_1 z_2) = \operatorname{Log} z_1 + \operatorname{Log} z_2 + 2N\pi i$$

where N has one of the values $0, \pm 1$. (Compare Exercise 9.)

11. Verify expression (4), Sec. 27, for $\log(z_1/z_2)$

    (*a*) by using the fact that $\arg(z_1/z_2) = \arg z_1 - \arg z_2$ (Sec. 6);

    (*b*) by first showing that $\log(1/z) = -\log z$ $(z \neq 0)$, in the sense that $\log(1/z)$ and $-\log z$ have the same set of values, and then referring to expression (1), Sec. 27, for $\log(z_1 z_2)$.

12. By choosing specific nonzero values of z_1 and z_2, show that expression (4), Sec. 27, for $\log(z_1/z_2)$ is not always valid when *log* is replaced by *Log*.

13. Show that

    (*a*) the function $\operatorname{Log}(z - i)$ is analytic everywhere except on the half line $y = 1$ $(x \leq 0)$;

    (*b*) the function

$$\frac{\operatorname{Log}(z + 4)}{z^2 + i}$$

    is analytic everywhere except at the points $\pm(1 - i)/\sqrt{2}$ and on the portion $x \leq -4$ of the real axis.

14. Show in two ways that the function $\ln(x^2 + y^2)$ is harmonic in every domain that does not contain the origin.

15. Show that

$$\operatorname{Re}[\log(z - 1)] = \frac{1}{2}\ln[(x - 1)^2 + y^2] \qquad (z \neq 1).$$

Why must this function satisfy Laplace's equation when $z \neq 1$?

16. Show that property (6), Sec. 27, also holds when n is a negative integer. Do this by writing $z^{1/n} = (z^{1/m})^{-1}(m = -n)$, where n has any one of the negative values $n = -1, -2, \ldots$ (see Exercise 3, Sec. 7), and using the fact that the property is already known to be valid for positive integers.

17. Let z denote any nonzero complex number, written $z = re^{i\Theta}(-\pi < \Theta \leq \pi)$, and let n denote any fixed positive integer $(n = 1, 2, \ldots)$. Show that all of the values of $\log(z^{1/n})$ are given by the equation

$$\log(z^{1/n}) = \frac{1}{n}\ln r + i\frac{\Theta + 2(pn + k)\pi}{n},$$

where $p = 0, \pm 1, \pm 2, \ldots$ and $k = 0, 1, 2, \ldots, n - 1$. Then, after writing

$$\frac{1}{n} \log z = \frac{1}{n} \ln r + i \frac{\Theta + 2q\pi}{n},$$

where $q = 0, \pm 1, \pm 2, \ldots$, show that the set of values of $\log(z^{1/n})$ is the same as the set of values of $(1/n) \log z$. Thus show that $\log(z^{1/n}) = (1/n) \log z$, where, corresponding to a value of $\log(z^{1/n})$ taken on the left, the appropriate value of $\log z$ is to be selected on the right, and conversely. [The result in Exercise 5(a) is a special case of this one.]

    Suggestion: Use the fact that the remainder upon dividing an integer by a positive integer n is always an integer between 0 and $n - 1$, inclusive; that is, when a positive integer n is specified, any integer q can be written $q = pn + k$, where p is an integer and k has one of the values $k = 0, 1, 2, \ldots, n - 1$.

28. COMPLEX EXPONENTS

When $z \neq 0$ and *the exponent c is any complex number,* the function z^c is defined by means of the equation

$$(1) \qquad\qquad z^c = e^{c \log z},$$

where $\log z$ denotes the multiple-valued logarithmic function. Equation (1) provides a consistent definition of z^c in the sense that the equation is already known (Sec. 27) to be valid when $c = n$ ($n = 0, \pm 1, \pm 2, \ldots$) and $c = 1/n$ ($n = \pm 1, \pm 2, \ldots$). The definition of z^c is, in fact, suggested by those particular choices of c.

    EXAMPLE 1. Powers of z are, in general, multiple-valued, as is illustrated by writing

$$i^{-2i} = \exp(-2i \log i) = \exp\left[-2i\left(2n + \frac{1}{2}\right)\pi i\right] = \exp[(4n + 1)\pi],$$

where $n = 0, \pm 1, \pm 2, \ldots$. Note that, since the exponential function has the property $1/e^z = e^{-z}$, the two sets of numbers $1/z^c$ and z^{-c} are the same. So one can write

$$(2) \qquad\qquad \frac{1}{z^c} = z^{-c};$$

and, in particular,

$$\frac{1}{i^{2i}} = \exp[(4n + 1)\pi] \qquad (n = 0, \pm 1, \pm 2, \ldots).$$

    If $z = re^{i\theta}$ and α is any real number, the branch

$$\log z = \ln r + i\theta \qquad (r > 0, \alpha < \theta < \alpha + 2\pi)$$

of the logarithmic function is single-valued and analytic in the indicated domain (Sec. 26). When that branch is used, it follows that the function $z^c = \exp(c \log z)$ is single-valued and analytic in the same domain. The derivative of such a *branch of* z^c is found by writing

$$\frac{d}{dz}z^c = \frac{d}{dz}\exp(c\log z) = [\exp(c\log z)]\frac{c}{z} = c\frac{\exp(c\log z)}{\exp(\log z)} = c\exp[(c-1)\log z]$$

and observing that the final term here is the single-valued function cz^{c-1}, which is also defined on the domain $r > 0, \alpha < \theta < \alpha + 2\pi$. That is,

(3) $$\frac{d}{dz}z^c = cz^{c-1} \qquad (|z| > 0, \alpha < \arg z < \alpha + 2\pi).$$

The *principal value* of z^c occurs when $\log z$ is replaced by $\mathrm{Log}\, z$ in definition (1):

(4) $$z^c = e^{c\,\mathrm{Log}\,z}.$$

Equation (4) also serves to define the *principal branch* of the function z^c on the domain $|z| > 0, -\pi < \mathrm{Arg}\, z < \pi$.

EXAMPLE 2. The principal value of $(-i)^i$ is

$$\exp[i\,\mathrm{Log}(-i)] = \exp\left[i\left(-\frac{\pi}{2}i\right)\right] = \exp\frac{\pi}{2}.$$

EXAMPLE 3. The principal branch of $z^{2/3}$ can be written

$$\exp\left(\frac{2}{3}\mathrm{Log}\,z\right) = \exp\left(\frac{2}{3}\ln r + \frac{2}{3}i\Theta\right) = \sqrt[3]{r^2}\exp\left(i\frac{2\Theta}{3}\right).$$

It is analytic in the domain $r > 0, -\pi < \Theta < \pi$, as one can also show directly from the theorem in Sec. 19.

According to definition (1), *the exponential function with base c*, where c is any nonzero complex constant, is written

(5) $$c^z = e^{z\log c}.$$

Note that although e^z is, in general, multiple-valued according to definition (5), the usual interpretation of e^z occurs when the principal value of the logarithm is taken. For the principal value of $\log e$ is unity.

When a value of $\log c$ is specified, c^z is an entire function of z. In fact,

$$\frac{d}{dz}c^z = \frac{d}{dz}e^{z\log c} = e^{z\log c}\log c;$$

and this shows that

(6) $$\frac{d}{dz}c^z = c^z \log c.$$

29. INVERSE TRIGONOMETRIC AND HYPERBOLIC FUNCTIONS

Inverses of the trigonometric and hyperbolic functions can be described in terms of logarithms.

To define the inverse sine function $\sin^{-1} z$, we write $w = \sin^{-1} z$ when $z = \sin w$. That is, $w = \sin^{-1} z$ when

$$z = \frac{e^{iw} - e^{-iw}}{2i}.$$

Let us put this equation in the form

$$(e^{iw})^2 - 2iz(e^{iw}) - 1 = 0,$$

which is quadratic in e^{iw}. Solving for e^{iw} [see Exercise 8(a), Sec. 7], we find that

$$(1) \qquad\qquad e^{iw} = iz + (1 - z^2)^{1/2},$$

where $(1 - z^2)^{1/2}$ is, of course, a double-valued function of z. Taking logarithms of each side of equation (1) and recalling that $w = \sin^{-1} z$, we arrive at the expression

$$(2) \qquad\qquad \sin^{-1} z = -i \log[iz + (1 - z^2)^{1/2}].$$

The following example illustrates the fact that $\sin^{-1} z$ is a multiple-valued function, with infinitely many values at each point z.

EXAMPLE.   Expression (2) tells us that

$$\sin^{-1}(-i) = -i \log(1 \pm \sqrt{2}).$$

But

$$\log(1 + \sqrt{2}) = \ln(1 + \sqrt{2}) + 2n\pi i \qquad (n = 0, \pm 1, \pm 2, \ldots)$$

and

$$\log(1 - \sqrt{2}) = \ln(\sqrt{2} - 1) + (2n + 1)\pi i \qquad (n = 0, \pm 1, \pm 2, \ldots).$$

Since

$$\ln(\sqrt{2} - 1) = \ln \frac{1}{1 + \sqrt{2}} = -\ln(1 + \sqrt{2}),$$

then, the numbers

$$(-1)^n \ln(1 + \sqrt{2}) + n\pi i \qquad (n = 0, \pm 1, \pm 2, \ldots)$$

constitute the set of values of $\log(1 \pm \sqrt{2})$. Thus

$$\sin^{-1}(-i) = n\pi + i(-1)^{n+1} \ln(1 + \sqrt{2}) \qquad (n = 0, \pm 1, \pm 2, \ldots).$$

One can apply the technique used to derive expression (2) for $\sin^{-1} z$ to show that

$$(3) \qquad \cos^{-1} z = -i \log[z + i(1 - z^2)^{1/2}]$$

and that

$$(4) \qquad \tan^{-1} z = \frac{i}{2} \log \frac{i + z}{i - z}.$$

The functions $\cos^{-1} z$ and $\tan^{-1} z$ are also multiple-valued. When specific branches of the square root and logarithmic functions are used, all three inverse functions become single-valued and analytic because they are then compositions of analytic functions.

The derivatives of these three functions are readily obtained from the above expressions. The derivatives of the first two depend on the values chosen for the square roots:

$$(5) \qquad \frac{d}{dz} \sin^{-1} z = \frac{1}{(1 - z^2)^{1/2}},$$

$$(6) \qquad \frac{d}{dz} \cos^{-1} z = \frac{-1}{(1 - z^2)^{1/2}}.$$

The derivative of the last one,

$$(7) \qquad \frac{d}{dz} \tan^{-1} z = \frac{1}{1 + z^2},$$

does not, however, depend on the manner in which the function is made single-valued.

Inverse hyperbolic functions can be treated in a corresponding manner. It turns out that

$$(8) \qquad \sinh^{-1} z = \log[z + (z^2 + 1)^{1/2}],$$

$$(9) \qquad \cosh^{-1} z = \log[z + (z^2 - 1)^{1/2}],$$

and

$$(10) \qquad \tanh^{-1} z = \frac{1}{2} \log \frac{1 + z}{1 - z}.$$

Finally, we remark that common alternative notation for all of these inverse functions is $\arcsin z$, etc.

EXERCISES

1. Show that when $n = 0, \pm 1, \pm 2, \ldots,$

(a) $(1 + i)^i = \exp\left(-\frac{\pi}{4} + 2n\pi\right) \exp\left(\frac{i}{2} \ln 2\right);$ (b) $(-1)^{1/\pi} = e^{(2n+1)i}.$

2. Find the principal value of

(a) i^i;   (b) $\left[\dfrac{e}{2}(-1 - \sqrt{3}i)\right]^{3\pi i}$;   (c) $(1 - i)^{4i}$.

    Ans. (a) $\exp(-\pi/2)$;   (b) $-\exp(2\pi^2)$;   (c) $e^{\pi}[\cos(2 \ln 2) + i \sin(2 \ln 2)]$.

3. Use definition (1), Sec. 28, of z^c to show that $(-1 + \sqrt{3}i)^{3/2} = \pm 2\sqrt{2}$.

4. Show that the result in Exercise 3 could have been obtained by writing

    (a) $(-1 + \sqrt{3}i)^{3/2} = [(-1 + \sqrt{3}i)^{1/2}]^3$ and first finding the square roots of $-1 + \sqrt{3}i$;

    (b) $(-1 + \sqrt{3}i)^{3/2} = [(-1 + \sqrt{3}i)^3]^{1/2}$ and first cubing $-1 + \sqrt{3}i$.

5. Show that the *principal* nth *root* of a nonzero complex number z_0, defined in Sec. 7, is the same as the principal value of $z_0^{1/n}$, defined in Sec. 28.

6. Show that if $z \neq 0$ and a is a real number, then $|z^a| = \exp(a \ln |z|) = |z|^a$, where the principal value of $|z|^a$ is to be taken.

7. Let $c = a + bi$ be a fixed complex number, where $c \neq 0, \pm 1, \pm 2, \ldots$, and note that i^c is multiple-valued. What restriction must be placed on the constant c so that the values of $|i^c|$ are all the same?

    Ans. c is real.

8. Let c, d, and z denote complex numbers, where $z \neq 0$. Prove that if all of the powers involved are principal values, then

    (a) $1/z^c = z^{-c}$;   (b) $(z^c)^n = z^{cn}$ ($n = 1, 2, \ldots$);

    (c) $z^c z^d = z^{c+d}$;   (d) $z^c/z^d = z^{c-d}$.

9. Assuming that $f'(z)$ exists, state the differentiation formula for $d[c^{f(z)}]/dz$.

10. Find all the values of

    (a) $\tan^{-1}(2i)$;   (b) $\tan^{-1}(1 + i)$;   (c) $\cosh^{-1}(-1)$;   (d) $\tanh^{-i} 0$.

$$\text{Ans.} \quad (a)\left(n + \frac{1}{2}\right)\pi + \frac{i}{2}\ln 3 \quad (n = 0, \pm 1, \pm 2, \ldots);$$

$$(d)\ n\pi i \quad (n = 0, \pm 1, \pm 2, \ldots).$$

11. Solve the equation $\sin z = 2$ for z

    (a) by equating real parts and imaginary parts in that equation;

    (b) using expression (2), Sec. 29, for $\sin^{-1} z$.

$$\text{Ans.}\ \left(2n + \frac{1}{2}\right)\pi \pm i \ln(2 + \sqrt{3}) \quad (n = 0, \pm 1, \pm 2, \ldots).$$

12. Solve the equation $\cos z = \sqrt{2}$ for z.

13. Derive formula (5), Sec. 29, for the derivative of $\sin^{-1} z$.

14. Derive expression (4), Sec. 29, for $\tan^{-1} z$.

15. Derive formula (7), Sec. 29, for the derivative of $\tan^{-1} z$.

16. Derive expression (9), Sec. 29, for $\cosh^{-1} z$.

CHAPTER
4

INTEGRALS

Integrals are extremely important in the study of functions of a complex variable. The theory of integration, to be developed in this chapter, is noted for its mathematical elegance. The theorems are generally concise and powerful, and most of the proofs are simple.

30. COMPLEX-VALUED FUNCTIONS $w(t)$

In order to introduce integrals of $f(z)$ in a fairly simple way, we first consider derivatives and definite integrals of complex-valued functions w of a real variable t. We write

(1)
$$w(t) = u(t) + iv(t),$$

where the functions u and v are *real-valued* functions of t.

The derivative $w'(t)$, or $d[w(t)]/dt$, of the function (1) at a point t is defined as

(2)
$$w'(t) = u'(t) + iv'(t),$$

provided each of the derivatives u' and v' exists at t.

From definition (2), it follows that, for each complex constant $z_0 = x_0 + iy_0$,

$$\frac{d}{dt}[z_0 w(t)] = \frac{d}{dt}[(x_0 + iy_0)(u + iv)]$$

$$= \frac{d}{dt}[(x_0 u - y_0 v) + i(y_0 u + x_0 v)]$$

$$= \frac{d}{dt}(x_0 u - y_0 v) + i\frac{d}{dt}(y_0 u + x_0 v)$$

$$= (x_0 u' - y_0 v') + i(y_0 u' + x_0 v')$$

$$= (x_0 + i y_0)(u' + i v').$$

That is,

(3)
$$\frac{d}{dt}[z_0 w(t)] = z_0 w'(t).$$

Various other rules learned in calculus, such as the ones for differentiating sums and products, apply just as they do for real-valued functions of t. As was the case with property (3), verifications may be based on the corresponding rules for real-valued functions. An expected differentiation formula, to be derived in the exercises, is

(4)
$$\frac{d}{dt} e^{z_0 t} = z_0 e^{z_0 t}.$$

It should be pointed out, however, that not every rule for derivatives in calculus carries over to functions of type (1). The following example illustrates this.

EXAMPLE 1.   Suppose that $w(t)$ is continuous on an interval $a \leq t \leq b$; that is, its component functions $u(t)$ and $v(t)$ are continuous there. Even if $w'(t)$ exists when $a < t < b$, the mean value theorem for derivatives no longer applies. To be precise, it is not necessarily true that there is a number c in the interval $a < t < b$ such that

$$w'(c) = \frac{w(b) - w(a)}{b - a}.$$

To see this, we need only consider the function $w(t) = e^{it}$ on the interval $0 \leq t \leq 2\pi$. For, when that function is used, $|w'(t)| = |i e^{it}| = 1$; and this means that $w'(t)$ is never zero, while $w(2\pi) - w(0) = 0$.

Definite integrals of functions of the type (1) over intervals $a \leq t \leq b$ are defined as

(5)
$$\int_a^b w(t)\, dt = \int_a^b u(t)\, dt + i \int_a^b v(t)\, dt$$

when the individual integrals on the right exist. Thus

(6) $\text{Re} \displaystyle\int_a^b w(t)\, dt = \int_a^b \text{Re}[w(t)]\, dt$ and $\text{Im} \displaystyle\int_a^b w(t)\, dt = \int_a^b \text{Im}[w(t)]\, dt.$

EXAMPLE 2.   As an illustration,

$$\int_0^1 (1 + it)^2\, dt = \int_0^1 (1 - t^2)\, dt + i \int_0^1 2t\, dt = \frac{2}{3} + i.$$

Improper integrals of $w(t)$ over unbounded intervals are defined in a similar way.

The existence of the integrals of u and v in definition (5) is ensured if those functions are *piecewise continuous* on the interval $a \le t \le b$. Such a function is continuous everywhere in the stated interval except possibly for a finite number of points where, although discontinuous, it has one-sided limits. Of course, only the right-hand limit is required at a; and only the left-hand limit is required at b. When both u and v are piecewise continuous, the function w is said to have that property.

Anticipated rules for integrating a complex constant times a function $w(t)$, for integrating sums of such functions, and for interchanging limits of integration are all valid. Those rules, as well as the property

$$\int_a^b w(t)\,dt = \int_a^c w(t)\,dt + \int_c^b w(t)\,dt,$$

are easy to verify by recalling corresponding results in calculus.

The fundamental theorem of calculus, involving antiderivatives, can, moreover, be extended so as to apply to integrals of the type (5). To be specific, suppose that the functions

$$w(t) = u(t) + iv(t) \qquad \text{and} \qquad W(t) = U(t) + iV(t)$$

are continuous on the interval $a \le t \le b$. If $W'(t) = w(t)$ when $a \le t \le b$, then $U'(t) = u(t)$ and $V'(t) = v(t)$. Hence, in view of definition (5),

$$\int_a^b w(t)\,dt = U(t)\Big]_a^b + iV(t)\Big]_a^b$$

$$= [U(b) + iV(b)] - [U(a) + iV(a)].$$

That is,

(7)
$$\int_a^b w(t)\,dt = W(t)\Big]_a^b = W(b) - W(a).$$

EXAMPLE 3.   Since $(e^{it})' = ie^{it}$,

$$\int_0^{\pi/4} e^{it}\,dt = -ie^{it}\Big]_0^{\pi/4} = -ie^{i\pi/4} + i$$

$$= -i\left(\frac{1}{\sqrt{2}} + \frac{i}{\sqrt{2}}\right) + i = \frac{1}{\sqrt{2}} + i\left(1 - \frac{1}{\sqrt{2}}\right).$$

We finish here with an important property of absolute values of integrals. Namely,

(8)
$$\left|\int_a^b w(t)\,dt\right| \le \int_a^b |w(t)|\,dt \qquad (a \le b).$$

This inequality clearly holds when the value of the integral on the left is zero, in particular when $a = b$. Thus, in the verification, we may assume that its value is a *nonzero* complex number. If r_0 is the modulus and θ_0 is an argument of that constant, then

$$\int_a^b w \, dt = r_0 e^{i\theta_0}.$$

Solving for r_0, we write

(9)
$$r_0 = \int_a^b e^{-i\theta_0} w \, dt.$$

Now the left-hand side of this equation is a real number, and so the right-hand side is too. Thus, using the fact that the real part of a real number is the number itself and referring to the first of properties (6), we see that the right-hand side of equation (9) can be rewritten in the following way:

$$\int_a^b e^{-i\theta_0} w \, dt = \text{Re} \int_a^b e^{-i\theta_0} w \, dt = \int_a^b \text{Re}(e^{-i\theta_0} w) \, dt.$$

Equation (9) then takes the form

(10)
$$r_0 = \int_a^b \text{Re}(e^{-i\theta_0} w) \, dt.$$

But

$$\text{Re}(e^{-i\theta_0} w) \le |e^{-i\theta_0} w| = |e^{-i\theta_0}||w| = |w|;$$

and so, according to equation (10),

$$r_0 \le \int_a^b |w| \, dt.$$

Because r_0 is, in fact, the left-hand side of inequality (8) when the value of the integral there is nonzero, the verification is now complete.

With only minor modifications, the above discussion yields inequalities such as

(11)
$$\left| \int_a^\infty w(t) \, dt \right| \le \int_a^\infty |w(t)| \, dt,$$

provided both improper integrals exist.

31. CONTOURS

Integrals of complex-valued functions of a *complex* variable are defined on curves in the complex plane, rather than on just intervals of the real line. Classes of curves that are adequate for the study of such integrals are introduced in this section.

A set of points $z = (x, y)$ in the complex plane is said to be an *arc* if

(1) $x = x(t), \qquad y = y(t) \qquad (a \le t \le b),$

where $x(t)$ and $y(t)$ are continuous functions of the real parameter t. This definition establishes a continuous mapping of the interval $a \le t \le b$ into the xy, or z, plane; and the image points are ordered according to increasing values of t. It is convenient to describe the points of C by means of the equation

(2) $z = z(t) \qquad (a \le t \le b),$

where

(3) $z(t) = x(t) + iy(t).$

The arc C is a *simple arc*, or a Jordan arc, if it does not cross itself; that is, C is simple if $z(t_1) \ne z(t_2)$ when $t_1 \ne t_2$. When the arc C is simple except for the fact that $z(b) = z(a)$, we say that C is a *simple closed curve*, or a Jordan curve.

The geometric nature of a particular arc often suggests different notation for the parameter t in equation (2). This is, in fact, the case in the examples below.

EXAMPLE 1.   The polygonal line

(4)
$$z = \begin{cases} x + ix & \text{when } 0 \le x \le 1, \\ x + i & \text{when } 1 \le x \le 2, \end{cases}$$

consisting of a line segment from 0 to $1 + i$ followed by one from $1 + i$ to $2 + i$ (Fig. 26), is a simple arc.

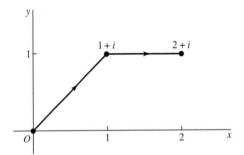

FIGURE 26

EXAMPLE 2.   The unit circle

(5) $z = e^{i\theta} \qquad (0 \le \theta \le 2\pi)$

about the origin is a simple closed curve, oriented in the counterclockwise direction. So is the circle

(6) $z = z_0 + Re^{i\theta} \qquad (0 \le \theta \le 2\pi),$

centered at the point z_0 and with radius R (see Sec. 5).

The same set of points can make up different arcs.

EXAMPLE 3.   The arc

(7) $z = e^{-i\theta}$ $(0 \le \theta \le 2\pi)$

is not the same as the arc described by equation (5). The set of points is the same, but now the circle is traversed in the *clockwise* direction.

EXAMPLE 4.   The points on the arc

(8) $z = e^{i2\theta}$ $(0 \le \theta \le 2\pi)$

are the same as those making up the arcs (5) and (7). The arc here differs, however, from each of those arcs since the circle is traversed *twice* in the counterclockwise direction.

Suppose that the derivatives $x'(t)$ and $y'(t)$ of the components of the function (3), used to describe an arc C, exist and are continuous throughout the entire interval $a \le t \le b$. Under these conditions, C is called a *differentiable arc*. It follows that, since the derivative of $z(t)$ is (Sec. 30)

(9) $z'(t) = x'(t) + iy'(t),$

the real-valued function

$$|z'(t)| = \sqrt{[x'(t)]^2 + [y'(t)]^2}$$

is integrable over the interval $a \le t \le b$, the length of the arc being

(10) $L = \int_a^b |z'(t)|\, dt.$

Expression (10) is the definition of arc length in calculus.

The parametric representation used for C is, of course, not unique; and the value of L given by expression (10) is invariant under certain changes in that representation, as one would expect. To be specific, suppose that

(11) $t = \phi(\tau)$ $(\alpha \le \tau \le \beta),$

where ϕ is a real-valued function mapping the interval $\alpha \le \tau \le \beta$ onto the interval $a \le t \le b$. We assume that ϕ is continuous with a continuous derivative. We also assume that $\phi'(\tau) > 0$ for each τ; this ensures that t increases with τ. With the change of variable indicated in equation (11), expression (10) for the length of C takes the form [see Exercise 6(b)]

$$L = \int_\alpha^\beta |z'[\phi(\tau)]||\phi'(\tau)\, d\tau.$$

So, if C has the representation

(12) $z = Z(\tau) = z[\phi(\tau)]$ $(\alpha \le \tau \le \beta),$

the fact (Exercise 10) that

(13) $$Z'(\tau) = z'[\phi(\tau)]\phi'(\tau)$$

enables us to write expression (10) as

$$L = \int_{\alpha}^{\beta} |Z'(\tau)| \, d\tau.$$

Thus the same length of C would be obtained if representation (12) were to be used.

If the equation $z = z(t)$ $(a \leq t \leq b)$ represents a differentiable arc and $z'(t) \neq 0$ anywhere in the interval $a < t < b$, then the unit tangent vector

$$\mathbf{T} = \frac{z'(t)}{|z'(t)|}$$

is well defined for all t in that open interval, with angle of inclination arg $z'(t)$. Also, when $\mathbf{T}$ turns, it does so continuously as the parameter t varies over the interval $a < t < b$. This expression for $\mathbf{T}$ is the one learned in calculus when $z(t)$ is interpreted as a radius vector. Such an arc is said to be *smooth*. In referring to a smooth arc $z = z(t)$ $(a \leq t \leq b)$, then, we agree that the derivative $z'(t)$ is continuous on the closed interval $a \leq t \leq b$ and nonzero on the open interval $a < t < b$.

A *contour*, or piecewise smooth arc, is an arc consisting of a finite number of smooth arcs joined end to end. Hence if equation (2) represents a contour, $z(t)$ is continuous, whereas its derivative $z'(t)$ is piecewise continuous. The polygonal path (4) is, for example, a contour. When only the initial and final values of $z(t)$ are the same, a contour C is called a *simple closed contour*. Examples are the circles (5) and (6), as well as the boundary of a triangle or a rectangle taken in a specific direction. The length of a contour or a simple closed contour is the sum of the lengths of the smooth arcs that make up the contour.

The points on any simple closed curve or simple closed contour C are boundary points of two distinct domains, one of which is the interior of C and is bounded. The other, which is the exterior of C, is unbounded. It will be convenient to accept this statement, known as the *Jordan curve theorem*, as geometrically evident; the proof is not easy.*

EXERCISES

1. Evaluate the following integrals:

(a) $\int_{1}^{2} \left(\frac{1}{t} - i \right)^2 dt$;   (b) $\int_{0}^{\pi/6} e^{i2t} \, dt$;   (c) $\int_{0}^{\infty} e^{-zt} \, dt$ (Re $z > 0$).

Ans. (a) $-\dfrac{1}{2} - (2 \ln 2)i$;   (b) $\dfrac{\sqrt{3}}{4} + \dfrac{i}{4}$;   (c) $\dfrac{1}{z}$.

---

*See pp. 115–116 of the book by Newman or Sec. 13 of the one by Thron, both of which are cited in Appendix 1. The special case in which C is a simple closed polygon is proved on pp. 281–285 of Vol. 1 of the work by Hille, also cited in Appendix 1.

*2. Show that if m and n are integers,

$$\int_0^{2\pi} e^{im\theta} e^{-in\theta} d\theta = \begin{cases} 0 & \text{when } m \neq n, \\ 2\pi & \text{when } m = n. \end{cases}$$

3. According to definition (5), Sec. 30, of integrals of complex-valued functions of a real variable,

$$\int_0^\pi e^{(1+i)x} dx = \int_0^\pi e^x \cos x \, dx + i \int_0^\pi e^x \sin x \, dx.$$

Evaluate the two integrals on the right here by evaluating the single integral on the left and then identifying the real and imaginary parts of the value found.

$\quad\quad$ *Ans.* $-(1 + e^\pi)/2, \quad (1 + e^\pi)/2.$

4. Verify the following differentiation rules in the manner indicated.

$\quad$ (*a*) Using the corresponding rule in calculus, show that

$$\frac{d}{dt}[w(t)]^2 = 2w(t)w'(t)$$

$\quad$ when $w(t) = u(t) + iv(t)$ is a complex-valued function of a real variable t and $w'(t)$ exists.

$\quad$ (*b*) Use the expression $e^{z_0 t} = e^{x_0 t} \cos y_0 t + i e^{x_0 t} \sin y_0 t$, where $z_0 = x_0 + iy_0$ is a fixed complex number, to show that

$$\frac{d}{dt} e^{z_0 t} = z_0 e^{z_0 t}.$$

5. Apply inequality (8), Sec. 30, to show that for all values of x in the interval $-1 \leq x \leq 1$, the functions*

$$P_n(x) = \frac{1}{\pi} \int_0^\pi (x + i\sqrt{1 - x^2} \cos \theta)^n d\theta \quad\quad (n = 0, 1, 2, \ldots)$$

satisfy the inequality $|P_n(x)| \leq 1$.

6. Show that if a function $w(t) = u(t) + iv(t)$ is continuous on an interval $a \leq t \leq b$, then

$\quad$ (*a*) $\displaystyle\int_{-b}^{-a} w(-t) \, dt = \int_a^b w(\tau) \, d\tau;$ $\quad$ (*b*) $\displaystyle\int_a^b w(t) \, dt = \int_\alpha^\beta w[\phi(\tau)]\phi'(\tau) \, d\tau,$

$\quad$ where $\phi(\tau)$ is the function in equation (11), Sec. 31.

$\quad\quad$ *Suggestion:* These identities can be obtained by noting that they are valid for *real-valued* functions of t.

7. Let $w(t) = u(t) + iv(t)$ denote a continuous complex-valued function defined on an interval $-a \leq t \leq a$.

$\quad$ (*a*) Suppose that $w(t)$ is *even*; that is, $w(-t) = w(t)$ for each point t in the given interval. Show that

$$\int_{-a}^a w(t) \, dt = 2 \int_0^a w(t) \, dt.$$

---

*These functions are actually polynomials in x. They are known as *Legendre polynomials* and are important in applied mathematics. See, for example, Chap. 4 of the book by Lebedev that is listed in Appendix 1.

(b) Show that if $w(t)$ is an *odd* function, one where $w(-t) = -w(t)$ for each point t in the interval, then

$$\int_{-a}^{a} w(t)\,dt = 0.$$

Suggestion: In each part of this exercise, use the corresponding property of integrals of *real-valued* functions of t, which is graphically evident.

8. Let $w(t)$ be a continuous complex-valued function of t defined on an interval $a \leq t \leq b$. By considering the special case $w(t) = e^{it}$ on the interval $0 \leq t \leq 2\pi$, show that it is *not* always true that there is a number c in the interval $a < t < b$ such that

$$\int_{a}^{b} w(t)\,dt = w(c)(b - a).$$

Thus show that the mean value theorem for definite integrals in calculus does not apply to such functions. (Compare Example 1 in Sec. 30.)

9. Let C denote the right-hand half of the circle $|z| = 2$, in the counterclockwise direction, and note that two parametric representations for C are

$$z = z(\theta) = 2e^{i\theta} \qquad \left(-\frac{\pi}{2} \leq \theta \leq \frac{\pi}{2}\right)$$

and

$$z = Z(y) = \sqrt{4 - y^2} + iy \qquad (-2 \leq y \leq 2).$$

Verify that $Z(y) = z[\phi(y)]$, where

$$\phi(y) = \arctan \frac{y}{\sqrt{4 - y^2}} \qquad \left(-\frac{\pi}{2} < \arctan t < \frac{\pi}{2}\right).$$

Also, show that this function ϕ has a positive derivative, as required in the conditions following equation (11), Sec. 31.

10. Verify expression (13), Sec. 31, for the derivative of $Z(\tau) = z[\phi(\tau)]$.
    Suggestion: Write $Z(\tau) = x[\phi(\tau)] + iy[\phi(\tau)]$ and apply the chain rule for real-valued functions of a real variable.

11. Suppose that a function $f(z)$ is analytic at a point $z_0 = z(t_0)$ on a smooth arc $z = z(t)$ $(a \leq t \leq b)$. Show that if $w(t) = f[z(t)]$, then

$$w'(t) = f'[z(t)]z'(t)$$

when $t = t_0$.
    Suggestion: Write $f(z) = u(x, y) + iv(x, y)$ and $z(t) = x(t) + iy(t)$, so that

$$w(t) = u[x(t), y(t)] + iv[x(t), y(t)].$$

Then apply the chain rule in calculus for functions of two variables to write

$$w' = (u_x x' + u_y y') + i(v_x x' + v_y y'),$$

and use the Cauchy-Riemann equations.

12. Let $y(x)$ be a real-valued function defined on the interval $0 \leq x \leq 1$ by means of the equations

$$y(x) = \begin{cases} x^3 \sin\left(\dfrac{\pi}{x}\right) & \text{when } 0 < x \leq 1, \\ 0 & \text{when } x = 0. \end{cases}$$

(*a*) Show that the equation

$$z = x + iy(x) \qquad (0 \le x \le 1)$$

represents an arc C_1 that intersects the real axis at the points $z = 1/n$ $(n = 1, 2, \ldots)$ and $z = 0$, as shown in Fig. 27.

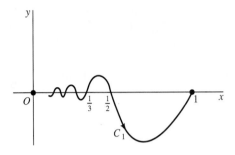

FIGURE 27

(*b*) Verify that the arc C_1 in part (*a*) is, in fact, a *smooth* arc.

Suggestion: To establish the continuity of $y(x)$ at $x = 0$, observe that

$$0 \le \left| x^3 \sin\left(\frac{\pi}{x}\right) \right| \le x^3$$

when $x > 0$. A similar remark applies in finding $y'(0)$ and showing that $y'(x)$ is continuous at $x = 0$.

32. CONTOUR INTEGRALS

We turn now to integrals of complex-valued functions f of the complex variable z. Such an integral is defined in terms of the values $f(z)$ along a given contour C, extending from a point $z = z_1$ to a point $z = z_2$ in the complex plane. It is, therefore, a line integral; and its value depends, in general, on the contour C as well as on the function f. It is written

$$\int_C f(z)\,dz \qquad \text{or} \qquad \int_{z_1}^{z_2} f(z)\,dz,$$

the latter notation often being used when the value of the integral is independent of the choice of the contour taken between the two end points. While the integral may be defined directly as the limit of a sum, we choose to define it in terms of a definite integral of the type introduced in Sec. 30.

Suppose that the equation

(1) $$z = z(t) \qquad (a \le t \le b)$$

represents a contour C, extending from a point $z_1 = z(a)$ to a point $z_2 = z(b)$. Let the function $f(z)$ be piecewise continuous on C; that is, $f[z(t)]$ is piecewise continuous on the interval $a \le t \le b$. We define the line integral, or *contour integral*, of f along

C as follows:

(2)
$$\int_C f(z)\,dz = \int_a^b f[z(t)]z'(t)\,dt.$$

Note that, since C is a contour, $z'(t)$ is also piecewise continuous on the interval $a \leq t \leq b$; and so the existence of integral (2) is ensured.

The value of a contour integral is invariant under a change in the representation of its contour when the change is of the type (12), Sec. 31. This can be seen by following the same general procedure that was used in Sec. 31 to show the invariance of arc length.

It follows immediately from definition (2) and properties of integrals of complex-valued functions $w(t)$ mentioned in Sec. 30 that

(3)
$$\int_C z_0 f(z)\,dz = z_0 \int_C f(z)\,dz,$$

for any complex constant z_0, and

(4)
$$\int_C [f(z) + g(z)]\,dz = \int_C f(z)\,dz + \int_C g(z)\,dz.$$

Associated with the contour C used in integral (2) is the contour $-C$, consisting of the same set of points but with the order reversed so that the new contour extends from the point z_2 to the point z_1. The contour $-C$ has parametric representation $z = z(-t)\,(-b \leq t \leq -a)$. Thus

$$\int_{-C} f(z)\,dz = \int_{-b}^{-a} f[z(-t)][-z'(-t)]\,dt,$$

where $z'(-t)$ denotes the derivative of $z(t)$ with respect to t, evaluated at $-t$. After a change of variable in this last integral [see Exercise 6(a), Sec. 31], we find that

(5)
$$\int_{-C} f(z)\,dz = -\int_C f(z)\,dz.$$

Suppose that C consists of a contour C_1 from z_1 to z_0 followed by a contour C_2 from z_0 to z_2, the initial point of C_2 being the final point of C_1. Then there is a real number c, where $z_0 = z(c)$, such that C_1 is represented by $z = z(t)\,(a \leq t \leq c)$ and C_2 is represented by $z = z(t)\,(c \leq t \leq b)$. Since

$$\int_C f(z)\,dz = \int_a^c f[z(t)]z'(t)\,dt + \int_c^b f[z(t)]z'(t)\,dt,$$

it is evident that

(6)
$$\int_C f(z)\,dz = \int_{C_1} f(z)\,dz + \int_{C_2} f(z)\,dz.$$

Sometimes the contour C is called the *sum* of its legs C_1 and C_2 and is denoted by $C_1 + C_2$. The sum of two contours C_1 and $-C_2$ is well defined when C_1 and C_2 have the same final points, and it is written $C_1 - C_2$.

Finally, according to definition (2) above and property (8), Sec. 30,

$$\left| \int_C f(z)\, dz \right| \leq \int_a^b |f[z(t)]z'(t)|\, dt.$$

So, for any nonnegative constant M such that the values of f on C satisfy the inequality $|f(z)| \leq M$,

$$\left| \int_C f(z)\, dz \right| \leq M \int_a^b |z'(t)|\, dt.$$

Since the integral on the right here represents the length L of the contour (see Sec. 31), it follows that the modulus of the value of the integral of f along C does not exceed ML:

(7)
$$\left| \int_C f(z)\, dz \right| \leq ML.$$

This is, of course, a strict inequality when the values of f on C are such that $|f(z)| < M$.

Note that since all of the paths of integration to be considered here are contours and the integrands are piecewise continuous functions defined on those contours, a number M such as the one appearing in inequality (7) will always exist. This is because the real-valued function $|f[z(t)]|$ is continuous on the closed bounded interval $a \leq t \leq b$ when f is *continuous* on C; and such a function always reaches a maximum value M on that interval.* Hence $|f(z)|$ has a maximum value on C when f is continuous on it. It now follows immediately that the same is true when f is *piecewise continuous* on C.

Definite integrals in calculus can be interpreted as areas, and they have other interpretations as well. Except in special cases, no corresponding helpful interpretation, geometric or physical, is available for integrals in the complex plane.

33. EXAMPLES

The purpose of this section is to provide examples of the definition in Sec. 32 of contour integrals and to illustrate various properties mentioned there. We defer development of the concept of antiderivatives of the integrands $f(z)$ in contour integrals until Sec. 34.

---

*See, for instance, A. E. Taylor and W. R. Mann, "Advanced Calculus," 3d ed., pp. 86–90, 1983.

EXAMPLE 1. Let us find the value of the integral

(1) $$I = \int_C \bar{z}\, dz$$

when C is the right-hand half

$$z = 2e^{i\theta} \qquad \left(-\frac{\pi}{2} \le \theta \le \frac{\pi}{2}\right)$$

of the circle $|z| = 2$, from $z = -2i$ to $z = 2i$ (Fig. 28). According to definition (2), Sec. 32,

$$I = \int_{-\pi/2}^{\pi/2} \overline{2e^{i\theta}}(2e^{i\theta})'\, d\theta;$$

and, since

$$\overline{e^{i\theta}} = e^{-i\theta} \qquad \text{and} \qquad (e^{i\theta})' = ie^{i\theta},$$

this means that

$$I = \int_{-\pi/2}^{\pi/2} 2e^{-i\theta}2ie^{i\theta}\, d\theta = 4i\int_{-\pi/2}^{\pi/2} d\theta = 4\pi i.$$

Note that when a point z is on the circle $|z| = 2$, it follows that $z\bar{z} = 4$, or $\bar{z} = 4/z$. Hence the result $I = 4\pi i$ can also be written

(2) $$\int_C \frac{dz}{z} = \pi i.$$

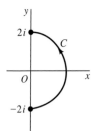

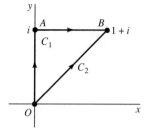

FIGURE 28   FIGURE 29

EXAMPLE 2. In this example, we first let C_1 denote the contour OAB shown in Fig. 29 and evaluate the integral

(3) $$\int_{C_1} f(z)\, dz = \int_{OA} f(z)\, dz + \int_{AB} f(z)\, dz,$$

where

$$f(z) = y - x - i3x^2 \qquad (z = x + iy).$$

The leg OA may be represented parametrically as $z = 0 + iy\ (0 \le y \le 1)$; and since $x = 0$ at points on that leg, the values of f there vary with the parameter y according to the equation $f(z) = y\ (0 \le y \le 1)$. Consequently,

$$\int_{OA} f(z)\,dz = \int_0^1 yi\,dy = i\int_0^1 y\,dy = \frac{i}{2}.$$

On the leg $AB, z = x + i\ (0 \le x \le 1)$; and so

$$\int_{AB} f(z)\,dz = \int_0^1 (1 - x - i3x^2)\cdot 1\,dx = \int_0^1 (1 - x)\,dx - 3i\int_0^1 x^2\,dx = \frac{1}{2} - i.$$

In view of equation (3), we now see that

(4) $$\int_{C_1} f(z)\,dz = \frac{1 - i}{2}.$$

If C_2 denotes the segment OB of the line $y = x$, with parametric representation $z = x + ix\ (0 \le x \le 1)$,

(5) $$\int_{C_2} f(z)\,dz = \int_0^1 -i3x^2(1 + i)\,dx = 3(1 - i)\int_0^1 x^2\,dx = 1 - i.$$

Evidently, then, the integrals of $f(z)$ along the two paths C_1 and C_2 have different values even though those paths have the same initial and the same final points.

Observe how it follows that the value of the integral of $f(z)$ over the simple closed contour $OABO$, or $C_1 - C_2$, is

$$\int_{C_1} f(z)\,dz - \int_{C_2} f(z)\,dz = \frac{-1 + i}{2}.$$

EXAMPLE 3. We begin here by letting C denote an arbitrary *smooth* arc $z = z(t)\ (a \le t \le b)$ from a fixed point z_1 to a fixed point z_2. In order to evaluate the integral

$$I = \int_C z\,dz = \int_a^b z(t)z'(t)\,dt,$$

we note that, according to Exercise 4(a), Sec. 31,

$$\frac{d}{dt}\frac{[z(t)]^2}{2} = z(t)z'(t).$$

Thus

$$I = \frac{[z(t)]^2}{2}\Bigg|_a^b = \frac{[z(b)]^2 - [z(a)]^2}{2}.$$

But $z(b) = z_2$ and $z(a) = z_1$; and so $I = (z_2^2 - z_1^2)/2$. Inasmuch as the value of I depends only on the end points of C, and is otherwise independent of the arc that is taken, we may write

(6)
$$\int_{z_1}^{z_2} z\, dz = \frac{z_2^2 - z_1^2}{2}.$$

(Compare Example 2, where the value of an integral from one fixed point to another depended on the path that was taken.)

Expression (6) is also valid when C is a contour that is not necessarily smooth since a contour consists of a finite number of smooth arcs C_k ($k = 1, 2, \ldots, n$), joined end to end. More precisely, suppose that each C_k extends from z_k to z_{k+1}. Then

(7)
$$\int_C z\, dz = \sum_{k=1}^{n} \int_{C_k} z\, dz = \sum_{k=1}^{n} \frac{z_{k+1}^2 - z_k^2}{2} = \frac{z_{n+1}^2 - z_1^2}{2},$$

z_1 being the initial point of C and z_{n+1} its final point.

It follows from expression (7) that the integral of the function $f(z) = z$ around each closed contour in the plane has value zero. (Once again, compare Example 2, where the value of the integral of a given function around a certain closed path was *not* zero.) The question of predicting when an integral around a closed contour has value zero will be discussed in the next several sections. That question is central to the theory of functions of a complex variable.

EXAMPLE 4. Let C denote the semicircular path

$$z = 3e^{i\theta} \qquad (0 \le \theta \le \pi)$$

from the point $z = 3$ to the point $z = -3$ (Fig. 30). Although the branch (Sec. 26)

(8)
$$f(z) = \sqrt{r}e^{i\theta/2} \qquad (r > 0, 0 < \theta < 2\pi)$$

of the multiple-valued function $z^{1/2}$ is not defined at the initial point $z = 3$ of the contour C, the integral

(9)
$$I = \int_C z^{1/2}\, dz$$

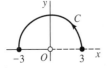

FIGURE 30

of that branch nevertheless exists. For the integrand is piecewise continuous on C. To see that this is so, we observe that when $z(\theta) = 3e^{i\theta}$, the right-hand limits of the real and imaginary components of the function

$$f[z(\theta)] = \sqrt{3}e^{i\theta/2} = \sqrt{3}\cos\frac{\theta}{2} + i\sqrt{3}\sin\frac{\theta}{2} \qquad (0 < \theta \le \pi)$$

at $\theta = 0$ are $\sqrt{3}$ and 0, respectively. Hence $f[z(\theta)]$ is continuous on the closed interval $0 \le \theta \le \pi$ when its value at $\theta = 0$ is defined as $\sqrt{3}$. Consequently,

$$I = \int_0^\pi \sqrt{3}e^{i\theta/2}3ie^{i\theta}\,d\theta = 3\sqrt{3}i\int_0^\pi e^{i3\theta/2}\,d\theta;$$

and

$$\int_0^\pi e^{i3\theta/2}\,d\theta = \frac{2}{3i}e^{i3\theta/2}\Big]_0^\pi = -\frac{2}{3i}(1+i).$$

Finally, then,

$$I = -2\sqrt{3}(1+i).$$

EXAMPLE 5.   Here C_R is the semicircular path

$$z = Re^{i\theta} \qquad (0 \le \theta \le \pi),$$

and $z^{1/2}$ denotes the branch (8) of the square root function that was used in Example 4. Without actually finding the value of the integral, we can easily show that

(10) $$\lim_{R\to\infty}\int_{C_R}\frac{z^{1/2}}{z^2+1}\,dz = 0.$$

For, when $|z| = R > 1$,

$$|z^{1/2}| = |\sqrt{R}e^{i\theta/2}| = \sqrt{R}$$

and

$$|z^2+1| \ge ||z^2| - 1| = R^2 - 1.$$

Consequently, at points on C_R where the integrand is defined,

$$\left|\frac{z^{1/2}}{z^2+1}\right| \le M_R \qquad \text{where} \qquad M_R = \frac{\sqrt{R}}{R^2-1}.$$

Since the length of C_R is the number $L = \pi R$, it follows from property (7), Sec. 32, that

$$\left|\int_{C_R}\frac{z^{1/2}}{z^2+1}\,dz\right| \le M_R L.$$

But

$$M_R L = \frac{\pi R \sqrt{R}}{R^2 - 1} \cdot \frac{1/R^2}{1/R^2} = \frac{\pi/\sqrt{R}}{1 - (1/R^2)},$$

and it is clear that the term on the far right here tends to zero as R tends to infinity. Limit (10) is, therefore, established.

EXERCISES

For the functions f and contours C in Exercises 1 through 6, use parametric representations for C, or legs of C, to evaluate

$$\int_C f(z)\, dz.$$

*1. $f(z) = (z + 2)/z$ and C is
   (a) the semicircle $z = 2e^{i\theta}$ $(0 \le \theta \le \pi)$;
   (b) the semicircle $z = 2e^{i\theta}$ $(\pi \le \theta \le 2\pi)$;
   (c) the circle $z = 2e^{i\theta}$ $(0 \le \theta \le 2\pi)$.
   Ans. (a) $-4 + 2\pi i$;   (b) $4 + 2\pi i$;   (c) $4\pi i$.

2. $f(z) = z - 1$ and C is the arc from $z = 0$ to $z = 2$ consisting of
   (a) the semicircle $z = 1 + e^{i\theta}$ $(\pi \le \theta \le 2\pi)$;
   (b) the segment $0 \le x \le 2$ of the real axis.
   Ans. (a) 0;   (b) 0.

3. $f(z) = \pi \exp(\pi \bar{z})$ and C is the boundary of the square with vertices at the points 0, 1, $1 + i$, and i, the orientation of C being in the counterclockwise direction.
   Ans. $4(e^{\pi} - 1)$.

4. $f(z)$ is defined by the equations

$$f(z) = \begin{cases} 1 & \text{when } y < 0, \\ 4y & \text{when } y > 0, \end{cases}$$

and C is the arc from $z = -1 - i$ to $z = 1 + i$ along the curve $y = x^3$.
   Ans. $2 + 3i$.

5. $f(z) = 1$ and C is an arbitrary contour from any fixed point z_1 to any fixed point z_2 in the plane.
   Ans. $z_2 - z_1$.

6. $f(z)$ is the branch

$$z^{-1+i} = \exp[(-1 + i) \log z] \qquad (|z| > 0, 0 < \arg z < 2\pi)$$

of the indicated power function, and C is the positively oriented unit circle $|z| = 1$.
   Ans. $i(1 - e^{-2\pi})$.

7. With the aid of the result in Exercise 2, Sec. 31, evaluate the integral

$$\int_C z^m \bar{z}^n\, dz,$$

where m and n are integers and C is the unit circle $|z| = 1$, taken counterclockwise.

8. Evaluate the integral I in Example 1, Sec. 33, using this representation for C:

$$z = \sqrt{4 - y^2} + iy \quad (-2 \le y \le 2).$$

(See Exercise 9, Sec. 31.)

9. Let C be the arc of the circle $|z| = 2$ from $z = 2$ to $z = 2i$ that lies in the first quadrant. Without evaluating the integral, show that

$$\left| \int_C \frac{dz}{z^2 - 1} \right| \le \frac{\pi}{3}.$$

10. Let C denote the line segment from $z = i$ to $z = 1$. By observing that, of all the points on that line segment, the midpoint is the closest to the origin, show that

$$\left| \int_C \frac{dz}{z^4} \right| \le 4\sqrt{2}$$

without evaluating the integral.

11. Show that if C is the boundary of the triangle with vertices at the points 0, $3i$, and -4, oriented in the counterclockwise direction, then

$$\left| \int_C (e^z - \bar{z}) \, dz \right| \le 60.$$

12. Let C and C_0 denote the circles $z = Re^{i\theta} (0 \le \theta \le 2\pi)$ and $z = z_0 + Re^{i\theta} (0 \le \theta \le 2\pi)$, respectively. Use these parametric representations to show that

$$\int_C f(z) \, dz = \int_{C_0} f(z - z_0) \, dz$$

when f is piecewise continuous on C.

*13. Let C_0 denote the circle $|z - z_0| = R$, taken counterclockwise. Use the parametric representation $z = z_0 + Re^{i\theta} (-\pi \le \theta \le \pi)$ for C_0 to derive the following integration formulas:

(a) $\displaystyle\int_{C_0} \frac{dz}{z - z_0} = 2\pi i;$ (b) $\displaystyle\int_{C_0} (z - z_0)^{n-1} \, dz = 0 \quad (n = \pm 1, \pm 2, \ldots);$

(c) $\displaystyle\int_{C_0} (z - z_0)^{a-1} \, dz = i\frac{2R^a}{a} \sin(a\pi)$, where a is any real number other than zero and

where the principal branch of the integrand and the principal value of R^a are taken.

14. Let C_R denote the upper half of the circle $|z| = R \ (R > 2)$, taken in the counterclockwise direction. Show that

$$\left| \int_{C_R} \frac{2z^2 - 1}{z^4 + 5z^2 + 4} \, dz \right| \le \frac{\pi R(2R^2 + 1)}{(R^2 - 1)(R^2 - 4)}.$$

Then, by dividing the numerator and denominator on the right here by R^4, show that the value of the integral tends to zero as R tends to infinity.

15. Let C_R be the circle $|z| = R \ (R > 1)$, described in the counterclockwise direction. Show that

$$\left| \int_{C_R} \frac{\text{Log } z}{z^2} \, dz \right| < 2\pi \left(\frac{\pi + \ln R}{R} \right),$$

and then use l'Hospital's rule to show that the value of this integral tends to zero as R tends to infinity.

16. Let C_ρ denote the circle $|z| = \rho$ $(0 < \rho < 1)$, oriented in the counterclockwise direction, and suppose that $f(z)$ is analytic in the disk $|z| \leq 1$. Show that if $z^{-1/2}$ represents any particular branch of that power of z, then there is a nonnegative constant M, *independent of* ρ, such that

$$\left| \int_{C_\rho} z^{-1/2} f(z)\, dz \right| \leq 2\pi M \sqrt{\rho}.$$

Thus show that the value of the integral here approaches 0 as ρ tends to 0.

   Suggestion: Note that since $f(z)$ is analytic, and therefore continuous, throughout the disk $|z| \leq 1$, it is bounded there (Sec. 14).

17. Let C_N denote the boundary of the square formed by the lines

$$x = \pm \left(N + \frac{1}{2} \right)\pi \qquad \text{and} \qquad y = \pm \left(N + \frac{1}{2} \right)\pi,$$

where N is a positive integer, and let the orientation of C_N be counterclockwise.

(a) With the aid of the inequalities

$$|\sin z| \geq |\sin x| \qquad \text{and} \qquad |\sin z| \geq |\sinh y|,$$

obtained in Exercises 8(a) and 9(a) of Sec. 24, show that $|\sin z| \geq 1$ on the vertical sides of the square and that $|\sin z| > \sinh(\pi/2)$ on the horizontal sides. Thus show that there is a positive constant A, *independent of* N, such that $|\sin z| \geq A$ for all points z lying on the contour C_N.

(b) Using the final result in part (a), show that

$$\left| \int_{C_N} \frac{dz}{z^2 \sin z} \right| \leq \frac{16}{(2N + 1)\pi A}$$

and hence that the value of this integral tends to zero as N tends to infinity.

18. (a) Suppose that a function $f(z)$ is continuous on a smooth arc C, which has a parametric representation $z = z(t)$ $(a \leq t \leq b)$; that is, $f[z(t)]$ is continuous on the interval $a \leq t \leq b$. Show that if $\phi(\tau)$ $(\alpha \leq \tau \leq \beta)$ is the function described in Sec. 31, then

$$\int_a^b f[z(t)]z'(t)\, dt = \int_\alpha^\beta f[Z(\tau)]Z'(\tau)\, d\tau,$$

where $Z(\tau) = z[\phi(\tau)]$.

(b) Point out how it follows that the identity obtained in part (a) remains valid when C is any contour, not necessarily a smooth one, and $f(z)$ is piecewise continuous on C. Thus show that the value of the integral of $f(z)$ along C is the same when the representation $z = Z(\tau)$ $(\alpha \leq \tau \leq \beta)$ is used, instead of the original one $z = z(t)$ $(a \leq t \leq b)$.

Suggestion: In part (a), use the result in Exercise 6(b), Sec. 31, and then refer to expression (13) in that section.

34. ANTIDERIVATIVES

Although the value of a contour integral of a function $f(z)$ from a fixed point z_1 to a fixed point z_2 depends, in general, on the path that is taken, there are certain functions whose integrals from z_1 and z_2 are independent of path. (Recall Examples 2 and 3 in Sec. 33.) The examples just cited also illustrate the fact that the values of integrals around closed paths are sometimes, but not always, zero. The theorem below is useful in determining when integration is independent of path and, moreover, when an integral around a closed path has value zero.

In proving the theorem, we shall discover an extension of the fundamental theorem of calculus that simplifies the evaluation of many contour integrals. That extension involves the concept of an *antiderivative* of a continuous function f in a domain D, or a function F such that $F'(z) = f(z)$ for all z in D. Note that an antiderivative is, of necessity, an analytic function. Note, too, that an antiderivative of a given function f is unique except for an additive complex constant. This is because the derivative of the difference $F(z) - G(z)$ of any two such antiderivatives $F(z)$ and $G(z)$ is zero; and, according to the theorem in Sec. 20, an analytic function is constant in a domain D when its derivative is zero throughout D.

Theorem. *Suppose that a function f is continuous on a domain D. If any one of the following statements is true, then so are the others:*

(a) f has an antiderivative F in D;

(b) the integrals of $f(z)$ along contours lying entirely in D and extending from any fixed point z_1 to any fixed point z_2 all have the same value;

(c) the integrals of $f(z)$ around closed contours lying entirely in D all have value zero.

Note that the theorem does *not* claim that any of these statements is true for a given function f and a given domain D. It tells us only that all of them are true or that none of them is true. To prove the theorem, it is sufficient to show that statement (a) implies statement (b), that statement (b) implies statement (c), and finally that statement (c) implies statement (a).

Let us assume that statement (a) is true. If a contour C from z_1 to z_2, lying in D, is actually a *smooth* arc, with parametric representation $z = z(t)\ (a \leq t \leq b)$, we know from Exercise 11, Sec. 31, that

$$\frac{d}{dt}F[z(t)] = F'[z(t)]z'(t) = f[z(t)]z'(t) \qquad (a \leq t \leq b).$$

Because the fundamental theorem of calculus can be extended so as to apply to complex-valued functions of a real variable (Sec. 30), it follows that

$$\int_C f(z)\,dz = \int_a^b f[z(t)]z'(t)\,dt = F[z(t)]\Big]_a^b = F[z(b)] - F[z(a)].$$

Since $z(b) = z_2$ and $z(a) = z_1$, the value of this contour integral is, then, $F(z_2) - F(z_1)$; and that value is evidently independent of the contour C as long as C extends from z_1 to z_2 and lies entirely in D. That is,

(1) $$\int_{z_1}^{z_2} f(z)\,dz = F(z)\Big]_{z_1}^{z_2} = F(z_2) - F(z_1).$$

This result is, of course, also valid when C is any contour, not necessarily a smooth one, that lies in D. Specifically, if C consists of a finite number of smooth arcs C_k ($k = 1, 2, \ldots, n$), each C_k extending from a point z_k to a point z_{k+1}, then

$$\int_C f(z)\,dz = \sum_{k=1}^{n} \int_{C_k} f(z)\,dz = \sum_{k=1}^{n} [F(z_{k+1}) - F(z_k)] = F(z_{n+1}) - F(z_1).$$

(Compare Example 3, Sec. 33.) The fact that statement (b) follows from statement (a) is now established.

To see that statement (b) implies statement (c), we let z_1 and z_2 denote any two points on a closed contour C lying in D and form two paths, each with initial point z_1 and final point z_2, such that $C = C_1 - C_2$ (Fig. 31). Assuming that statement (b) is true, one can write

$$(2) \qquad\qquad \int_{C_1} f(z)\,dz = \int_{C_2} f(z)\,dz,$$

or

$$(3) \qquad\qquad \int_{C_1} f(z)\,dz + \int_{-C_2} f(z)\,dz = 0.$$

That is, the integral of $f(z)$ around the closed contour $C = C_1 - C_2$ has value zero.

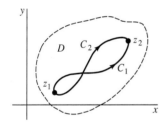

FIGURE 31

It remains to show that statement (c) implies statement (a). We do this by assuming that statement (c) is true, establishing the validity of statement (b), and then arriving at statement (a). To see that statement (b) is true, we let C_1 and C_2 denote any two contours, lying in D, from a point z_1 to a point z_2 and observe that, in view of statement (c), equation (3) holds (see Fig. 31). Thus equation (2) holds. Integration is, therefore, independent of path in D; and we can define the function

$$F(z) = \int_{z_0}^{z} f(s)\,ds$$

on D. The proof of the theorem is complete once we show that $F'(z) = f(z)$ everywhere in D. We do this by letting $z + \Delta z$ be any point, distinct from z, lying in some neighborhood of z that is sufficiently small to be contained in D. Then

$$F(z + \Delta z) - F(z) = \int_{z_0}^{z+\Delta z} f(s)\, ds - \int_{z_0}^{z} f(s)\, ds = \int_{z}^{z+\Delta z} f(s)\, ds,$$

where the path of integration from z to $z + \Delta z$ may be selected as a line segment (Fig. 32). Since

$$\int_{z}^{z+\Delta z} ds = \Delta z$$

(see Exercise 5, Sec. 33), we can write

$$f(z) = \frac{1}{\Delta z} \int_{z}^{z+\Delta z} f(z)\, ds;$$

and it follows that

$$\frac{F(z + \Delta z) - F(z)}{\Delta z} - f(z) = \frac{1}{\Delta z} \int_{z}^{z+\Delta z} [f(s) - f(z)]\, ds.$$

But f is continuous at the point z. Hence, for each positive number ε, a positive number δ exists such that

$$|f(s) - f(z)| < \varepsilon \qquad \text{whenever} \qquad |s - z| < \delta.$$

Consequently, if the point $z + \Delta z$ is close enough to z so that $|\Delta z| < \delta$, then

$$\left| \frac{F(z + \Delta z) - F(z)}{\Delta z} - f(z) \right| < \frac{1}{|\Delta z|} \varepsilon |\Delta z| = \varepsilon;$$

that is,

$$\lim_{\Delta z \to 0} \frac{F(z + \Delta z) - F(z)}{\Delta z} = f(z),$$

or $F'(z) = f(z)$.

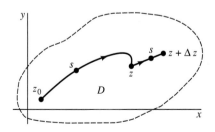

FIGURE 32

35. EXAMPLES

The following examples illustrate the theorem in Sec. 34 and, in particular, the use of formula (1) there, which is an extension of the fundamental theorem of calculus.

EXAMPLE 1. The continuous function $f(z) = z^2$ has an antiderivative $F(z) = z^3/3$ throughout the plane. Hence

$$\int_0^{1+i} z^2 \, dz = \frac{z^3}{3}\Bigg]_0^{1+i} = \frac{1}{3}(1+i)^3 = \frac{2}{3}(-1+i)$$

for every contour from $z = 0$ to $z = 1 + i$.

EXAMPLE 2. The function $1/z^2$, which is continuous everywhere except at the origin, has an antiderivative $-1/z$ in the domain $|z| > 0$. Consequently,

$$\int_{z_1}^{z_2} \frac{dz}{z^2} = -\frac{1}{z}\Bigg]_{z_1}^{z_2} = \frac{1}{z_1} - \frac{1}{z_2} \qquad (z_1 \neq 0, z_2 \neq 0)$$

for any contour from z_1 to z_2 that does not pass through the origin. In particular,

$$\int_C \frac{dz}{z^2} = 0$$

when C is the circle $z = 2e^{i\theta}$ ($-\pi \leq \theta \leq \pi$) about the origin.

Note that the integral of the function $f(z) = 1/z$ around the same circle *cannot* be evaluated in a similar way. For, although the derivative of any branch $F(z)$ of $\log z$ is $1/z$ (Sec. 26), $F(z)$ is not differentiable, or even defined, along its branch cut. In particular, if a ray $\theta = \alpha$ from the origin is used to form the branch cut, $F'(z)$ fails to exist at the point where that ray intersects the circle C. So C does not lie in a domain throughout which $F'(z) = 1/z$, and we cannot make direct use of an antiderivative.

EXAMPLE 3. Let D be the domain $|z| > 0, -\pi < \operatorname{Arg} z < \pi$, consisting of the entire plane with the origin and the negative real axis excluded. The principal branch $\operatorname{Log} z$ of the logarithmic function serves as an antiderivative of the continuous function $1/z$ throughout D. Hence we can write

$$\int_{-2i}^{2i} \frac{dz}{z} = \operatorname{Log} z\Bigg]_{-2i}^{2i} = \operatorname{Log}(2i) - \operatorname{Log}(-2i) = \left(\ln 2 + i\frac{\pi}{2}\right) - \left(\ln 2 - i\frac{\pi}{2}\right) = \pi i$$

when the path of integration from $-2i$ to $2i$ is, for instance, the arc

$$z = 2e^{i\theta} \qquad \left(-\frac{\pi}{2} \leq \theta \leq \frac{\pi}{2}\right)$$

of the circle in Example 2. (Compare Example 1, Sec. 33, where this integral was evaluated using the above parametric representation for the arc.)

EXAMPLE 4. Let us use an antiderivative to evaluate the integral

$$(1) \qquad\qquad\qquad \int_{C_1} z^{1/2} \, dz,$$

where the integrand is the branch

$$(2) \qquad\qquad z^{1/2} = \sqrt{r}e^{i\theta/2} \qquad (r > 0, 0 < \theta < 2\pi)$$

of the square root function and where C_1 is any contour from $z = -3$ to $z = 3$ that, except for its end points, lies above the x axis (Fig. 33). Although the integrand is piecewise continuous on C_1, and the integral therefore exists, the branch (2) of $z^{1/2}$ is not defined on the ray $\theta = 0$, in particular at the point $z = 3$. But another branch,

$$f_1(z) = \sqrt{r}e^{i\theta/2} \qquad \left(r > 0, -\frac{\pi}{2} < \theta < \frac{3\pi}{2}\right),$$

is defined and continuous everywhere on C_1. The values of $f_1(z)$ at all points on C_1 except $z = 3$ coincide with those of our integrand (2); so the integrand can be replaced by $f_1(z)$. Since an antiderivative of $f_1(z)$ is the function

$$F_1(z) = \frac{2}{3}z^{3/2} = \frac{2}{3}r\sqrt{r}e^{i3\theta/2} \qquad \left(r > 0, -\frac{\pi}{2} < \theta < \frac{3\pi}{2}\right),$$

we can now write

$$\int_{C_1} z^{1/2}\,dz = \int_{-3}^{3} f_1(z)\,dz = F_1(z)\Big]_{-3}^{3} = 2\sqrt{3}(e^{i0} - e^{i3\pi/2}) = 2\sqrt{3}(1+i).$$

(Compare Example 4 in Sec. 33.)

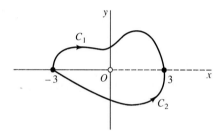

FIGURE 33

Integral (1) over any contour C_2 that extends from $z = -3$ to $z = 3$ *below* the real axis has another value. In this case, we can replace the integrand by the branch

$$f_2(z) = \sqrt{r}e^{i\theta/2} \qquad \left(r > 0, \frac{\pi}{2} < \theta < \frac{5\pi}{2}\right),$$

whose values coincide with those of the branch (2) in the lower half plane. The analytic function

$$F_2(z) = \frac{2}{3}z^{3/2} = \frac{2}{3}r\sqrt{r}e^{i3\theta/2} \qquad \left(r > 0, \frac{\pi}{2} < \theta < \frac{5\pi}{2}\right)$$

is an antiderivative of $f_2(z)$. Thus

$$\int_{C_2} z^{1/2}\,dz = \int_{-3}^{3} f_2(z)\,dz = F_2(z)\Big]_{-3}^{3} = 2\sqrt{3}(e^{i3\pi} - e^{i3\pi/2}) = 2\sqrt{3}(-1+i).$$

Note how it follows that the integral of the function (2) around the closed contour $C_2 - C_1$ has the value

$$2\sqrt{3}(-1+i) - 2\sqrt{3}(1+i) = -4\sqrt{3}.$$

36.   CAUCHY-GOURSAT THEOREM

In Sec. 34, we saw that when a continuous function f has an antiderivative in a domain D, the integral of $f(z)$ around any given closed contour C lying entirely in D has value zero. In this section, we present a theorem giving other conditions on a function f ensuring that the value of the integral of $f(z)$ around a *simple* closed contour (Sec. 31) is zero. The theorem is central to the theory of functions of a complex variable; and some extensions of it, involving certain special types of domains, will be given in Sec. 38.

We let C denote a simple closed contour $z = z(t)$ $(a \leq t \leq b)$, described in the *positive sense* (counterclockwise), and we assume that f is analytic at each point interior to and on C. According to Sec. 32,

$$(1) \qquad \int_C f(z)\, dz = \int_a^b f[z(t)]z'(t)\, dt;$$

and if $f(z) = u(x, y) + iv(x, y)$ and $z(t) = x(t) + iy(t)$, the integrand on the right in expression (1) is the product of the functions

$$u[x(t), y(t)] + iv[x(t), y(t)], \qquad x'(t) + iy'(t)$$

of the real variable t. Thus

$$(2) \qquad \int_C f(z)\, dz = \int_a^b (ux' - vy')\, dt + i \int_a^b (vx' + uy')\, dt.$$

In terms of line integrals of real-valued functions of two real variables, then,

$$(3) \qquad \int_C f(z)\, dz = \int_C u\, dx - v\, dy + i \int_C v\, dx + u\, dy.$$

Observe that expression (3) can be obtained formally by replacing $f(z)$ by $u + iv$ and dz by $dx + i\, dy$ on the left and expanding the product. Expression (3) is, of course, also valid when C is any contour, not necessarily a simple closed one, and $f[z(t)]$ is only piecewise continuous on it.

We next recall a result from advanced calculus that allows us to express the line integrals on the right in equation (3) as double integrals. Specifically, if two real-valued functions $P(x, y)$ and $Q(x, y)$, together with their first-order partial derivatives, are continuous throughout the closed region R consisting of all points interior to and on C, *Green's theorem* tells us that*

$$\int_C P\, dx + Q\, dy = \iint_R (Q_x - P_y)\, dA.$$

---

*See, for example, A. E. Taylor and W. R. Mann, "Advanced Calculus," 3d ed., p. 457, 1983.

Now f is continuous in R, since it is analytic there. Hence the functions u and v are also continuous in R. Likewise, if the derivative f' of f is continuous in R, so are the first-order partial derivatives of u and v. Green's theorem then enables us to rewrite equation (3) as

$$(4) \qquad \int_C f(z)\,dz = \int\int_R (-v_x - u_y)\,dA + i\int\int_R (u_x - v_y)\,dA.$$

But, in view of the Cauchy-Riemann equations

$$u_x = v_y, \qquad u_y = -v_x,$$

the integrands of these two double integrals are zero throughout R. So, *when f is analytic in R and f' is continuous there,*

$$(5) \qquad \int_C f(z)\,dz = 0.$$

This result was obtained by Cauchy in the early part of the nineteenth century.

Note that, once it has been established that the value of this integral is zero, the orientation of C is immaterial. That is, statement (5) is also true if C is taken in the clockwise direction, since then

$$\int_C f(z)\,dz = -\int_{-C} f(z)\,dz = 0.$$

Goursat* was the first to prove that *the condition of continuity on f' can be omitted.* Its removal is important and will allow us to show, for example, that the derivative f' of an analytic function f is analytic without having to assume the continuity of f', which follows as a consequence. We now state the revised form of Cauchy's result, known as the *Cauchy-Goursat theorem.*

Theorem. *If a function f is analytic at all points interior to and on a simple closed contour C, then*

$$\int_C f(z)\,dz = 0.$$

The proof is presented in the next section, where, to be specific, we assume that C is positively oriented. The reader who wishes to accept this theorem without proof may pass directly to Sec. 38.

---

*E. Goursat (1858–1936), pronounced *gour-sah'*.

37. PROOF OF THE THEOREM

We preface the proof of the Cauchy-Goursat theorem with a lemma. We start by forming subsets of the region R which consists of the points on a positively oriented simple closed contour C together with the points interior to C. To do this, we draw equally spaced lines parallel to the real and imaginary axes such that the distance between adjacent vertical lines is the same as that between adjacent horizontal lines. We thus form a finite number of closed square subregions, where each point of R lies in at least one such subregion and each subregion contains points of R. We refer to these square subregions simply as *squares*, always keeping in mind that by a square we mean a boundary together with the points interior to it. If a particular square contains points that are not in R, we remove those points and call what remains a *partial square*. We thus *cover* the region R with a finite number of squares and partial squares (Fig. 34), and our proof of the following lemma starts with this covering.

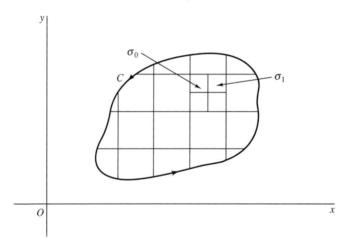

FIGURE 34

> **Lemma.** *Let f be analytic throughout a closed region R consisting of the points interior to a positively oriented simple closed contour C together with the points on C itself. For any positive number ε, the region R can be covered with a finite number of squares and partial squares, indexed by $j = 1, 2, \ldots, n$, such that in each one there is a fixed point z_j for which the inequality*

(1)
$$\left| \frac{f(z) - f(z_j)}{z - z_j} - f'(z_j) \right| < \varepsilon \qquad (z \neq z_j)$$

> *is satisfied by all other points in that square or partial square.*

To start the proof, we consider the possibility that, in the covering constructed just prior to the statement of the lemma, there is some square or partial square in which no point z_j exists such that inequality (1) holds for all other points z in it. If that subregion is a square, we construct four smaller squares by drawing line segments joining the midpoints of its opposite sides (Fig. 34). If the subregion is a partial square, we treat the whole square in the same manner and then let the portions that lie outside R be discarded. If, in any one of these smaller subregions, no point z_j exists such that inequality (1) holds for all other points z in it, we construct still smaller squares and partial squares, etc. When this is done to each of the original subregions that requires it, it turns out that, *after a finite number of steps*, the region R can be covered with a finite number of squares and partial squares such that the lemma is true.

To verify this, we suppose that the needed points z_j do *not* exist after subdividing one of the original subregions a finite number of times and reach a contradiction. We let σ_0 denote that subregion if it is a square; if it is a partial square, we let σ_0 denote the entire square of which it is a part. After we subdivide σ_0, at least one of the four smaller squares, denoted by σ_1, must contain points of R but no appropriate point z_j. We then subdivide σ_1 and continue in this manner. It may be that after a square σ_{k-1} ($k = 1, 2, \ldots$) has been subdivided, more than one of the four smaller squares constructed from it can be chosen. To make a specific choice, we take σ_k to be the one lowest and then farthest to the left.

In view of the manner in which the nested infinite sequence

(2) $$\sigma_0, \sigma_1, \sigma_2, \ldots, \sigma_{k-1}, \sigma_k, \ldots$$

of squares is constructed, it is easily shown (Exercise 11, Sec. 38) that there is a point z_0 common to each σ_k; also, each of these squares contain points of R other than possibly z_0. Recall how the sizes of the squares in the sequence are decreasing, and note that any δ neighborhood $|z - z_0| < \delta$ of z_0 contains such squares when their diagonals have lengths less than δ. Every δ neighborhood $|z - z_0| < \delta$ therefore contains points of R distinct from z_0, and this means that z_0 is an accumulation point of R. Since the region R is a closed set, it follows that z_0 is a point in R. (See Sec. 8.)

Now the function f is analytic throughout R and, in particular, at z_0. Consequently, $f'(z_0)$ exists. According to the definition of derivative (Sec. 15), there is, for each positive number ε, a δ neighborhood $|z - z_0| < \delta$ such that the inequality

$$\left| \frac{f(z) - f(z_0)}{z - z_0} - f'(z_0) \right| < \varepsilon$$

is satisfied by all points distinct from z_0 in that neighborhood. But the neighborhood $|z - z_0| < \delta$ contains a square σ_K when the integer K is large enough that the length of a diagonal of that square is less than δ (Fig. 35). Consequently, z_0 serves as the point z_j in inequality (1) for the subregion consisting of the square σ_K or a part of σ_K. Contrary to the way in which sequence (2) was formed, then, it is not necessary to subdivide σ_K. We thus arrive at a contradiction, and the proof of the lemma is complete.

Continuing with a function f which is analytic throughout a region R consisting of a positively oriented simple closed contour C and points interior to it, we are now ready to prove the Cauchy-Goursat theorem, namely that

(3) $$\int_C f(z)\, dz = 0.$$

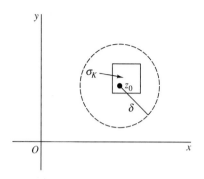

FIGURE 35

Given an arbitrary positive number ε, we consider the covering of R in the statement of the lemma. Let us define on the jth square or partial square the following function, where z_j is the fixed point in that subregion for which inequality (1) holds:

$$(4) \qquad \delta_j(z) = \begin{cases} \dfrac{f(z) - f(z_j)}{z - z_j} - f'(z_j) & \text{when } z \neq z_j, \\ 0 & \text{when } z = z_j. \end{cases}$$

According to inequality (1),

$$(5) \qquad |\delta_j(z)| < \varepsilon$$

at all points z in the subregion on which $\delta_j(z)$ is defined. Also, the function $\delta_j(z)$ is continuous throughout the subregion since $f(z)$ is continuous there and

$$\lim_{z \to z_j} \delta_j(z) = f'(z_j) - f'(z_j) = 0.$$

Next, let C_j ($j = 1, 2, \ldots, n$) denote the positively oriented boundaries of the above squares or partial squares covering R. In view of definition (4), the value of f at a point z on any particular C_j can be written

$$f(z) = f(z_j) - z_j f'(z_j) + f'(z_j)z + (z - z_j)\delta_j(z);$$

and this means that

$$(6) \int_{C_j} f(z)\,dz = [f(z_j) - z_j f'(z_j)] \int_{C_j} dz + f'(z_j) \int_{C_j} z\,dz + \int_{C_j} (z - z_j)\delta_j(z)\,dz.$$

But

$$\int_{C_j} dz = 0 \qquad \text{and} \qquad \int_{C_j} z\,dz = 0$$

since the functions 1 and z possess antiderivatives everywhere in the finite plane. So equation (6) reduces to

$$(7) \qquad \int_{C_j} f(z)\,dz = \int_{C_j} (z - z_j)\delta_j(z)\,dz \qquad (j = 1, 2, \ldots, n).$$

The sum of all n integrals on the left in equations (7) can be written

$$\sum_{j=1}^{n}\int_{C_j}f(z)\,dz = \int_{C}f(z)\,dz$$

since the two integrals along the common boundary of every pair of adjacent subregions cancel each other, the integral being taken in one sense along that line segment in one subregion and in the opposite sense in the other (Fig. 36). Only the integrals along the arcs that are parts of C remain. Thus, in view of equations (7),

$$\int_{C}f(z)\,dz = \sum_{j=1}^{n}\int_{C_j}(z-z_j)\delta_j(z)\,dz;$$

and so

(8)
$$\left|\int_{C}f(z)\,dz\right| \le \sum_{j=1}^{n}\left|\int_{C_j}(z-z_j)\delta_j(z)\,dz\right|.$$

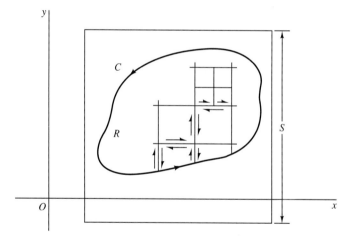

FIGURE 36

Let us now use property (7), Sec. 32, to find an upper bound for each absolute value on the right in inequality (8). To do this, we first recall that each C_j coincides either entirely or partially with the boundary of a square. In either case, we let s_j denote the length of a side of the square. Since, in the jth integral, both the variable z and the point z_j lie in that square,

$$|z - z_j| \le \sqrt{2}s_j.$$

In view of inequality (5), then, we know that each integrand on the right in inequality (8) satisfies the condition

(9)
$$|(z - z_j)\delta_j(z)| < \sqrt{2}s_j\varepsilon.$$

As for the length of the path C_j, it is $4s_j$ if C_j is the boundary of a square. In that case, we let A_j denote the area of the square and observe that

$$(10) \qquad \left| \int_{C_j} (z - z_j) \delta_j(z) \, dz \right| < \sqrt{2} s_j \varepsilon 4 s_j = 4\sqrt{2} A_j \varepsilon.$$

If C_j is the boundary of a partial square, its length does not exceed $4s_j + L_j$, where L_j is the length of that part of C_j which is also a part of C. Again letting A_j denote the area of the full square, we find that

$$(11) \qquad \left| \int_{C_j} (z - z_j) \delta_j(z) \, dz \right| < \sqrt{2} s_j \varepsilon (4 s_j + L_j) < 4\sqrt{2} A_j \varepsilon + \sqrt{2} S L_j \varepsilon,$$

where S is the length of a side of some square that encloses the entire contour C as well as all of the squares originally used in covering R (Fig. 36). Note that the sum of all the A_j's does not exceed S^2.

If L denotes the length of C, it now follows from inequalities (8), (10), and (11) that

$$\left| \int_C f(z) \, dz \right| < (4\sqrt{2} S^2 + \sqrt{2} S L) \varepsilon.$$

Since the value of the positive number ε is arbitrary, we can choose it so that the right-hand side of this last inequality is as small as we please. The left-hand side, which is independent of ε, must therefore be equal to zero; and statement (3) follows. This completes the proof of the Cauchy-Goursat theorem.

38. SIMPLY AND MULTIPLY CONNECTED DOMAINS

A *simply connected* domain D is a domain such that every simple closed contour within it encloses only points of D. The set of points interior to a simple closed contour is an example. The annular domain between two concentric circles is, however, not simply connected. A domain that is not simply connected is said to be *multiply connected*.

The Cauchy-Goursat theorem can be extended in the following way, involving a simply connected domain.

Theorem 1. *If a function f is analytic throughout a simply connected domain D, then*

$$(1) \qquad \int_C f(z) \, dz = 0$$

for every closed contour C lying in D.

The proof is easy if C is a *simple* closed contour or if it is a closed contour that intersects itself a *finite* number of times. For, if C is simple and lies in D, the function f is analytic at each point interior to and on C; and the Cauchy-Goursat theorem ensures that equation (1) holds. Furthermore, if C is closed but intersects itself a finite number of times, it consists of a finite number of simple closed contours, as illustrated in Fig. 37. By applying the Cauchy-Goursat theorem to each of those simple closed contours, we obtain the desired result for C. Subtleties arise if the closed contour has an infinite number of self-intersection points. One method that can sometimes be used to show that the theorem still applies is illustrated in Exercise 13 below.*

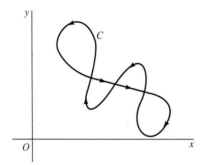

FIGURE 37

Corollary 1. *A function f which is analytic throughout a simply connected domain D must have an antiderivative in D.*

This corollary follows immediately from Theorem 1 because of the theorem in Sec. 34, which tells us that a continuous function f always has an antiderivative in a given domain when equation (1) holds for each closed contour C in that domain. Note that, since the finite plane is simply connected, Corollary 1 tells us that *entire functions always possess antiderivatives.*

The Cauchy-Goursat theorem can also be extended in a way that involves integrals along the boundary of a multiply connected domain. The following theorem is such an extension. In this theorem,

    (*a*) C is a simple closed contour, described in the counterclockwise direction;

    (*b*) C_k ($k = 1, 2, \ldots, n$) are simple closed contours, all described in the clockwise direction, that are interior to C and whose interiors have no points in common.

Theorem 2. *Let C and C_k ($k = 1, 2, \ldots, n$) be the simple closed contours in statements (a) and (b) above. If a function f is analytic throughout the closed region (Fig. 38) consisting of all points within and on C except for points interior to any*

---

*For a proof of the theorem involving more general paths of finite length, see, for example, Secs. 63–65 in Vol. 1 of the book by Markushevich, cited in Appendix 1.

C_k, *then*

(2) $$\int_C f(z)\,dz + \sum_{k=1}^{n} \int_{C_k} f(z)\,dz = 0.$$

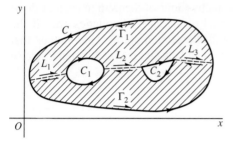

FIGURE 38

Note that, in equation (2), the direction of each path of integration is such that the interior of the closed region lies to the *left* of that path.

To prove the theorem, we introduce a polygonal path L_1, consisting of a finite number of line segments joined end to end, to connect the outer contour C to the inner contour C_1. We introduce another polygonal path L_2 which connects C_1 to C_2; and we continue in this manner, with L_{n+1} connecting C_n to C. As indicated by the single-barbed arrows in Fig. 38, two simple closed contours Γ_1 and Γ_2 can be formed, each consisting of polygonal paths L_k or $-L_k$ and pieces of C and C_k and each described in such a direction that the points enclosed by them lie to the left. The Cauchy-Goursat theorem can now be applied to f on Γ_1 and Γ_2, and the sum of the values of the integrals over those contours is found to be zero. Since the integrals in opposite directions along each path L_k cancel, only the integrals along C and C_k remain; and we arrive at statement (2).

The following corollary is an especially important consequence of Theorem 2.

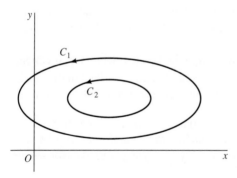

FIGURE 39

Corollary 2. *Let C_1 and C_2 denote positively oriented simple closed contours, where C_2 is interior to C_1 (Fig. 39). If a function f is analytic in the closed region* *consisting of those contours and all points between them, then*

(3)
$$\int_{C_1} f(z)\, dz = \int_{C_2} f(z)\, dz.$$

For a verification, we use Theorem 2 to write

$$\int_{C_1} f(z)\, dz + \int_{-C_2} f(z)\, dz = 0;$$

and we note that this is just a different form of equation (3).

Corollary 2 is known as the *principle of deformation of paths* since it tells us that if C_1 is continuously deformed into C_2, always passing through points at which f is analytic, then the value of the integral of f over C_1 never changes.

EXAMPLE. When C is any positively oriented simple closed contour surrounding the origin, Corollary 2 can be used to show that

$$\int_C \frac{dz}{z} = 2\pi i.$$

To accomplish this, we need only construct a positively oriented circle C_0 with center at the origin and radius so small that C_0 lies entirely inside C (Fig. 40). Since [Exercise 13(*a*), Sec. 33]

$$\int_{C_0} \frac{dz}{z} = 2\pi i$$

and since $1/z$ is analytic everywhere except at $z = 0$, the desired result follows.

Note that the radius of C_0 could equally well have been so large that C lies entirely inside C_0.

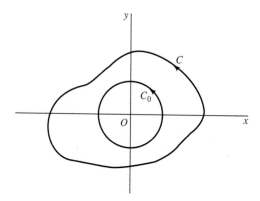

FIGURE 40

EXERCISES

1. Use an antiderivative to show that, for every contour C extending from a point z_1 to a point z_2,

$$\int_C z^n\, dz = \frac{1}{n+1}(z_2^{n+1} - z_1^{n+1}) \qquad (n = 0, 1, 2, \dots).$$

2. By finding an antiderivative, evaluate each of these integrals, where the path is an arbitrary contour between the indicated limits of integration:

(a) $\displaystyle\int_{i}^{i/2} e^{\pi z}\,dz;$ (b) $\displaystyle\int_{0}^{\pi+2i} \cos\left(\frac{z}{2}\right) dz;$ (c) $\displaystyle\int_{1}^{3} (z-2)^3\,dz.$

Ans. (a) $(1+i)/\pi$;   (b) $e + (1/e)$;   (c) 0.

3. Use the theorem in Sec. 34 to show that

$$\int_{C_0} (z-z_0)^{n-1}\,dz = 0 \qquad (n = \pm 1, \pm 2, \ldots)$$

when C_0 is any closed contour which does not pass through the point z_0. [Compare Exercise 13(b), Sec. 33].

4. Apply the Cauchy-Goursat theorem to show that

$$\int_{C} f(z)\,dz = 0$$

when the contour C is the circle $|z| = 1$, in either direction, and when

(a) $f(z) = \dfrac{z^2}{z-3}$;   (b) $f(z) = ze^{-z}$;   (c) $f(z) = \dfrac{1}{z^2+2z+2}$;

(d) $f(z) = \operatorname{sech} z$;   (e) $f(z) = \tan z$;   (f) $f(z) = \operatorname{Log}(z+2)$.

5. Let C_1 denote the positively oriented circle $|z| = 4$ and C_2 the positively oriented boundary of the square whose sides lie along the lines $x = \pm 1, y = \pm 1$ (Fig. 41). With the aid of Corollary 2 in Sec. 38, point out why

$$\int_{C_1} f(z)\,dz = \int_{C_2} f(z)\,dz$$

when

(a) $f(z) = \dfrac{1}{3z^2+1}$;   (b) $f(z) = \dfrac{z+2}{\sin(z/2)}$;   (c) $f(z) = \dfrac{z}{1-e^z}$.

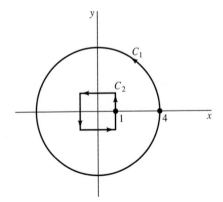

FIGURE 41

6. If C_0 denotes a positively oriented circle $|z - z_0| = R$, then

$$\int_{C_0} (z-z_0)^{n-1}\,dz = \begin{cases} 0 & \text{when } n = \pm 1, \pm 2, \ldots, \\ 2\pi i & \text{when } n = 0, \end{cases}$$

according to Exercise 13, Sec. 33. Use that result and Corollary 2 in Sec. 38 to show that if C is the boundary of the rectangle $0 \le x \le 3, 0 \le y \le 2$, described in the positive sense, then

$$\int_C (z - 2 - i)^{n-1}\, dz = \begin{cases} 0 & \text{when } n = \pm 1, \pm 2, \ldots, \\ 2\pi i & \text{when } n = 0. \end{cases}$$

7. (a) By using the branch

$$\log z = \ln r + i\theta \qquad (r > 0, 0 < \theta < 2\pi)$$

of the logarithmic function as an antiderivative of $1/z$, show that

$$\int_{-2i}^{2i} \frac{dz}{z} = -\pi i$$

when the path of integration from $-2i$ to $2i$ is the left-hand half of the circle $|z| = 2$.

(b) Point out how if follows from the results in part (a) and Example 3, Sec. 35, that

$$\int_C \frac{dz}{z} = 2\pi i$$

when C is the entire positively oriented circle $|z| = 2$. [Compare Exercise 13(a), Sec. 33.]

8. Show that

$$\int_{-1}^{1} z^i\, dz = \frac{1 + e^{-\pi}}{2}(1 - i),$$

where z^i denotes the principal branch

$$z^i = \exp(i \, \text{Log } z) \qquad (|z| > 0, -\pi < \text{Arg } z < \pi)$$

and where the path of integration is any contour from $z = -1$ to $z = 1$ that, except for its end points, lies above the real axis.

Suggestion: Use an antiderivative of the branch

$$z^i = \exp(i \log z) \qquad \left(|z| > 0, -\frac{\pi}{2} < \arg z < \frac{3\pi}{2}\right)$$

of the same power function.

9. Let C denote the entire positively oriented boundary of the half disk $0 \le r \le 1, 0 \le \theta \le \pi$, and let $f(z)$ be a continuous function defined on that half disk by writing $f(0) = 0$ and using the branch

$$f(z) = \sqrt{r}e^{i\theta/2} \qquad \left(r > 0, -\frac{\pi}{2} < \theta < \frac{3\pi}{2}\right)$$

of the multiple-valued function $z^{1/2}$. Show that

$$\int_C f(z)\, dz = 0$$

by evaluating separately the integrals of $f(z)$ over the semicircle and the two radii which constitute C. Why does the Cauchy-Goursat theorem not apply here?

10. *Nested Intervals.* An infinite sequence of closed intervals $a_n \leq x \leq b_n$ ($n = 0, 1, 2, \ldots$) is formed in the following way. The interval $a_1 \leq x \leq b_1$ is either the left-hand or right-hand half of the first interval $a_0 \leq x \leq b_0$, and the interval $a_2 \leq x \leq b_2$ is then one of the two halves of $a_1 \leq x \leq b_1$, etc. Prove that there is a point x_0 which belongs to every one of the closed intervals $a_n \leq x \leq b_n$.

   Suggestion: Note that the left-hand end points a_n represent a bounded nondecreasing sequence of numbers, since $a_0 \leq a_n \leq a_{n+1} < b_0$; hence they have a limit A as n tends to infinity. Show that the end points b_n also have a limit B. Then show that $A = B$, and write $x_0 = A = B$.

11. *Nested Squares.* A square $\sigma_0 \colon a_0 \leq x \leq b_0, c_0 \leq y \leq d_0$ is divided into four equal squares by line segments parallel to the coordinate axes. One of those four smaller squares $\sigma_1 \colon a_1 \leq x \leq b_1, c_1 \leq y \leq d_1$ is selected according to some rule. It, in turn, is divided into four equal squares one of which, called σ_2, is selected, etc. (see Sec. 37). Prove that there is a point (x_0, y_0) which belongs to each of the closed regions of the infinite sequence $\sigma_0, \sigma_1, \sigma_2, \ldots$.

   Suggestion: Apply the result in Exercise 10 to each of the sequences of closed intervals $a_n \leq x \leq b_n$ and $c_n \leq y \leq d_n$ ($n = 0, 1, 2, \ldots$).

12. Show that if C is a positively oriented simple closed contour, then the area of the region enclosed by C can be written

$$\frac{1}{2i} \int_C \bar{z} \, dz.$$

   Suggestion: Note that expression (4), Sec. 36, can be used here even though the function $f(z) = \bar{z}$ is not analytic.

13. According to Exercise 12, Sec. 31, the path C_1 from the origin to the point $z = 1$ along the graph of the function defined by means of the equations

$$y(x) = \begin{cases} x^3 \sin\left(\dfrac{\pi}{x}\right) & \text{when } 0 < x \leq 1, \\ 0 & \text{when } x = 0 \end{cases}$$

is a smooth arc that intersects the real axis an infinite number of times. Let C_2 denote the line segment along the real axis from $z = 1$ back to the origin, and let C_3 denote any smooth arc from the origin to $z = 1$ that does not intersect itself and has only its end points in common with the arcs C_1 and C_2 (Fig. 42). Apply the Cauchy-Goursat theorem

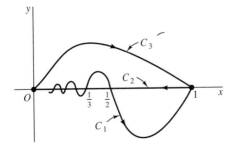

FIGURE 42

to show that if a function f is entire, then

$$\int_{C_1} f(z)\, dz = \int_{C_3} f(z)\, dz \quad \text{and} \quad \int_{C_2} f(z)\, dz = -\int_{C_3} f(z)\, dz.$$

Conclude that, even though the closed contour $C = C_1 + C_2$ intersects itself an infinite number of times,

$$\int_C f(z)\, dz = 0.$$

39. CAUCHY INTEGRAL FORMULA

Another fundamental result will now be established.

Theorem. *Let f be analytic everywhere within and on a simple closed contour C, taken in the positive sense. If z_0 is any point interior to C, then*

(1)
$$f(z_0) = \frac{1}{2\pi i} \int_C \frac{f(z)\, dz}{z - z_0}.$$

Formula (1) is called the *Cauchy integral formula*. It tells us that if a function f is to be analytic within and on a simple closed contour C, then the values of f interior to C are completely determined by the values of f on C.

When the Cauchy integral formula is written

(2)
$$\int_C \frac{f(z)\, dz}{z - z_0} = 2\pi i f(z_0),$$

it can be used to evaluate certain integrals along simple closed contours.

EXAMPLE.   Let C be the positively oriented circle $|z| = 2$. Since the function

$$f(z) = \frac{z}{9 - z^2}$$

is analytic within and on C and since the point $z_0 = -i$ is interior to C, formula (2) tells us that

$$\int_C \frac{z\, dz}{(9 - z^2)(z + i)} = \int_C \frac{z/(9 - z^2)}{z - (-i)}\, dz = 2\pi i \left(\frac{-i}{10}\right) = \frac{\pi}{5}.$$

Our proof of the theorem begins with the fact (Sec. 14) that since f is continuous at z_0, there corresponds to any positive number ε, however small, a positive number δ such that

$$|f(z) - f(z_0)| < \varepsilon \quad \text{whenever} \quad |z - z_0| < \delta.$$

Let us now choose a positive number ρ that is less than δ and is so small that the positively oriented circle $|z - z_0| = \rho$, denoted by C_0 in Fig. 43, is interior to C. Then

(3) $$|f(z) - f(z_0)| < \varepsilon \qquad \text{whenever} \qquad |z - z_0| = \rho.$$

This statement plays an important role in what follows.

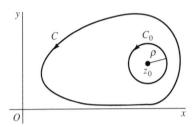

FIGURE 43

Since the function $f(z)/(z - z_0)$ is analytic in the closed region consisting of the contours C and C_0 and all points between them, we know from Corollary 2 in Sec. 38 that

$$\int_C \frac{f(z)\, dz}{z - z_0} = \int_{C_0} \frac{f(z)\, dz}{z - z_0}.$$

This enables us to write

(4) $$\int_C \frac{f(z)\, dz}{z - z_0} - f(z_0) \int_{C_0} \frac{dz}{z - z_0} = \int_{C_0} \frac{f(z) - f(z_0)}{z - z_0}\, dz.$$

But [see Exercise 13(a), Sec. 33]

$$\int_{C_0} \frac{dz}{z - z_0} = 2\pi i;$$

and so equation (4) becomes

(5) $$\int_C \frac{f(z)\, dz}{z - z_0} - 2\pi i f(z_0) = \int_{C_0} \frac{f(z) - f(z_0)}{z - z_0}\, dz.$$

Referring now to statement (3) and noting that the length of C_0 is $2\pi\rho$, we may apply property (7), Sec. 32, of integrals:

$$\left| \int_{C_0} \frac{f(z) - f(z_0)}{z - z_0}\, dz \right| < \frac{\varepsilon}{\rho} 2\pi\rho = 2\pi\varepsilon.$$

In view of equation (5), then,

$$\left| \int_C \frac{f(z)\, dz}{z - z_0} - 2\pi i f(z_0) \right| < 2\pi\varepsilon.$$

Since the left-hand side of this inequality is a nonnegative constant that is less than an arbitrarily small positive number, it must equal to zero. Hence equation (2) is valid, and the theorem is proved.

40. DERIVATIVES OF ANALYTIC FUNCTIONS

We are now ready to prove that if a function is analytic at a point, then its derivatives of all orders exist at that point and are themselves analytic there.

We begin by assuming that f is analytic within and on a positively oriented simple closed contour C, and we let z be any point interior to C. Letting s denote points on C and using the Cauchy integral formula (Sec. 39)

$$(1) \qquad\qquad f(z) = \frac{1}{2\pi i} \int_C \frac{f(s)\,ds}{s-z},$$

we shall show that the derivative of f at z exists and has the integral representation

$$(2) \qquad\qquad f'(z) = \frac{1}{2\pi i} \int_C \frac{f(s)\,ds}{(s-z)^2}.$$

Note that expression (2) can be obtained formally by differentiating the integrand in equation (1) with respect to z.

To verify expression (2), we observe that, according to formula (1),

$$\frac{f(z+\Delta z) - f(z)}{\Delta z} = \frac{1}{2\pi i} \int_C \left(\frac{1}{s - z - \Delta z} - \frac{1}{s-z} \right) \frac{f(s)}{\Delta z}\,ds$$

$$= \frac{1}{2\pi i} \int_C \frac{f(s)\,ds}{(s - z - \Delta z)(s - z)}$$

when $0 < |\Delta z| < d$, where d is the shortest distance from z to points s on C. We then use the fact that f is continuous on C to show that the value of the last integral here approaches the value of the integral

$$\int_C \frac{f(s)\,ds}{(s-z)^2}$$

as Δz approaches zero. To accomplish this, we first write the difference

$$\int_C \left[\frac{1}{(s - z - \Delta z)(s - z)} - \frac{1}{(s-z)^2} \right] f(s)\,ds$$

of those two integrals in the form

$$\Delta z \int_C \frac{f(s)\,ds}{(s - z - \Delta z)(s - z)^2}.$$

Next we let M denote the maximum value of $|f(s)|$ on C, and we let L be the length of C. After noting that $|s - z| \geq d$ and

$$|s - z - \Delta z| \geq \|s - z| - |\Delta z\| \geq d - |\Delta z|,$$

we readily obtain the inequality

$$\left| \Delta z \int_C \frac{f(s)\, ds}{(s - z - \Delta z)(s - z)^2} \right| \leq \frac{|\Delta z| M L}{(d - |\Delta z|) d^2},$$

where the last fraction approaches zero as Δz tends to zero. Consequently,

$$\lim_{\Delta z \to 0} \frac{f(z + \Delta z) - f(z)}{\Delta z} = \frac{1}{2\pi i} \int_C \frac{f(s)\, ds}{(s - z)^2};$$

and expression (2) is now verified.

If we apply the same technique to expression (2), we find that

$$(3) \qquad f''(z) = \frac{1}{\pi i} \int_C \frac{f(s)\, ds}{(s - z)^3}.$$

More precisely, when $0 < |\Delta z| < d$,

$$\frac{f'(z + \Delta z) - f'(z)}{\Delta z} = \frac{1}{2\pi i} \int_C \left[\frac{1}{(s - z - \Delta z)^2} - \frac{1}{(s - z)^2} \right] \frac{f(s)\, ds}{\Delta z}$$

$$= \frac{1}{2\pi i} \int_C \frac{2(s - z) - \Delta z}{(s - z - \Delta z)^2 (s - z)^2} f(s)\, ds;$$

and, again since f is continuous on C, it can be shown that the value of the integral

$$\int_C \left[\frac{2(s - z) - \Delta z}{(s - z - \Delta z)^2 (s - z)^2} - \frac{2}{(s - z)^3} \right] f(s)\, ds = \int_C \frac{3(s - z)\Delta z - 2(\Delta z)^2}{(s - z - \Delta z)^2 (s - z)^3} f(s)\, ds$$

approaches zero as Δz tends to zero. (See Exercise 7.)

Expression (3) establishes the existence of the second derivative of the function f at each point z interior to C. In fact, it shows that if a function f is analytic at a point, then its derivative f' is also analytic at that point. For if f is analytic at a point z, there must exist a circle about z such that f is analytic within and on the circle. Then, in view of expression (3), $f''(z)$ exists at each point interior to the circle; and the derivative f' is, therefore, analytic at z. One can apply the same argument to the analytic function f' to conclude that its derivative f'' is analytic, etc. The following fundamental result is thus obtained.

Theorem 1. *If a function f is analytic at a point, then its derivatives of all orders are also analytic functions at that point.*

In particular, when a function

$$f(z) = u(x, y) + iv(x, y)$$

is analytic at a point $z = (x, y)$, the analyticity of f' ensures the continuity of f' there. Then, since

$$f'(z) = u_x(x, y) + iv_x(x, y) = v_y(x, y) - iu_y(x, y),$$

we may conclude that the first-order partial derivatives of u and v are continuous at the point. Furthermore, since f'' is analytic and continuous at z and since

$$f''(z) = u_{xx}(x, y) + iv_{xx}(x, y) = v_{yx}(x, y) - iu_{yx}(x, y),$$

etc., we arrive at a corollary that was anticipated in Sec. 22, where harmonic functions were introduced.

Corollary. *If a function $f(z) = u(x, y) + iv(x, y)$ is analytic at a point $z = x + iy$, then the component functions u and v have continuous partial derivatives of all orders at that point.*

If we agree that $f^{(0)}(z)$ denotes $f(z)$ and that $0! = 1$, we can use mathematical induction to verify the remarkable formula

(4) $$f^{(n)}(z) = \frac{n!}{2\pi i} \int_C \frac{f(s)\, ds}{(s - z)^{n+1}} \qquad (n = 0, 1, 2, \ldots).$$

When $n = 0$, this is just the Cauchy integral formula (1); if we assume that formula (4) holds for any particular nonnegative integer $n = m$, we can, by proceeding as we did in obtaining formulas (2) and (3), show that it is valid when $n = m + 1$. The details of the proof are left as an exercise.

When written in the form

(5) $$f^{(n)}(z_0) = \frac{n!}{2\pi i} \int_C \frac{f(z)\, dz}{(z - z_0)^{n+1}} \qquad (n = 0, 1, 2, \ldots),$$

where z_0 is a fixed point interior to C, expression (4) is an extension of the Cauchy integral formula in the notation of Sec. 39, where that formula was derived. This provides the useful formula

(6) $$\int_C \frac{f(z)\, dz}{(z - z_0)^{n+1}} = \frac{2\pi i}{n!} f^{(n)}(z_0) \qquad (n = 0, 1, 2, \ldots)$$

for evaluating certain integrals when f is analytic inside and on a simple closed contour C, taken in the positive sense, and z_0 is any point interior to C. It has already been illustrated in Sec. 39 when $n = 0$.

EXAMPLE 1. If C is the positively oriented unit circle $|z| = 1$, and $f(z) = \exp(2z)$, then

$$\int_C \frac{\exp(2z)\, dz}{z^4} = \int_C \frac{f(z)\, dz}{(z - 0)^{3+1}} = \frac{2\pi i}{3!} f'''(0) = \frac{8\pi i}{3}.$$

EXAMPLE 2. Let z_0 be any point interior to a positively oriented simple closed contour C. When $f(z) = 1$, formula (6) shows that

$$\int_C \frac{dz}{z - z_0} = 2\pi i, \qquad \int_C \frac{dz}{(z - z_0)^{n+1}} = 0 \quad (n = 1, 2, \ldots).$$

(Compare Exercise 13, Sec. 33.)

We conclude this section with a theorem due to E. Morera (1856–1909). The proof here depends on the fact that the derivative of an analytic function is itself analytic, as stated in Theorem 1.

Theorem 2. *If a function f is continuous throughout a domain D and if*

(7)
$$\int_C f(z)\, dz = 0$$

for every closed contour C lying in D, then f is analytic throughout D.

In particular, when D is *simply connected*, we have for the class of continuous functions on D a converse of Theorem 1 in Sec. 38, which is the extension of the Cauchy-Goursat theorem involving such domains.

When the hypothesis of Theorem 2 is satisfied, we need only refer to the theorem in Sec. 34 to see that f has an antiderivative in D; that is, there exists an analytic function F such that $F'(z) = f(z)$ at each point in D. Since f is the derivative of F, it then follows from Theorem 1 above that f is analytic in D.

EXERCISES

1. Let C denote the positively oriented boundary of the square whose sides lie along the lines $x = \pm 2$ and $y = \pm 2$. Evaluate each of these integrals:

(a) $\displaystyle\int_C \frac{e^{-z}\, dz}{z - (\pi i/2)}$;   (b) $\displaystyle\int_C \frac{\cos z}{z(z^2 + 8)}\, dz$;   (c) $\displaystyle\int_C \frac{z\, dz}{2z + 1}$;

(d) $\displaystyle\int_C \frac{\tan(z/2)}{(z - x_0)^2}\, dz \quad (-2 < x_0 < 2)$;   (e) $\displaystyle\int_C \frac{\cosh z}{z^4}\, dz$.

Ans. (a) 2π;   (b) $\pi i/4$;   (c) $-\pi i/2$;   (d) $i\pi \sec^2(x_0/2)$;   (e) 0.

*2. Find the value of the integral of $g(z)$ around the circle $|z - i| = 2$ in the positive sense when

(a) $g(z) = \dfrac{1}{z^2 + 4}$;   (b) $g(z) = \dfrac{1}{(z^2 + 4)^2}$.

Ans. (a) $\pi/2$;   (b) $\pi/16$.

3. Let C be the circle $|z| = 3$, described in the positive sense. Show that if

$$g(w) = \int_C \frac{2z^2 - z - 2}{z - w}\, dz \qquad (|w| \neq 3),$$

then $g(2) = 8\pi i$. What is the value of $g(w)$ when $|w| > 3$?

4. Let C be any simple closed contour, described in the positive sense in the z plane, and write

$$g(w) = \int_C \frac{z^3 + 2z}{(z - w)^3} \, dz.$$

Show that $g(w) = 6\pi i w$ when w is inside C and that $g(w) = 0$ when w is outside C.

5. Show that if f is analytic within and on a simple closed contour C and z_0 is not on C, then

$$\int_C \frac{f'(z) \, dz}{z - z_0} = \int_C \frac{f(z) \, dz}{(z - z_0)^2}.$$

6. Let f denote a function that is *continuous* on a simple closed contour C. Following the procedure used in Sec. 40, prove that the function

$$g(z) = \frac{1}{2\pi i} \int_C \frac{f(s) \, ds}{s - z}$$

is *analytic* at each point z interior to C and that

$$g'(z) = \frac{1}{2\pi i} \int_C \frac{f(s) \, ds}{(s - z)^2}$$

at such a point.

7. Give details in the derivation of integral representation (3), Sec. 40, for $f''(z)$.

   Suggestion: In the algebraic simplifications, retain the difference $s - z$ as a single term. Also, let D denote the *largest* distance from z to points s on C.

8. Carry out the induction argument used to establish integral representation (4), Sec. 40, for $f^{(n)}(z)$ $(n = 0, 1, 2, \ldots)$.

   Suggestion: Use the binomial formula (Exercise 14, Sec. 6) and the suggestion in Exercise 7.

*9. Let C be the unit circle $z = e^{i\theta}$ $(-\pi \le \theta \le \pi)$. First show that, for any real constant a,

$$\int_C \frac{e^{az}}{z} \, dz = 2\pi i.$$

Then write the integral in terms of θ to derive the integration formula

$$\int_0^\pi e^{a \cos \theta} \cos(a \sin \theta) \, d\theta = \pi.$$

10. (*a*) With the aid of the binomial formula (Exercise 14, Sec. 6), show that, for each value of n, the function

$$P_n(z) = \frac{1}{n! 2^n} \frac{d^n}{dz^n} (z^2 - 1)^n \qquad (n = 0, 1, 2, \ldots)$$

is a polynomial of degree n.*

---

*These are the Legendre polynomials which appear in Exercise 5, Sec. 31, when $z = x$. See the footnote to that exercise.

(*b*) Let C denote any positively oriented simple closed contour surrounding a fixed point z. With the aid of the integral representation (4), Sec. 40, for the nth derivative of an analytic function, show that the polynomials in part (*a*) can be expressed in the form

$$P_n(z) = \frac{1}{2^{n+1}\pi i} \int_C \frac{(s^2 - 1)^n}{(s - z)^{n+1}} \, ds \qquad (n = 0, 1, 2, \ldots).$$

(*c*) Point out how the integrand in the representation for $P_n(z)$ in part (*b*) can be written $(s+1)^n/(s-1)$ if $z = 1$. Then apply the Cauchy integral formula to show that $P_n(1) = 1$ $(n = 0, 1, 2, \ldots)$. Similarly, show that $P_n(-1) = (-1)^n$ $(n = 0, 1, 2, \ldots)$.

41. LIOUVILLE'S THEOREM AND THE FUNDAMENTAL THEOREM OF ALGEBRA

Let z_0 be a fixed complex number. If a function f is analytic within and on a circle $|z - z_0| = R$, taken in the positive sense and denoted by C (Fig. 44), we know from Sec. 40 that

$$f^{(n)}(z_0) = \frac{n!}{2\pi i} \int_C \frac{f(z)\, dz}{(z - z_0)^{n+1}} \qquad (n = 1, 2, \ldots).$$

Now the maximum value of $|f(z)|$ on the circle C depends, in general, on the radius of C; and if we let M_R denote that maximum value, an application of inequality (7), Sec. 32, tells us that

$$\left| f^{(n)}(z_0) \right| \leq \frac{n!}{2\pi} \cdot \frac{M_R}{R^{n+1}} 2\pi R \qquad (n = 1, 2, \ldots).$$

We thus have *Cauchy's inequality*:

(1) $$\left| f^{(n)}(z_0) \right| \leq \frac{n! M_R}{R^n} \qquad (n = 1, 2, \ldots).$$

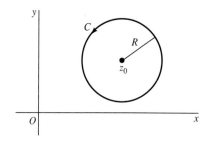

FIGURE 44

In particular, when $n = 1$,

(2) $$\left| f'(z_0) \right| \leq \frac{M_R}{R}.$$

From this it is easy to show that no entire function except a constant is bounded in the complex plane, as stated below in *Liouville's theorem* in a slightly different way.

Theorem 1. *If f is entire and bounded in the complex plane, then $f(z)$ is constant throughout the plane.*

To start the proof, we note that since f is entire, inequality (2) holds for any choices of z_0 and R. The boundedness condition in the theorem tells us that a non-negative constant M exists such that $|f(z)| \leq M$ for all z; and, because the constant M_R in inequality (2) is always less than or equal to M, it follows that

$$(3) \qquad\qquad |f'(z_0)| \leq \frac{M}{R},$$

where z_0 is any fixed point in the plane and R is arbitrarily large. Now the number M in inequality (3) is independent of the value of R that is taken. Hence that inequality can hold for arbitrarily large values of R only if $f'(z_0) = 0$. Since the choice of z_0 was arbitrary, this means that $f'(z) = 0$ everywhere in the complex plane. Consequently, f is a constant function (Sec. 20).

The following theorem, known as the *fundamental theorem of algebra*, follows readily from Liouville's theorem.

Theorem 2. *Any polynomial*

$$P(z) = a_0 + a_1 z + a_2 z^2 + \cdots + a_n z^n \qquad (a_n \neq 0)$$

of degree n ($n \geq 1$) has at least one zero. That is, there exists at least one point z_0 such that $P(z_0) = 0$.

The proof here is by contradiction. Suppose that $P(z)$ is *not* zero for any value of z. Then the reciprocal

$$f(z) = \frac{1}{P(z)}$$

is clearly entire, and it is also bounded in the complex plane.

To show that it is bounded, we first write

$$(4) \qquad\qquad w = \frac{a_0}{z^n} + \frac{a_1}{z^{n-1}} + \frac{a_2}{z^{n-2}} + \cdots + \frac{a_{n-1}}{z},$$

so that $P(z) = (a_n + w)z^n$. We then observe that a sufficiently large positive number R can be found such that the modulus of each of the quotients in expression (4) is less than the number $|a_n|/(2n)$ when $|z| \geq R$. The generalized triangle inequality, applied to n complex numbers, thus shows that $|w| < |a_n|/2$ for such values of z. Consequently, when $|z| \geq R$,

$$|a_n + w| \geq \big| \, |a_n| - |w| \, \big| > \frac{|a_n|}{2};$$

and this enables us to write

$$(5) \quad |P(z)| = |a_n + w| \, |z^n| > \frac{|a_n|}{2}|z|^n \geq \frac{|a_n|}{2}R^n \qquad \text{whenever} \qquad |z| \geq R.$$

Evidently, then,

$$|f(z)| = \frac{1}{|P(z)|} < \frac{2}{|a_n|R^n} \qquad \text{whenever} \qquad |z| > R.$$

So f is bounded in the region *exterior to* the disk $|z| \leq R$. But f is continuous in that closed disk, and this means that f is bounded there too. Hence f is bounded in the entire plane.

It now follows from Liouville's theorem that $f(z)$, and consequently $P(z)$, is constant. But $P(z)$ is not constant, and we have reached a contradiction.*

The fundamental theorem tells us that any polynomial $P(z)$ of degree n $(n \geq 1)$ can be expressed as a product of linear factors:

$$(6) \qquad P(z) = c(z - z_1)(z - z_2) \cdots (z - z_n),$$

where c and z_k $(k = 1, 2, \ldots, n)$ are complex constants. Specifically, the theorem ensures that $P(z)$ has a zero z_1. Then, according to Exercise 10, Sec. 42,

$$P(z) = (z - z_1)Q_1(z),$$

where $Q_1(z)$ is a polynomial of degree $n - 1$. The same argument, applied to $Q_1(z)$, reveals that there is a number z_2 such that

$$P(z) = (z - z_1)(z - z_2)Q_2(z),$$

where $Q_2(z)$ is a polynomial of degree $n - 2$. Continuing in this way, we arrive at expression (6). Some of the constants z_k in expression (6) may, of course, appear more than once, and it is clear that $P(z)$ can have no more than n *distinct* zeros.

42. MAXIMUM MODULI OF FUNCTIONS

In this section, we shall derive some important results involving maximum values of the moduli of analytic functions.

Lemma. *Suppose that $f(z)$ is analytic throughout a neighborhood $|z - z_0| < \varepsilon$ of a point z_0. If $|f(z)| \leq |f(z_0)|$ for each point z in that neighborhood, then $f(z)$ has the constant value $f(z_0)$ throughout the neighborhood.*

To prove this, we assume that f satisfies the stated conditions and let z_1 be any point other than z_0 in the given neighborhood. We then let ρ be the distance between z_1 and z_0. If C_ρ denotes the positively oriented circle $|z - z_0| = \rho$, centered at z_0 and passing through z_1 (Fig. 45), the Cauchy integral formula tells us that

$$(1) \qquad f(z_0) = \frac{1}{2\pi i} \int_{C_\rho} \frac{f(z)\,dz}{z - z_0};$$

---

*For an interesting proof of the fundamental theorem using the Cauchy-Goursat theorem, see R. P. Boas, Jr., *Amer. Math. Monthly*, Vol. 71, No. 2, p. 180, 1964.

and the parametric representation

$$z = z_0 + \rho e^{i\theta} \qquad (0 \le \theta \le 2\pi)$$

for C_ρ enables us to write equation (1) as

(2) $$f(z_0) = \frac{1}{2\pi} \int_0^{2\pi} f(z_0 + \rho e^{i\theta})\, d\theta.$$

We note from expression (2) that when a function is analytic within and on a given circle, its value at the center is the arithmetic mean of its values on the circle. This result is called *Gauss's mean value theorem*.

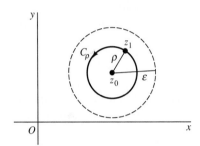

FIGURE 45

From equation (2), we obtain the inequality

(3) $$|f(z_0)| \le \frac{1}{2\pi} \int_0^{2\pi} |f(z_0 + \rho e^{i\theta})|\, d\theta.$$

On the other hand, since

(4) $$|f(z_0 + \rho e^{i\theta})| \le |f(z_0)| \qquad (0 \le \theta \le 2\pi),$$

we find that

$$\int_0^{2\pi} |f(z_0 + \rho e^{i\theta})|\, d\theta \le \int_0^{2\pi} |f(z_0)|\, d\theta = 2\pi |f(z_0)|.$$

Thus

(5) $$|f(z_0)| \ge \frac{1}{2\pi} \int_0^{2\pi} |f(z_0 + \rho e^{i\theta})|\, d\theta.$$

It is now evident from inequalities (3) and (5) that

$$|f(z_0)| = \frac{1}{2\pi} \int_0^{2\pi} |f(z_0 + \rho e^{i\theta})|\, d\theta,$$

or

$$\int_0^{2\pi} [|f(z_0)| - |f(z_0 + \rho e^{i\theta})|]\, d\theta = 0.$$

The integrand in this last integral is continuous in the variable θ; and, in view of condition (4), it is greater than or equal to zero on the entire interval $0 \le \theta \le 2\pi$. Because the value of the integral is zero, then, the integrand must be identically equal to zero. That is,

$$(6) \qquad |f(z_0 + \rho e^{i\theta})| = |f(z_0)| \qquad (0 \le \theta \le 2\pi).$$

This shows that $|f(z)| = |f(z_0)|$ *for all points z on the circle* $|z - z_0| = \rho$.

Finally, since z_1 is an arbitrarily chosen point in the deleted neighborhood $0 < |z - z_0| < \varepsilon$, we see that the equation $|f(z)| = |f(z_0)|$ is, in fact, satisfied by all points z lying on *any* circle $|z - z_0| = \rho$, where $0 < \rho < \varepsilon$. Consequently, $|f(z)| = |f(z_0)|$ everywhere in the neighborhood $|z - z_0| < \varepsilon$. But we know from Exercise 7(c), Sec. 22, that when the modulus of an analytic function is constant in a domain, the function itself is constant there. Thus $f(z) = f(z_0)$ for each point z in the neighborhood, and the proof of the lemma is complete.

This lemma can be used to prove the following theorem, which is known as the *maximum modulus principle*.

Theorem. *If a function f is analytic and not constant in a given domain D, then* $|f(z)|$ *has no maximum value in D. That is, there is no point z_0 in the domain such that* $|f(z)| \le |f(z_0)|$ *for all points z in it.*

Given that f is analytic in D, we shall prove the theorem by assuming that $|f(z)|$ *does* have a maximum value at some point z_0 in D and then showing that $f(z)$ must be constant throughout D.

We begin by observing that, since D is a *connected* open set (Sec. 8), there is a polygonal line L, consisting of a finite number of line segments joined end to end and lying entirely in D, that extends from z_0 to any other point P in D. We let d be the shortest distance from points on L to the boundary of D, unless D is the entire plane; in that case, d may be any positive number. We then form a finite sequence of points $z_0, z_1, z_2, \ldots, z_n$ along L, where the point z_n coincides with P (Fig. 46) and where each point is sufficiently close to the adjacent ones that

$$|z_k - z_{k-1}| < d \qquad (k = 1, 2, \ldots, n).$$

Finally, we construct a finite sequence of neighborhoods $N_0, N_1, N_2, \ldots, N_n$, where each neighborhood N_k is centered at z_k and has radius d. Note that these neighborhoods are all contained in D and that the center z_k of each neighborhood N_k ($k = 1, 2, \ldots, n$) lies in the preceding neighborhood N_{k-1}.

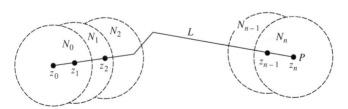

FIGURE 46

Since $|f(z)|$ was assumed to have a maximum value in D at z_0, it also has a maximum value in N_0 at that point. Hence, according to the preceding lemma, $f(z)$ has the constant value $f(z_0)$ throughout N_0. In particular, $f(z_1) = f(z_0)$. This means that $|f(z)| \leq |f(z_1)|$ for each point z in N_1; and the lemma can be applied again, this time telling us that

$$f(z) = f(z_1) = f(z_0)$$

when z is in N_1. Since z_2 is in N_1, then, $f(z_2) = f(z_0)$. Hence $|f(z)| \leq |f(z_2)|$ when z is in N_2; and the lemma is once again applicable, showing that

$$f(z) = f(z_2) = f(z_0)$$

when z is in N_2. Continuing in this manner, we eventually reach the neighborhood N_n and arrive at the fact that $f(z_n) = f(z_0)$.

Since z_n coincides with the point P, which is any point other than z_0 in D, we may conclude that $f(z) = f(z_0)$ for *every* point z in D. Inasmuch as $f(z)$ has now been shown to be constant throughout D, the theorem is proved.

If a function f that is analytic at each point in the interior of a closed bounded region R is also continuous throughout R, then the modulus $|f(z)|$ has a maximum value somewhere in R (Sec. 14). That is, there exists a nonnegative constant M such that $|f(z)| \leq M$ for all points z in R, and equality holds for at least one such point. If f is a constant function, then $|f(z)| = M$ for all z in R. If, however, $f(z)$ is not constant, then, according to the maximum modulus principle, $|f(z)| \neq M$ for any point z in the interior of R. We thus arrive at an important corollary of the maximum modulus principle.

Corollary. *Suppose that a function f is continuous in a closed bounded region R and that it is analytic and not constant in the interior of R. Then the maximum value of $|f(z)|$ in R, which is always reached, occurs somewhere on the boundary of R and never in the interior.*

EXAMPLE. Let R denote the rectangular region $0 \leq x \leq \pi, 0 \leq y \leq 1$. The corollary tells us that the modulus of the entire function $f(z) = \sin z$ has a maximum value in R that occurs somewhere on the boundary, and not in the interior, of R. This can be verified directly by writing (see Sec. 24)

$$|f(z)| = \sqrt{\sin^2 x + \sinh^2 y}$$

and noting that, in R, $\sin^2 x$ is greatest when $x = \pi/2$ and that the increasing function $\sinh^2 y$ is greatest when $y = 1$. Thus the maximum value of $|f(z)|$ in R occurs at the boundary point $z = (\pi/2) + i$ and at no other point in R.

When the function f in the corollary is written $f(z) = u(x, y) + iv(x, y)$, *the component function $u(x, y)$ also has a maximum value in R which is assumed on the boundary of R and never in the interior*, where it is harmonic. For the composite function $g(z) = \exp[f(z)]$ is continuous in R and analytic and not constant in the interior. Consequently, its modulus $|g(z)| = \exp[u(x, y)]$, which is continuous in R, must assume its maximum value in R on the boundary. Because of the increasing nature of the exponential function, it follows that the maximum value of $u(x, y)$ also occurs on the boundary.

Properties of *minimum* values of $|f(z)|$ and $u(x, y)$ are treated in the exercises.

EXERCISES

1. Let f be an entire function such that $|f(z)| \leq A|z|$ for all z, where A is a fixed positive number. Show that $f(z) = a_1 z$, where a_1 is a complex constant.

   Suggestion: Use Cauchy's inequality (Sec. 41) to show that the second derivative $f''(z)$ is zero everywhere in the plane. Note that the constant M_R in Cauchy's inequality is less than or equal to $A(|z_0| + R)$.

2. Suppose that $f(z)$ is entire and that the harmonic function $u(x, y) = \text{Re}[f(z)]$ has an upper bound; that is, $u(x, y) \leq u_0$ for all points (x, y) in the xy plane. Show that $u(x, y)$ must be constant throughout the plane.

   Suggestion: Apply Liouville's theorem (Sec. 41) to the function $g(z) = \exp[f(z)]$.

3. Show that, for R sufficiently large, the polynomial $P(z)$ in Theorem 2, Sec. 41, satisfies the inequality

$$|P(z)| < 2|a_n||z|^n \qquad \text{whenever} \qquad |z| \geq R.$$

   [Compare the first of inequalities (5), Sec. 41.]

   Suggestion: Observe that there is a positive number R such that the modulus of each quotient in expression (4), Sec. 41, is less than $|a_n|/n$ when $|z| \geq R$.

4. Let a function f be continuous in a closed bounded region R, and let it be analytic and not constant throughout the interior of R. Assuming that $f(z) \neq 0$ anywhere in R, prove that $|f(z)|$ has a *minimum value m* in R which occurs on the boundary of R and never in the interior. Do this by applying the corresponding result for maximum values (Sec. 42) to the function $g(z) = 1/f(z)$.

5. Use the function $f(z) = z$ to show that in Exercise 4 the condition $f(z) \neq 0$ anywhere in R is necessary in order to obtain the result of that exercise. That is, show that $|f(z)|$ *can* reach its minimum value at an interior point when that minimum value is zero.

6. Consider the function $f(z) = (z + 1)^2$ and the closed triangular region R with vertices at the points $z = 0, z = 2$, and $z = i$. Find points in R where $|f(z)|$ has its maximum and minimum values, thus illustrating results in Sec. 42 and Exercise 4.

   Suggestion: Interpret $|f(z)|$ as the square of the distance between z and -1.

   Ans. $z = 2, z = 0$.

7. Let $f(z) = u(x, y) + iv(x, y)$ be a function that is continuous in a closed bounded region R and analytic and not constant throughout the interior of R. Prove that the component function $u(x, y)$ has a minimum value in R which occurs on the boundary of R and never in the interior. (See Exercise 4.)

8. Let f be the function $f(z) = e^z$ and R the rectangular region $0 \leq x \leq 1, 0 \leq y \leq \pi$. Illustrate results in Sec. 42 and Exercise 7 by finding points in R where the component function $u(x, y) = \text{Re}[f(z)]$ reaches its maximum and minimum values.

   Ans. $z = 1, z = 1 + \pi i$.

9. Let the function $f(z) = u(x, y) + iv(x, y)$ be continuous in a closed bounded region R, and suppose that it is analytic and not constant in the interior of R. Show that the

component function $v(x, y)$ has maximum and minimum values in R which are reached on the boundary of R and never in the interior, where it is harmonic.

   Suggestion: Apply results in Sec. 42 and Exercise 7 to the function $g(z) = -if(z)$.

10. Let z_0 be a zero of the polynomial

$$P(z) = a_0 + a_1 z + a_2 z^2 + \cdots + a_n z^n \qquad (a_n \neq 0)$$

of degree n ($n \geq 1$). Show in the following way that

$$P(z) = (z - z_0)Q(z),$$

where $Q(z)$ is a polynomial of degree $n - 1$.

(*a*) Verify that

$$z^k - z_0^k = (z - z_0)(z^{k-1} + z^{k-2}z_0 + \cdots + z z_0^{k-2} + z_0^{k-1}) \qquad (k = 2, 3, \ldots).$$

(*b*) Use the factorization in part (*a*) to show that

$$P(z) - P(z_0) = (z - z_0)Q(z),$$

where $Q(z)$ is a polynomial of degree $n - 1$, and deduce the desired result from this.

CHAPTER
5

SERIES

This chapter is devoted mainly to series representations of analytic functions. We present theorems that guarantee the existence of such representations, and we develop some facility in manipulating series.

43. CONVERGENCE OF SEQUENCES AND SERIES

An infinite *sequence*

(1)
$$z_1, z_2, \ldots, z_n, \ldots$$

of complex numbers has a *limit z* if, for each positive number ε, there exists a positive integer n_0 such that

(2)
$$|z_n - z| < \varepsilon \quad \text{whenever} \quad n > n_0.$$

Geometrically, this means that, for sufficiently large values of n, the points z_n lie in any given ε neighborhood of z (Fig. 47). Since we can choose ε as small as we please, it follows that the points z_n become arbitrarily close to z as their subscripts increase. Note that the value of n_0 that is needed will, in general, depend on the value of ε.

The sequence (1) can have at most one limit. That is, a limit z is unique if it exists (Exercise 6). When that limit exists, the sequence is said to *converge* to z; and we write

(3)
$$\lim_{n \to \infty} z_n = z.$$

If the sequence has no limit, it *diverges*.

138

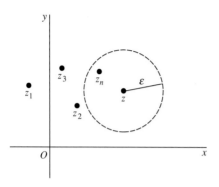

FIGURE 47

Theorem 1. *Suppose that* $z_n = x_n + iy_n$ $(n = 1, 2, \ldots)$ *and* $z = x + iy$. *Then*

(4)
$$\lim_{n \to \infty} z_n = z$$

if and only if

(5)
$$\lim_{n \to \infty} x_n = x \qquad and \qquad \lim_{n \to \infty} y_n = y.$$

To prove this theorem, we first assume that conditions (5) hold and obtain condition (4) from it. According to conditions (5), there exist, for each positive number ε, positive integers n_1 and n_2 such that

$$|x_n - x| < \frac{\varepsilon}{2} \qquad \text{whenever} \qquad n > n_1$$

and

$$|y_n - y| < \frac{\varepsilon}{2} \qquad \text{whenever} \qquad n > n_2.$$

Hence, if n_0 is the larger of the two integers n_1 and n_2,

$$|x_n - x| < \frac{\varepsilon}{2} \quad \text{and} \quad |y_n - y| < \frac{\varepsilon}{2} \qquad \text{whenever} \qquad n > n_0.$$

Since

$$|(x_n + iy_n) - (x + iy)| = |(x_n - x) + i(y_n - y)| \le |x_n - x| + |y_n - y|,$$

then,

$$|z_n - z| < \frac{\varepsilon}{2} + \frac{\varepsilon}{2} = \varepsilon \qquad \text{whenever} \qquad n > n_0.$$

Condition (4) thus holds.

Conversely, if we start with condition (4), we know that, for each positive number ε, there exists a positive integer n_0 such that

$$|(x_n + iy_n) - (x + iy)| < \varepsilon \qquad \text{whenever} \qquad n > n_0.$$

But

$$|x_n - x| \le |(x_n - x) + i(y_n - y)| = |(x_n + iy_n) - (x + iy)|$$

and

$$|y_n - y| \le |(x_n - x) + i(y_n - y)| = |(x_n + iy_n) - (x + iy)|;$$

and this means that

$$|x_n - x| < \varepsilon \quad \text{and} \quad |y_n - y| < \varepsilon \qquad \text{whenever} \qquad n > n_0.$$

That is, conditions (5) are satisfied.

An infinite *series*

(6)
$$\sum_{n=1}^{\infty} z_n = z_1 + z_2 + \cdots + z_n + \cdots$$

of complex numbers *converges* to the *sum S* if the sequence

$$S_N = \sum_{n=1}^{N} z_n = z_1 + z_2 + \cdots + z_N \qquad (N = 1, 2, \ldots)$$

of *partial sums* converges to S; we then write

$$\sum_{n=1}^{\infty} z_n = S.$$

Note that, since a sequence can have at most one limit, a series can have at most one sum. When a series does not converge, we say that it *diverges*.

Theorem 2. *Suppose that $z_n = x_n + iy_n$ $(n = 1, 2, \ldots)$ and $S = X + iY$. Then*

(7)
$$\sum_{n=1}^{\infty} z_n = S$$

if and only if

(8)
$$\sum_{n=1}^{\infty} x_n = X \qquad and \qquad \sum_{n=1}^{\infty} y_n = Y.$$

The proof is based on Theorem 1. Let S_N denote the partial sum consisting of the first N terms of the series in condition (7), and observe that

(9)
$$S_N = X_N + iY_N,$$

where

$$X_N = \sum_{n=1}^{N} x_n \qquad and \qquad Y_N = \sum_{n=1}^{N} y_n.$$

Now condition (7) holds if and only if

$$\lim_{N\to\infty} S_N = S;$$

and, in view of relation (9) and Theorem 1, this condition holds if and only if

(10) $$\lim_{N\to\infty} X_N = X \quad \text{and} \quad \lim_{N\to\infty} Y_N = Y.$$

Conditions (10) therefore imply condition (7), and conversely. Since X_N and Y_N are the partial sums of the series in conditions (8), Theorem 2 is proved.

By recalling from calculus that the nth term of a convergent series of real numbers approaches zero as n tends to infinity, we can see immediately from Theorems 1 and 2 that the same is true of a convergent series of complex numbers. That is, *a necessary condition for the convergence of series (6) is that*

(11) $$\lim_{n\to\infty} z_n = 0.$$

The terms of a convergent series of complex numbers are, therefore, *bounded*. More precisely, there exists a positive constant M such that $|z_n| \leq M$ for each positive integer n. (See Exercise 10.)

For another important property of series of complex numbers, we assume that series (6) is *absolutely convergent*. That is, when $z_n = x_n + iy_n$, the series

$$\sum_{n=1}^{\infty} |z_n| = \sum_{n=1}^{\infty} \sqrt{x_n^2 + y_n^2}$$

of real numbers $\sqrt{x_n^2 + y_n^2}$ converges. Since

$$|x_n| \leq \sqrt{x_n^2 + y_n^2} \quad \text{and} \quad |y_n| \leq \sqrt{x_n^2 + y_n^2},$$

we know from the comparison test in calculus that the two series

$$\sum_{n=1}^{\infty} |x_n| \quad \text{and} \quad \sum_{n=1}^{\infty} |y_n|$$

converge. Moreover, since the absolute convergence of a series of real numbers implies the convergence of the series itself, it follows that there are real numbers X and Y such that conditions (8) hold. According to Theorem 2, then, series (6) converges. Consequently, *absolute convergence of a series of complex numbers implies convergence of that series.*

In establishing the fact that the sum of a series is a given number S, it is often convenient to define the *remainder* ρ_N after N terms:

(12) $$\rho_N = S - S_N.$$

Thus $S = S_N + \rho_N$; and, since $|S_N - S| = |\rho_N - 0|$, we see that *a series converges to a number S if and only if the sequence of remainders tends to zero.* We shall make

considerable use of this observation in our treatment of *power series*. They are series
of the form

$$\sum_{n=0}^{\infty} a_n(z - z_0)^n = a_0 + a_1(z - z_0) + a_2(z - z_0)^2 + \cdots + a_n(z - z_0)^n + \cdots,$$

where z_0 and the coefficients a_n are complex constants and z may be any point in a
stated region containing z_0. In such series, involving a variable z, we shall denote
sums, partial sums, and remainders by $S(z), S_N(z),$ and $\rho_N(z)$, respectively.

EXERCISES

1. Show in two ways that the sequence

$$z_n = -2 + i\frac{(-1)^n}{n^2} \qquad (n = 1, 2, \ldots)$$

   converges to -2.

2. Let r_n denote the moduli and Θ_n the principal values of the arguments of the complex
   numbers z_n in Exercise 1. Show that the sequence r_n $(n = 1, 2, \ldots)$ converges but that
   the sequence Θ_n $(n = 1, 2, \ldots)$ does not.

3. Show that

$$\text{if} \qquad \lim_{n \to \infty} z_n = z, \qquad \text{then} \qquad \lim_{n \to \infty} |z_n| = |z|.$$

4. By considering the remainders $\rho_N(z)$, verify that

$$\sum_{n=1}^{\infty} z^n = \frac{z}{1 - z} \qquad \text{when} \qquad |z| < 1.$$

   Suggestion: Use the identity (Exercise 13, Sec. 6)

$$1 + z + z^2 + \cdots + z^N = \frac{1 - z^{N+1}}{1 - z} \qquad (z \neq 1)$$

   to show that $\rho_N(z) = z^{N+1}/(1 - z)$.

5. Write $z = re^{i\theta}$, where $0 < r < 1$, in the summation formula obtained in Exercise 4. Then,
   with the aid of Theorem 2 in Sec. 43, show that

$$\sum_{n=1}^{\infty} r^n \cos n\theta = \frac{r \cos \theta - r^2}{1 - 2r \cos \theta + r^2} \qquad \text{and} \qquad \sum_{n=1}^{\infty} r^n \sin n\theta = \frac{r \sin \theta}{1 - 2r \cos \theta + r^2}$$

   when $0 < r < 1$. (Note that these formulas are also valid when $r = 0$.)

6. Show that a limit of a convergent sequence of complex numbers is unique by appealing
   to the corresponding result for a sequence of real numbers.

7. Show that

$$\text{if} \qquad \sum_{n=1}^{\infty} z_n = S, \qquad \text{then} \qquad \sum_{n=1}^{\infty} \overline{z}_n = \overline{S}.$$

8. Let c denote any complex number and show that

$$\text{if} \quad \sum_{n=1}^{\infty} z_n = S, \quad \text{then} \quad \sum_{n=1}^{\infty} c z_n = cS.$$

9. By recalling the corresponding result for series of real numbers and referring to Theorem 2 in Sec. 43, show that

$$\text{if} \quad \sum_{n=1}^{\infty} z_n = S \quad \text{and} \quad \sum_{n=1}^{\infty} w_n = T, \quad \text{then} \quad \sum_{n=1}^{\infty} (z_n + w_n) = S + T.$$

10. Let a sequence z_n $(n = 1, 2, \ldots)$ converge to a number z. Show that there exists a positive number M such that the inequality $|z_n| \leq M$ holds for all n. Do this in each of the ways indicated below.

(*a*) Note that there is a positive integer n_0 such that

$$|z_n| = |z + (z_n - z)| < |z| + 1$$

whenever $n > n_0$.

(*b*) Write $z_n = x_n + i y_n$ and recall from the theory of sequences of real numbers that the convergence of x_n and y_n $(n = 1, 2, \ldots)$ implies that $|x_n| \leq M_1$ and $|y_n| \leq M_2$ $(n = 1, 2, \ldots)$ for some positive numbers M_1 and M_2.

44. TAYLOR SERIES

We turn now to *Taylor's theorem*, which is one of the most important results of the chapter.

Theorem. *Suppose that a function* f *is analytic throughout an open disk* $|z - z_0| < R_0$, *centered at* z_0 *and with radius* R_0 *(Fig. 48). Then, at each point* z *in that disk,* $f(z)$ *has the series representation*

(1) $$f(z) = \sum_{n=0}^{\infty} a_n (z - z_0)^n \qquad (|z - z_0| < R_0),$$

where

(2) $$a_n = \frac{f^{(n)}(z_0)}{n!} \qquad (n = 0, 1, 2, \ldots).$$

That is, the power series here converges to $f(z)$ *when* $|z - z_0| < R_0$.

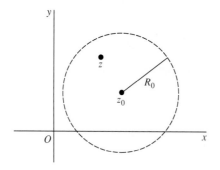

FIGURE 48

This is the expansion of $f(z)$ into a *Taylor series* about the point z_0. It is the familiar Taylor series from calculus, adapted to functions of a complex variable. Note that, with the agreement that $f^{(0)}(z_0) = f(z_0)$ and $0! = 1$, series (1) can be written

(3) $\quad f(z) = f(z_0) + \dfrac{f'(z_0)}{1!}(z - z_0) + \dfrac{f''(z_0)}{2!}(z - z_0)^2 + \cdots \qquad (|z - z_0| < R_0).$

Any function that is known to be analytic at a point z_0 must have a Taylor series about that point. For, if f is analytic at z_0, it is analytic in some neighborhood $|z - z_0| < \varepsilon$ of z_0 (Sec. 20); and ε may serve as the value of R_0 in the statement of Taylor's theorem. If, on the other hand, f is entire, R_0 can be chosen arbitrarily large; and the condition of validity becomes $|z - z_0| < \infty$. The series then converges to $f(z)$ at each point in the finite plane.

We first prove the theorem when $z_0 = 0$, in which case series (1) becomes

(4) $$f(z) = \sum_{n=0}^{\infty} \frac{f^{(n)}(0)}{n!} z^n \qquad (|z| < R_0)$$

and is called a *Maclaurin series*. The proof when z_0 is arbitrary will follow as an immediate consequence.

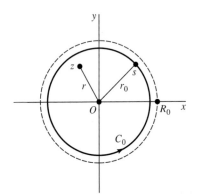

FIGURE 49

To begin, we let C_0 denote any positively oriented circle $|z| = r_0$ that is contained in the disk $|z| < R_0$ but is large enough so that the point z is interior to it (Fig. 49). The Cauchy integral formula then applies:

(5) $$f(z) = \frac{1}{2\pi i} \int_{C_0} \frac{f(s)\,ds}{s - z}.$$

Now the factor $1/(s - z)$ in the integrand here can be put in the form

(6) $$\frac{1}{s - z} = \frac{1}{s} \cdot \frac{1}{1 - (z/s)};$$

and we know from Exercise 13, Sec. 6, that

(7) $$\frac{1}{1 - z} = \sum_{n=0}^{N-1} z^n + \frac{z^N}{1 - z}.$$

when z is any complex number other than unity. Replacing z by z/s in expression (7), then, we can rewrite equation (6) as

(8) $$\frac{1}{s-z} = \sum_{n=0}^{N-1} \frac{1}{s^{n+1}} z^n + z^N \frac{1}{(s-z)s^N}.$$

Multiplying through this equation by $f(s)$ and then integrating each side with respect to s around C_0, we find that

$$\int_{C_0} \frac{f(s)\,ds}{s-z} = \sum_{n=0}^{N-1} \int_{C_0} \frac{f(s)\,ds}{s^{n+1}} z^n + z^N \int_{C_0} \frac{f(s)\,ds}{(s-z)s^N}.$$

In view of expression (5) and the fact (Sec. 40) that

$$\frac{1}{2\pi i} \int_{C_0} \frac{f(s)\,ds}{s^{n+1}} = \frac{f^{(n)}(0)}{n!} \qquad (n = 0,1,2,\ldots),$$

this reduces, after we multiply through by $1/(2\pi i)$, to

(9) $$f(z) = \sum_{n=0}^{N-1} \frac{f^{(n)}(0)}{n!} z^n + \rho_n(z),$$

where

(10) $$\rho_N(z) = \frac{z^N}{2\pi i} \int_{C_0} \frac{f(s)\,ds}{(s-z)s^N}.$$

Representation (4) now follows once it is shown that

(11) $$\lim_{N\to\infty} \rho_N(z) = 0.$$

To accomplish this, suppose that $|z| = r$. Then, if s is a point on C_0,

$$|s-z| \geq ||s| - |z|| = r_0 - r.$$

Hence, if M denotes the maximum value of $|f(s)|$ on C_0,

$$|\rho_N(z)| \leq \frac{r^N}{2\pi} \cdot \frac{M}{(r_0-r)r_0^N} 2\pi r_0 = \frac{Mr_0}{r_0-r}\left(\frac{r}{r_0}\right)^N.$$

But $(r/r_0) < 1$ since z is interior to C_0, and limit (11) clearly holds.

To verify the theorem when the disk of radius R_0 is centered at an arbitrary point z_0, we suppose that f is analytic when $|z - z_0| < R_0$ and note that the composite function $f(z + z_0)$ must be analytic when $|(z + z_0) - z_0| < R_0$. This last inequality is, of course, just $|z| < R_0$; and, if we write $g(z) = f(z + z_0)$, the analyticity of g in the disk $|z| < R_0$ ensures the existence of a Maclaurin series representation:

$$g(z) = \sum_{n=0}^{\infty} \frac{g^{(n)}(0)}{n!} z^n \qquad (|z| < R_0).$$

That is,

$$f(z + z_0) = \sum_{n=0}^{\infty} \frac{f^{(n)}(z_0)}{n!} z^n \qquad (|z| < R_0).$$

After replacing z by $z - z_0$ in this equation and its condition of validity, we have the desired Taylor series expansion (1).

45. EXAMPLES

When it is known that f is analytic at all points within a circle centered at z_0, convergence of its Taylor series about z_0 to $f(z)$ for each point z within that circle is ensured; no test for the convergence of the series is required. In fact, according to Taylor's theorem, the series converges to $f(z)$ within the circle about z_0 whose radius is the distance from z_0 to the nearest point z_1 where f fails to be analytic. In Sec. 49, we shall find that this is actually the largest circle centered at z_0 such that the series converges to $f(z)$ for all z interior to it.

Also, in Sec. 50, we shall see that if there are constants a_n $(n = 0, 1, 2, \ldots)$ such that

$$f(z) = \sum_{n=0}^{\infty} a_n (z - z_0)^n$$

for all points z interior to some circle centered at z_0, then the power series here must be *the* Taylor series for f about z_0, regardless of how those constants arise. This observation often allows us to find the coefficients a_n in Taylor series in more efficient ways than by appealing directly to the formula $a_n = f^{(n)}(z_0)/n!$ in Taylor's theorem.

In the following examples, we use the formula in Taylor's theorem to find the Maclaurin series expansions of some fairly simple functions, and we emphasize the use of those expansions in finding other representations.

EXAMPLE 1.   Since the function $f(z) = e^z$ is entire, it has a Maclaurin series representation which is valid for all z. Here $f^{(n)}(z) = e^z$; and, because $f^{(n)}(0) = 1$, it follows that

$$(1) \qquad e^z = \sum_{n=0}^{\infty} \frac{z^n}{n!} \qquad (|z| < \infty).$$

Note that if $z = x + i0$, expansion (1) becomes

$$e^x = \sum_{n=0}^{\infty} \frac{x^n}{n!} \qquad (-\infty < x < \infty).$$

The entire function $z^2 e^{3z}$ also has a Maclaurin series expansion. The simplest way to obtain it is to replace z by $3z$ on each side of equation (1) and then multiply through the resulting equation by z^2 (see Exercise 8, Sec. 43):

$$z^2 e^{3z} = \sum_{n=0}^{\infty} \frac{3^n}{n!} z^{n+2} \qquad (|z| < \infty).$$

Finally, if we replace n by $n - 2$ here, we have

$$z^2 e^{3z} = \sum_{n=2}^{\infty} \frac{3^{n-2}}{(n-2)!} z^n \qquad (|z| < \infty).$$

EXAMPLE 2. If $f(z) = \sin z$, then

$$f^{(2n)}(0) = 0 \qquad \text{and} \qquad f^{(2n+1)}(0) = (-1)^n \qquad (n = 0, 1, 2, \ldots).$$

Hence

$$(2) \qquad \sin z = \sum_{n=0}^{\infty} (-1)^n \frac{z^{2n+1}}{(2n+1)!} \qquad (|z| < \infty).$$

The condition $|z| < \infty$ once again follows from the fact that the function is entire.

Because $\sinh z = -i \sin(iz)$ (Sec. 25), we need only replace z by iz on each side of equation (2) and multiply through the resulting equation by $-i$ to see that

$$(3) \qquad \sinh z = \sum_{n=0}^{\infty} \frac{z^{2n+1}}{(2n+1)!} \qquad (|z| < \infty).$$

EXAMPLE 3. When $f(z) = \cos z$,

$$f^{(2n)}(0) = (-1)^n \qquad \text{and} \qquad f^{(2n+1)}(0) = 0 \qquad (n = 0, 1, 2, \ldots).$$

So this entire function has the Maclaurin series representation

$$(4) \qquad \cos z = \sum_{n=0}^{\infty} (-1)^n \frac{z^{2n}}{(2n)!} \qquad (|z| < \infty);$$

and, since $\cosh z = \cos(iz)$,

$$(5) \qquad \cosh z = \sum_{n=0}^{\infty} \frac{z^{2n}}{(2n)!} \qquad (|z| < \infty).$$

Observe that the Taylor series for $\cosh z$ about the point $z_0 = -2\pi i$ is obtained by replacing the variable z by $z + 2\pi i$ on each side of equation (5) and then recalling that $\cosh(z + 2\pi i) = \cosh z$ for all z:

$$\cosh z = \sum_{n=0}^{\infty} \frac{(z + 2\pi i)^{2n}}{(2n)!} \qquad (|z| < \infty).$$

EXAMPLE 4. Another Maclaurin series representation is

$$(6) \qquad \frac{1}{1-z} = \sum_{n=0}^{\infty} z^n \qquad (|z| < 1).$$

The derivatives of the function $f(z) = 1/(1-z)$, which fails to be analytic at $z = 1$, are

$$f^{(n)}(z) = \frac{n!}{(1-z)^{n+1}} \qquad (n = 0,1,2,\ldots);$$

and, in particular, $f^{(n)}(0) = n!$. Note that expansion (6) gives us the sum of an infinite *geometric series*, where z is the common ratio of adjacent terms:

$$1 + z + z^2 + z^3 + \cdots = \frac{1}{1-z} \qquad (|z| < 1).$$

This is essentially the summation formula that was found in another way in Exercise 4, Sec. 43.

If we substitute $-z$ for z in equation (6) and its condition of validity, and note that $|z| < 1$ when $|-z| < 1$, we see that

$$\frac{1}{1+z} = \sum_{n=0}^{\infty} (-1)^n z^n \qquad (|z| < 1).$$

If, on the other hand, we replace the variable z in equation (6) by $1 - z$, we have the Taylor series representation

$$\frac{1}{z} = \sum_{n=0}^{\infty} (-1)^n (z-1)^n \qquad (|z-1| < 1).$$

This condition of validity follows from the one associated with expansion (6) since $|1 - z| < 1$ is the same as $|z - 1| < 1$.

EXAMPLE 5. For our final example, let us expand the function

$$f(z) = \frac{1 + 2z^2}{z^3 + z^5} = \frac{1}{z^3} \cdot \frac{2(1+z^2) - 1}{1+z^2} = \frac{1}{z^3}\left(2 - \frac{1}{1+z^2}\right)$$

into a series involving powers of z. We cannot find a Maclaurin series for $f(z)$ since it is not analytic at $z = 0$. But we do know from expansion (6) that

$$\frac{1}{1+z^2} = 1 - z^2 + z^4 - z^6 + z^8 - \cdots \qquad (|z| < 1).$$

Hence, when $0 < |z| < 1$,

$$f(z) = \frac{1}{z^3}(2 - 1 + z^2 - z^4 + z^6 - z^8 + \cdots) = \frac{1}{z^3} + \frac{1}{z} - z + z^3 - z^5 + \cdots.$$

We call such terms as $1/z^3$ and $1/z$ *negative* powers of z since they can be written z^{-3} and z^{-1}, respectively. The theory of expansions involving negative powers of $z - z_0$ will be discussed in the next section.

EXERCISES*

1. Obtain the Maclaurin series representation

$$z \cosh(z^2) = \sum_{n=0}^{\infty} \frac{z^{4n+1}}{(2n)!} \qquad (|z| < \infty).$$

*2. Show that

$$e^z = e \sum_{n=0}^{\infty} \frac{(z-1)^n}{n!} \qquad (|z| < \infty).$$

3. Find the Maclaurin series expansion of the function

$$f(z) = \frac{z}{z^4 + 9} = \frac{z}{9} \cdot \frac{1}{1 + (z^4/9)}.$$

Ans. $\displaystyle \sum_{n=0}^{\infty} \frac{(-1)^n}{3^{2n+2}} z^{4n+1} \quad (|z| < \sqrt{3}).$

4. Write the Maclaurin series representation of the function $f(z) = \sin(z^2)$, and point out how it follows that

$$f^{(4n)}(0) = 0 \qquad \text{and} \qquad f^{(2n+1)}(0) = 0 \qquad (n = 0, 1, 2, \ldots).$$

5. Derive the Taylor series representation

$$\frac{1}{1 - z} = \sum_{n=0}^{\infty} \frac{(z - i)^n}{(1 - i)^{n+1}} \qquad (|z - i| < \sqrt{2}).$$

Suggestion: Start by writing

$$\frac{1}{1 - z} = \frac{1}{(1 - i) - (z - i)} = \frac{1}{1 - i} \cdot \frac{1}{1 - (z - i)/(1 - i)}.$$

6. Expand $\cos z$ into a Taylor series about the point $z = \pi/2$.

7. Expand $\sinh z$ into a Taylor series about the point $z = \pi i$.

Ans. $\displaystyle -\sum_{n=0}^{\infty} \frac{(z - \pi i)^{2n+1}}{(2n + 1)!} \quad (|z - \pi i| < \infty).$

8. What is the largest circle within which the Maclaurin series for the function $\tanh z$ converges to $\tanh z$? Write the first two nonzero terms of that series.

*9. Use the relation (Sec. 24) $\sin z = (e^{iz} - e^{-iz})/(2i)$, along with the results in Exercises 8 and 9 of Sec. 43 to justify certain steps, to derive the Maclaurin series for $\sin z$ from the one for e^z.

*10. Show that when $z \neq 0$,

(a) $\displaystyle \frac{e^z}{z^2} = \frac{1}{z^2} + \frac{1}{z} + \frac{1}{2!} + \frac{z}{3!} + \frac{z^2}{4!} + \cdots;$ (b) $\displaystyle \frac{\sin(z^2)}{z^4} = \frac{1}{z^2} - \frac{z^2}{3!} + \frac{z^6}{5!} - \frac{z^{10}}{7!} + \cdots;$

(c) $\displaystyle \frac{\sinh z}{z^2} = \frac{1}{z} + \sum_{n=0}^{\infty} \frac{z^{2n+1}}{(2n + 3)!};$ (d) $\displaystyle z^3 \cosh\left(\frac{1}{z}\right) = \frac{z}{2} + z^3 + \sum_{n=1}^{\infty} \frac{1}{(2n + 2)!} \cdot \frac{1}{z^{2n-1}}.$

---

*In these and subsequent exercises on series expansions, it is recommended that the reader use, when possible, representations (1) through (6) in Sec. 45.

11. Show that when $0 < |z| < 4$,

$$\frac{1}{4z - z^2} = \frac{1}{4z} + \sum_{n=0}^{\infty} \frac{z^n}{4^{n+2}}.$$

46. LAURENT SERIES

If a function f fails to be analytic at a point z_0, we cannot apply Taylor's theorem at that point. It is often possible, however, to find a series representation for $f(z)$ involving both positive and negative powers of $z - z_0$ (see Example 5, Sec. 45). We now present the theory of such representations, and we begin with *Laurent's theorem.*

   Theorem. *Suppose that a function f is analytic throughout an annular domain $R_1 < |z - z_0| < R_2$, and let C denote any positively oriented simple closed contour around z_0 and lying in that domain (Fig. 50). Then, at each point z in the domain, $f(z)$ has the series representation*

$$(1) \qquad f(z) = \sum_{n=0}^{\infty} a_n(z - z_0)^n + \sum_{n=1}^{\infty} \frac{b_n}{(z - z_0)^n} \qquad (R_1 < |z - z_0| < R_2),$$

where

$$(2) \qquad\qquad a_n = \frac{1}{2\pi i} \int_C \frac{f(z)\, dz}{(z - z_0)^{n+1}} \qquad (n = 0, 1, 2, \ldots)$$

and

$$(3) \qquad\qquad b_n = \frac{1}{2\pi i} \int_C \frac{f(z)\, dz}{(z - z_0)^{-n+1}} \qquad (n = 1, 2, \ldots).$$

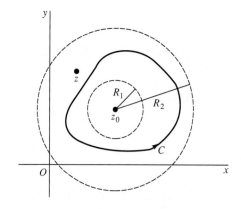

FIGURE 50

   Expansion (1) is often written

$$(4) \qquad\qquad f(z) = \sum_{n=-\infty}^{\infty} c_n(z - z_0)^n \qquad (R_1 < |z - z_0| < R_2),$$

where

(5) $$c_n = \frac{1}{2\pi i}\int_C \frac{f(z)\,dz}{(z-z_0)^{n+1}} \qquad (n = 0, \pm 1, \pm 2, \ldots).$$

In either of the forms (1) and (4), it is called a *Laurent series*.

Observe that the integrand in expression (3) can be written $f(z)(z - z_0)^{n-1}$. Thus it is clear that when f is actually analytic throughout the disk $|z - z_0| < R_2$, this integrand is too. Hence all of the coefficients b_n are zero; and, because (Sec. 40)

$$\frac{1}{2\pi i}\int_C \frac{f(z)\,dz}{(z-z_0)^{n+1}} = \frac{f^{(n)}(z_0)}{n!} \qquad (n = 0, 1, 2, \ldots),$$

expansion (1) reduces to a Taylor series about z_0.

If, however, f fails to be analytic at z_0 but is otherwise analytic in the disk $|z - z_0| < R_2$, the radius R_1 can be chosen arbitrarily small. Representation (1) is then valid when $0 < |z - z_0| < R_2$. Similarly, if f is analytic at each point in the finite plane exterior to the circle $|z - z_0| = R_1$, the condition of validity is $R_1 < |z - z_0| < \infty$. Observe that if f is analytic *everywhere* in the finite plane except at z_0, series (1) is valid at each point of analyticity, or when $0 < |z - z_0| < \infty$.

We shall prove Laurent's theorem first when $z_0 = 0$, in which case the annulus is centered at the origin. The verification of the theorem when z_0 is arbitrary will follow readily.

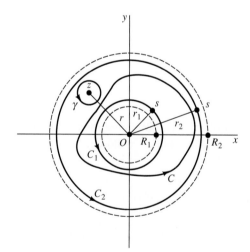

FIGURE 51

We start the proof by forming a closed annular region $r_1 \le |z| \le r_2$ that is contained in the domain $R_1 < |z| < R_2$ and whose interior contains both the point z and the contour C (Fig. 51). We let C_1 and C_2 denote the circles $|z| = r_1$ and $|z| = r_2$, respectively, and we assign those two circles a positive orientation. Observe that f is analytic on C_1 and C_2, as well as in the annular domain between them.

Next, we construct a positively oriented circle γ with center at z and small enough to be completely contained in the interior of the annular region $r_1 \le |z| \le r_2$, as shown in Fig. 51. It then follows from the extension of the Cauchy-Goursat theorem to integrals of analytic functions around the oriented boundaries of multiply connected domains (Theorem 2, Sec. 38) that

$$\int_{C_2} \frac{f(s)\,ds}{s-z} - \int_{C_1} \frac{f(s)\,ds}{s-z} - \int_{\gamma} \frac{f(s)\,ds}{s-z} = 0.$$

But, according to the Cauchy integral formula, the value of the third integral here is $2\pi i f(z)$. Hence

$$(6) \qquad f(z) = \frac{1}{2\pi i}\int_{C_2} \frac{f(s)\,ds}{s-z} + \frac{1}{2\pi i}\int_{C_1} \frac{f(s)\,ds}{z-s}.$$

Now the factor $1/(s-z)$ in the first of these integrals is the same as in expression (5), Sec. 44, where Taylor's theorem was proved; and we shall need here the expansion

$$(7) \qquad \frac{1}{s-z} = \sum_{n=0}^{N-1} \frac{1}{s^{n+1}} z^n + z^N \frac{1}{(s-z)s^N},$$

which was used in that earlier section. As for the factor $1/(z-s)$ in the second integral, an interchange of s and z in equation (7) reveals that

$$\frac{1}{z-s} = \sum_{n=0}^{N-1} \frac{1}{s^{-n}} \cdot \frac{1}{z^{n+1}} + \frac{1}{z^N} \cdot \frac{s^N}{z-s}.$$

If we replace the index of summation n here by $n-1$, this expansion takes the form

$$(8) \qquad \frac{1}{z-s} = \sum_{n=1}^{N} \frac{1}{s^{-n+1}} \cdot \frac{1}{z^n} + \frac{1}{z^N} \cdot \frac{s^N}{z-s},$$

which is to be used in what follows.

Multiplying through equations (7) and (8) by $f(s)/(2\pi i)$ and then integrating each side of the resulting equations with respect to s around C_2 and C_1, respectively, we find from expression (6) that

$$(9) \qquad f(z) = \sum_{n=0}^{N-1} a_n z^n + \rho_N(z) + \sum_{n=1}^{N} \frac{b_n}{z^n} + \sigma_N(z),$$

where the numbers a_n ($n = 0,1,2,\ldots,N-1$) and b_n ($n = 1,2,\ldots,N$) are given by the equations

$$(10) \qquad a_n = \frac{1}{2\pi i}\int_{C_2} \frac{f(s)\,ds}{s^{n+1}}, \qquad b_n = \frac{1}{2\pi i}\int_{C_1} \frac{f(s)\,ds}{s^{-n+1}}$$

and where

$$\rho_N(z) = \frac{z^N}{2\pi i}\int_{C_2} \frac{f(s)\,ds}{(s-z)s^N}, \qquad \sigma_N(z) = \frac{1}{2\pi i z^N}\int_{C_1} \frac{s^N f(s)\,ds}{z-s}.$$

As N tends to ∞, expression (9) evidently takes the proper form of a Laurent series in the domain $R_1 < |z| < R_2$, provided that

(11) $$\lim_{N \to \infty} \rho_N(z) = 0 \quad \text{and} \quad \lim_{N \to \infty} \sigma_N(z) = 0.$$

These limits are readily established by a method already used in the proof of Taylor's theorem in Sec. 44. We write $|z| = r$, so that $r_1 < r < r_2$, and let M denote the maximum value of $|f(s)|$ on C_1 and C_2. We also note that if s is a point on C_2, then $|s - z| \geq r_2 - r$; and if s is on $C_1, |z - s| \geq r - r_1$. This enables us to write

$$|\rho_N(z)| \leq \frac{Mr_2}{r_2 - r} \left(\frac{r}{r_2} \right)^N \quad \text{and} \quad |\sigma_N(z)| \leq \frac{Mr_1}{r - r_1} \left(\frac{r_1}{r} \right)^N.$$

Since $(r/r_2) < 1$ and $(r_1/r) < 1$, it is now clear that both $\rho_N(z)$ and $\sigma_N(z)$ have the desired property.

Finally, we need only recall Corollary 2 in Sec. 38 to see that the contours used in integrals (10) may be replaced by the contour C. This completes the proof of Laurent's theorem when $z_0 = 0$ since, if z is used instead of s as the variable of integration, expressions (10) for the coefficients a_n and b_n are the same as expressions (2) and (3) when $z_0 = 0$ there.

To extend the proof to the general case in which z_0 is an arbitrary point in the finite plane, we let f be a function satisfying the conditions in the theorem; and, just as we did in the proof of Taylor's theorem, we write $g(z) = f(z + z_0)$. Since $f(z)$ is analytic in the annulus $R_1 < |z - z_0| < R_2$, the function $f(z + z_0)$ is analytic when $R_1 < |(z + z_0) - z_0| < R_2$. That is, g is analytic in the annulus $R_1 < |z| < R_2$, which is centered at the origin. Now the simple closed contour C in the statement of the theorem has some parametric representation $z = z(t)$ $(a \leq t \leq b)$, where

(12) $$R_1 < |z(t) - z_0| < R_2$$

for all t in the interval $a \leq t \leq b$. Hence if Γ denotes the path

(13) $$z = z(t) - z_0 \quad (a \leq t \leq b),$$

Γ is not only a simple closed contour but, in view of inequalities (12), it lies in the domain $R_1 < |z| < R_2$. Consequently, $g(z)$ has a Laurent series representation

(14) $$g(z) = \sum_{n=0}^{\infty} a_n z^n + \sum_{n=1}^{\infty} \frac{b_n}{z^n} \quad (R_1 < |z| < R_2),$$

where

(15) $$a_n = \frac{1}{2\pi i} \int_{\Gamma} \frac{g(z)\,dz}{z^{n+1}} \quad (n = 0, 1, 2, \ldots),$$

(16) $$b_n = \frac{1}{2\pi i} \int_{\Gamma} \frac{g(z)\,dz}{z^{-n+1}} \quad (n = 1, 2, \ldots).$$

Representation (1) is obtained if we write $f(z + z_0)$ instead of $g(z)$ in equation (14) and then replace z by $z - z_0$ in the resulting equation, as well as in the condition of validity $R_1 < |z| < R_2$. Expression (15) for the coefficients a_n is, moreover, the same as expression (2) since

$$\int_{\Gamma} \frac{g(z)\,dz}{z^{n+1}} = \int_a^b \frac{f[z(t)]z'(t)}{[z(t) - z_0]^{n+1}}\,dt = \int_C \frac{f(z)\,dz}{(z - z_0)^{n+1}}.$$

Similarly, the coefficients b_n in expression (16) are the same as those in expression (3).

47. EXAMPLES

The coefficients in a Laurent series are generally found by means other than by appealing directly to their integral representations. This is illustrated in the examples below, where it is always assumed that, when the annular domain is specified, a Laurent series for a given function is unique. As was the case with Taylor series, we defer the proof of such uniqueness until Sec. 50.

  EXAMPLE 1.   Replacing z by $1/z$ in the Maclaurin series expansion

$$e^z = \sum_{n=0}^{\infty} \frac{z^n}{n!} = 1 + \frac{z}{1!} + \frac{z^2}{2!} + \frac{z^3}{3!} + \cdots \qquad (|z| < \infty),$$

we have the Laurent series representation

$$e^{1/z} = \sum_{n=0}^{\infty} \frac{1}{n!z^n} = 1 + \frac{1}{1!z} + \frac{1}{2!z^2} + \frac{1}{3!z^3} + \cdots \qquad (0 < |z| < \infty).$$

  Note that no positive powers of z appear here, the coefficients of the positive powers being zero. Note, too, that the coefficient of $1/z$ is unity; and, according to Laurent's theorem in Sec. 46, that coefficient is the number

$$b_1 = \frac{1}{2\pi i} \int_C e^{1/z}\,dz,$$

where C is any positively oriented simple closed contour around the origin. Since $b_1 = 1$, then,

$$\int_C e^{1/z}\,dz = 2\pi i.$$

This method of evaluating certain integrals around simple closed contours will be developed in considerable detail in Chap. 6.

  EXAMPLE 2.   The function $f(z) = 1/(z - i)^2$ is already in the form of a Laurent series, where $z_0 = i$. That is,

$$f(z) = \sum_{n=-\infty}^{\infty} c_n(z - i)^n \qquad (0 < |z - i| < \infty),$$

where $c_{-2} = 1$ and all of the other coefficients are zero. From formula (5), Sec. 46, for the coefficients in a Laurent series, we know that

$$c_n = \frac{1}{2\pi i} \int_C \frac{dz}{(z-i)^{n+3}} \qquad (n = 0, \pm 1, \pm 2, \ldots),$$

where C is, for instance, any positively oriented circle $|z - i| = R$ about the point $z_0 = i$. Thus (compare Exercise 13, Sec. 33)

$$\int_C \frac{dz}{(z-i)^{n+3}} = \begin{cases} 0 & \text{when } n \neq -2, \\ 2\pi i & \text{when } n = -2. \end{cases}$$

EXAMPLE 3. The function

(1)
$$f(z) = \frac{-1}{(z-1)(z-2)} = \frac{1}{z-1} - \frac{1}{z-2},$$

which has the two singular points $z = 1$ and $z = 2$, is analytic in the domains

$$|z| < 1, \quad 1 < |z| < 2, \quad \text{and} \quad 2 < |z| < \infty.$$

In each of those domains, denoted by $D_1, D_2,$ and D_3, respectively, in Fig. 52, $f(z)$ has series representations in powers of z. They can all be found by recalling from Example 4, Sec. 45, that

$$\frac{1}{1-z} = \sum_{n=0}^{\infty} z^n \qquad (|z| < 1).$$

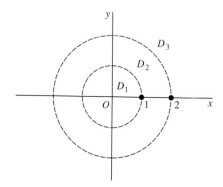

FIGURE 52

The representation in D_1 is a Maclaurin series. To find it, we write

$$f(z) = -\frac{1}{1-z} + \frac{1}{2} \cdot \frac{1}{1-(z/2)}$$

and observe that, since $|z| < 1$ and $|z/2| < 1$ in D_1,

(2)
$$f(z) = -\sum_{n=0}^{\infty} z^n + \sum_{n=0}^{\infty} \frac{z^n}{2^{n+1}} = \sum_{n=0}^{\infty} (2^{-n-1} - 1)z^n \qquad (|z| < 1).$$

As for the representation in D_2, we write

$$f(z) = \frac{1}{z} \cdot \frac{1}{1-(1/z)} + \frac{1}{2} \cdot \frac{1}{1-(z/2)}.$$

Since $|1/z| < 1$ and $|z/2| < 1$ when $1 < |z| < 2$, it follows that

$$f(z) = \sum_{n=0}^{\infty} \frac{1}{z^{n+1}} + \sum_{n=0}^{\infty} \frac{z^n}{2^{n+1}} \qquad (1 < |z| < 2).$$

If we replace the index of summation n in the first of these series by $n-1$ and then interchange the two series, we arrive at an expansion having the same form as the one in the statement of Laurent's theorem (Sec. 46):

$$(3) \qquad f(z) = \sum_{n=0}^{\infty} \frac{z^n}{2^{n+1}} + \sum_{n=1}^{\infty} \frac{1}{z^n} \qquad (1 < |z| < 2).$$

Since there is only one such representation for $f(z)$ in the annulus $1 < |z| < 2$, expansion (3) is, in fact, the Laurent series for $f(z)$ there.

The representation of $f(z)$ in the unbounded domain D_3 is also a Laurent series. If we put expression (1) in the form

$$f(z) = \frac{1}{z} \cdot \frac{1}{1-(1/z)} - \frac{1}{z} \cdot \frac{1}{1-(2/z)}$$

and observe that $|1/z| < 1$ and $|2/z| < 1$ when $2 < |z| < \infty$, we find that

$$f(z) = \sum_{n=0}^{\infty} \frac{1}{z^{n+1}} - \sum_{n=0}^{\infty} \frac{2^n}{z^{n+1}} = \sum_{n=0}^{\infty} \frac{1-2^n}{z^{n+1}} \qquad (2 < |z| < \infty).$$

That is,

$$(4) \qquad f(z) = \sum_{n=1}^{\infty} \frac{1-2^{n-1}}{z^n} \qquad (2 < |z| < \infty).$$

EXERCISES

1. Find the Laurent series that represents the function

$$f(z) = z^2 \sin\left(\frac{1}{z^2}\right)$$

in the domain $0 < |z| < \infty$.

$$Ans. \ 1 + \sum_{n=1}^{\infty} \frac{(-1)^n}{(2n+1)!} \cdot \frac{1}{z^{4n}}.$$

2. Derive the Laurent series representation

$$\frac{e^z}{(z+1)^2} = \frac{1}{e}\left[\sum_{n=0}^{\infty} \frac{(z+1)^n}{(n+2)!} + \frac{1}{z+1} + \frac{1}{(z+1)^2}\right] \qquad (0 < |z+1| < \infty).$$

*3. Find a representation for $1/(1 + z)$ in negative powers of z that is valid in the domain $1 < |z| < \infty$.

$$Ans. \sum_{n=1}^{\infty} \frac{(-1)^{n+1}}{z^n}.$$

4. Give two Laurent series expansions in powers of z for the function

$$f(z) = \frac{1}{z^2(1 - z)},$$

and specify the regions in which those expansions are valid.

$$Ans. \sum_{n=0}^{\infty} z^n + \frac{1}{z} + \frac{1}{z^2} \quad (0 < |z| < 1); \quad -\sum_{n=3}^{\infty} \frac{1}{z^n} \quad (1 < |z| < \infty).$$

5. Represent the function $f(z) = (z + 1)/(z - 1)$ by
(a) its Maclaurin series, and give the region of validity for the representation;
(b) its Laurent series for the domain $1 < |z| < \infty$.

$$Ans. (a) -1 - 2\sum_{n=1}^{\infty} z^n \quad (|z| < 1); \quad (b) \ 1 + 2\sum_{n=1}^{\infty} \frac{1}{z^n}.$$

6. Show that when $0 < |z - 1| < 2$,

$$\frac{z}{(z - 1)(z - 3)} = -3\sum_{n=0}^{\infty} \frac{(z - 1)^n}{2^{n+2}} - \frac{1}{2(z - 1)}.$$

7. Write the two Laurent series in powers of z that represent the function

$$f(z) = \frac{1}{z(1 + z^2)}$$

in certain domains, and specify those domains.

$$Ans. \sum_{n=0}^{\infty} (-1)^{n+1} z^{2n+1} + \frac{1}{z} \quad (0 < |z| < 1); \quad \sum_{n=1}^{\infty} \frac{(-1)^{n+1}}{z^{2n+1}} \quad (1 < |z| < \infty).$$

8. (a) Let a denote a real number, where $-1 < a < 1$, and derive the Laurent series representation

$$\frac{a}{z - a} = \sum_{n=1}^{\infty} \frac{a^n}{z^n} \quad (|a| < |z| < \infty).$$

(b) Write $z = e^{i\theta}$ in the equation obtained in part (a) and then equate real parts and imaginary parts on each side of the result to derive the summation formulas

$$\sum_{n=1}^{\infty} a^n \cos n\theta = \frac{a\cos\theta - a^2}{1 - 2a\cos\theta + a^2} \quad \text{and} \quad \sum_{n=1}^{\infty} a^n \sin n\theta = \frac{a\sin\theta}{1 - 2a\cos\theta + a^2},$$

where $-1 < a < 1$. (Compare Exercise 5, Sec. 43.)

9. Suppose that a series

$$\sum_{n=-\infty}^{\infty} x[n]z^{-n}$$

converges to an analytic function $X(z)$ in some annulus $R_1 < |z| < R_2$. That sum $X(z)$ is called the z-*transform* of $x[n]$ $(n = 0, \pm 1, \pm 2, \ldots)$.* Use expression (5), Sec. 46, for the coefficients in a Laurent series to show that if the annulus contains the unit circle $|z| = 1$, then the *inverse* z-transform of $X(z)$ can be written

$$x[n] = \frac{1}{2\pi} \int_{-\pi}^{\pi} X(e^{i\theta}) e^{in\theta} \, d\theta \qquad (n = 0, \pm 1, \pm 2, \ldots).$$

10. (*a*) Let z be any complex number, and let C denote the unit circle

$$w = e^{i\phi} \qquad (-\pi \le \phi \le \pi)$$

in the w plane. Then use that contour in expression (5), Sec. 46, for the coefficients in a Laurent series, adapted to such series about the origin in the w plane, to show that

$$\exp\left[\frac{z}{2}\left(w - \frac{1}{w}\right)\right] = \sum_{n=-\infty}^{\infty} J_n(z) w^n \qquad (0 < |w| < \infty),$$

where

$$J_n(z) = \frac{1}{2\pi} \int_{-\pi}^{\pi} \exp[-i(n\phi - z \sin \phi)] \, d\phi \qquad (n = 0, \pm 1, \pm 2, \ldots).$$

(*b*) With the aid of Exercise 7, Sec. 31, regarding certain definite integrals of even and odd complex-valued functions of a real variable, show that the coefficients in part (*a*) can be written†

$$J_n(z) = \frac{1}{\pi} \int_0^{\pi} \cos(n\phi - z \sin \phi) \, d\phi \qquad (n = 0, \pm 1, \pm 2, \ldots).$$

11. (*a*) Let $f(z)$ denote a function which is analytic in some annular domain about the origin that includes the unit circle $z = e^{i\phi}(-\pi \le \phi \le \pi)$. By taking that circle as the path of integration in expressions (2) and (3), Sec. 46, for the coefficients a_n and b_n in a Laurent series in powers of z, show that

$$f(z) = \frac{1}{2\pi} \int_{-\pi}^{\pi} f(e^{i\phi}) \, d\phi + \frac{1}{2\pi} \sum_{n=1}^{\infty} \int_{-\pi}^{\pi} f(e^{i\phi}) \left[\left(\frac{z}{e^{i\phi}}\right)^n + \left(\frac{e^{i\phi}}{z}\right)^n\right] d\phi$$

when z is any point in the annular domain.

(*b*) Write $u(\theta) = \mathrm{Re}[f(e^{i\theta})]$, and show how it follows from the expansion in part (*a*) that

$$u(\theta) = \frac{1}{2\pi} \int_{-\pi}^{\pi} u(\phi) \, d\phi + \frac{1}{\pi} \sum_{n=1}^{\infty} \int_{-\pi}^{\pi} u(\phi) \cos[n(\theta - \phi)] \, d\phi.$$

---

*The z-transform arises in studies of discrete-time linear systems. See, for instance, Chap. 4 of the book by Oppenheim and Schafer that is listed in Appendix 1.

†These coefficients $J_n(z)$ are called *Bessel functions* of the first kind. They play a prominent role in certain areas of applied mathematics. See, for example, the authors' "Fourier Series and Boundary Value Problems," 5th ed., Chap 7, 1993.

This is one form of the *Fourier series* expansion of the real-valued function $u(\theta)$ on the interval $-\pi \le \theta \le \pi$. The restriction of $u(\theta)$ is more severe than is necessary in order for it to be represented by a Fourier series.*

48. ABSOLUTE AND UNIFORM CONVERGENCE OF POWER SERIES

This section and the three following it are devoted mainly to various properties of power series.

We recall from Sec. 43 that a series of complex numbers converges *absolutely* if the series of absolute values of those numbers converges. The following theorem concerns the absolute convergence of power series.

Theorem 1. *If a power series*

$$(1) \qquad\qquad \sum_{n=0}^{\infty} a_n(z - z_0)^n$$

converges when $z = z_1$ $(z_1 \ne z_0)$*, then it is absolutely convergent at each point z in the open disk* $|z - z_0| < R_1$*, where* $R_1 = |z_1 - z_0|$ *(Fig. 53).*

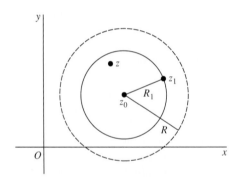

FIGURE 53

We first prove the theorem when $z_0 = 0$, and we assume that the series

$$\sum_{n=0}^{\infty} a_n z_1^n \qquad (z_1 \ne 0)$$

converges. The terms $a_n z_1^n$ are thus bounded; that is,

$$|a_n z_1^n| \le M \qquad (n = 0, 1, 2, \ldots)$$

---

*For other sufficient conditions, see Secs. 19 and 20 of the book cited in the footnote to Exercise 10.

for some positive constant M (see Sec. 43). If $|z| < |z_1|$ and we let ρ denote the modulus $|z/z_1|$, we can see that

$$|a_n z^n| = |a_n z_1^n| \left| \frac{z}{z_1} \right|^n \leq M\rho^n \qquad (n = 0, 1, 2, \ldots),$$

where $\rho < 1$. Now the series whose terms are the real numbers $M\rho^n$ $(n = 0, 1, 2, \ldots)$ is a geometric series, which converges when $\rho < 1$. Hence, by the comparison test for series of real numbers, the series

$$\sum_{n=0}^{\infty} |a_n z^n|$$

converges in the open disk $|z| < |z_1|$; and the theorem is proved when $z_0 = 0$.

When z_0 is any nonzero number, we assume that series (1) converges at $z = z_1$ $(z_1 \neq z_0)$. If we write $w = z - z_0$, series (1) becomes

(2) $$\sum_{n=0}^{\infty} a_n w^n ;$$

and this series converges at $w = z_1 - z_0$. Consequently, since the theorem is known to be true when $z_0 = 0$, we see that series (2) is absolutely convergent in the open disk $|w| < |z_1 - z_0|$. Finally, by replacing w by $z - z_0$ in series (2) and this condition of validity, as well as writing $R_1 = |z_1 - z_0|$, we arrive at the proof of the theorem as it is stated.

The theorem tells us that the set of all points inside some circle centered at z_0 is a region of convergence for the power series (1), provided it converges at some point other than z_0. The greatest circle centered at z_0 such that series (1) converges at each point inside is called the *circle of convergence* of series (1). The series cannot converge at any point z_2 outside that circle, according to the theorem; for if it did, it would converge everywhere inside the circle centered at z_0 and passing through z_2. The first circle could not then be the circle of convergence.

Our next theorem involves terminology that we must first define. Suppose that the power series (1) has circle of convergence $|z - z_0| = R$, and let $S(z)$ and $S_N(z)$ represent the sum and partial sums, respectively, of that series:

$$S(z) = \sum_{n=0}^{\infty} a_n (z - z_0)^n, \quad S_N(z) = \sum_{n=0}^{N-1} a_n (z - z_0)^n \qquad (|z - z_0| < R).$$

Then write the remainder function

(3) $$\rho_N(z) = S(z) - S_N(z) \qquad (|z - z_0| < R).$$

Since the power series converges for any fixed value of z when $|z - z_0| < R$, we know that the remainder $\rho_N(z)$ approaches zero for any such z as N tends to infinity. According to definition (2), Sec. 43, of the limit of a sequence, this means that, corresponding to each positive number ε, there is a positive integer N_ε such that

(4) $$|\rho_N(z)| < \varepsilon \qquad \text{whenever} \qquad N > N_\varepsilon.$$

When the choice of N_ε depends only on the value of ε and is independent of the point z taken in a specified region within the circle of convergence, the convergence is said to be *uniform* in that region.

Theorem 2. *If z_1 is a point inside the circle of convergence $|z - z_0| = R$ of a power series*

$$
(5) \qquad \sum_{n=0}^{\infty} a_n(z - z_0)^n,
$$

then that series is uniformly convergent in the closed disk $|z - z_0| \le R_1$, where $R_1 = |z_1 - z_0|$ (Fig. 54).

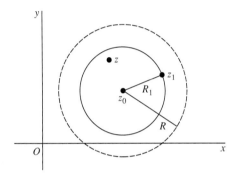

FIGURE 54

As in the proof of Theorem 1, we first treat the case in which $z_0 = 0$. Given that z_1 is a point lying inside the circle of convergence of the series

$$
(6) \qquad \sum_{n=0}^{\infty} a_n z^n,
$$

we note that there are points with modulus greater than $|z_1|$ for which it converges. According to Theorem 1, then, the series

$$
(7) \qquad \sum_{n=0}^{\infty} |a_n z_1^n|
$$

converges. Letting m and N denote positive integers, where $m > N$, we can write the remainders of series (6) and (7) as

$$
(8) \qquad \rho_N(z) = \lim_{m \to \infty} \sum_{n=N}^{m} a_n z^n
$$

and

$$
(9) \qquad \sigma_N = \lim_{m \to \infty} \sum_{n=N}^{m} |a_n z_1^n|,
$$

respectively.

Now, in view of Exercise 3, Sec. 43,

$$|\rho_N(z)| = \lim_{m \to \infty} \left| \sum_{n=N}^{m} a_n z^n \right|;$$

and, when $|z| \leq |z_1|$,

$$\left| \sum_{n=N}^{m} a_n z^n \right| \leq \sum_{n=N}^{m} |a_n||z|^n \leq \sum_{n=N}^{m} |a_n||z_1|^n = \sum_{n=N}^{m} |a_n z_1^n|.$$

Hence

(10) $$|\rho_N(z)| \leq \sigma_N \qquad \text{when} \qquad |z| \leq |z_1|.$$

Since σ_N are the remainders of a convergent series, they tend to zero as N tends to infinity. That is, for each positive number ε, an integer N_ε exists such that

(11) $$\sigma_N < \varepsilon \qquad \text{whenever} \qquad N > N_\varepsilon.$$

Because of conditions (10) and (11), then, condition (4) holds for all points z in the disk $|z| \leq |z_1|$; and the value of N_ε is independent of the choice of z. Hence the convergence of series (6) is uniform in that disk.

The extension of the proof to the case in which z_0 is arbitrary is, of course, accomplished by writing $w = z - z_0$ in series (5). For then the hypothesis of the theorem is that $z_1 - z_0$ is a point inside the circle of convergence $|w| = R$ of the series

$$\sum_{n=0}^{\infty} a_n w^n.$$

Since we now know that this series must converge uniformly in the closed disk $|w| \leq |z_1 - z_0|$, the conclusion in the statement of the theorem is evident.

The following corollary is an important consequence of Theorem 2.

Corollary. *A power series*

(12) $$\sum_{n=0}^{\infty} a_n (z - z_0)^n$$

represents a continuous function $S(z)$ at each point inside its circle of convergence $|z - z_0| = R$.

Another way to state the corollary is that if $S(z)$ denotes the sum of series (12) within its circle of convergence $|z - z_0| = R$ and if z_1 is a point inside that circle, then, for each positive number ε, there is a positive number δ such that

(13) $$|S(z) - S(z_1)| < \varepsilon \qquad \text{whenever} \qquad |z - z_1| < \delta,$$

the number δ being small enough so that z lies in the domain of definition $|z - z_0| < R$ of $S(z)$.

To show this, we let $S_N(z)$ denote the sum of the first N terms of series (12) and write

$$S(z) = S_N(z) + \rho_N(z),$$

where $\rho_N(z)$ is the remainder function (3). Then

$$|S(z) - S(z_1)| = |S_N(z) - S_N(z_1) + \rho_N(z) - \rho_N(z_1)|,$$

or

(14) $$|S(z) - S(z_1)| \leq |S_N(z) - S_N(z_1)| + |\rho_N(z)| + |\rho_N(z_1)|.$$

If z is any point lying in some closed disk $|z - z_0| \leq R_0$ whose radius R_0 is greater than $|z_1 - z_0|$ but less than the radius R of the circle of convergence of series (12) (see Fig. 55), the uniform convergence stated in Theorem 2 ensures that there is a positive integer N_ε such that

(15) $$|\rho_N(z)| < \frac{\varepsilon}{3} \qquad \text{whenever} \qquad N > N_\varepsilon.$$

In particular, condition (15) holds for each point z in some neighborhood $|z - z_1| < \delta$ of z_1 that is small enough to be contained in the disk $|z - z_0| \leq R_0$.

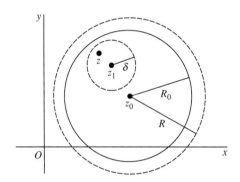

FIGURE 55

Now the partial sum $S_N(z)$ is a polynomial and is, therefore, continuous at z_1 for each value of N. In particular, when $N = N_\varepsilon + 1$, we can choose our δ so small that

(16) $$|S_N(z) - S_N(z_1)| < \frac{\varepsilon}{3} \qquad \text{whenever} \qquad |z - z_1| < \delta.$$

By writing $N = N_\varepsilon + 1$ in inequality (14) and using the fact that statements (15) and (16) are true when $N = N_\varepsilon + 1$, we now find that

$$|S(z) - S(z_1)| < \frac{\varepsilon}{3} + \frac{\varepsilon}{3} + \frac{\varepsilon}{3} \qquad \text{whenever} \qquad |z - z_1| < \delta.$$

This is statement (13), and the corollary is now established.

By writing $w = 1/(z - z_0)$, one can modify the two theorems and corollary in this section so as to apply to series of the type

$$\sum_{n=1}^{\infty} \frac{b_n}{(z - z_0)^n}.$$

For instance, if this series converges at a point z_1 ($z_1 \neq z_0$), then it must converge absolutely to a continuous function in the domain *exterior to* the circle $|z - z_0| = R_1$, where $R_1 = |z_1 - z_0|$. Also, we know that if a Laurent series representation

$$f(z) = \sum_{n=0}^{\infty} a_n (z - z_0)^n + \sum_{n=1}^{\infty} \frac{b_n}{(z - z_0)^n}$$

is valid in an annulus $R_1 < |z - z_0| < R_2$, then *both* of the series on the right converge uniformly in any closed annulus which is concentric to and interior to that region of validity.

49. INTEGRATION AND DIFFERENTIATION OF POWER SERIES

We have just seen that a power series

$$(1) \qquad\qquad\qquad S(z) = \sum_{n=0}^{\infty} a_n (z - z_0)^n$$

represents a continuous function at each point interior to its circle of convergence. We prove in this section that the sum $S(z)$ is actually analytic within that circle. Our proof depends on the following theorem, which is of interest in itself.

Theorem 1. Let C denote any contour interior to the circle of convergence of the power series (1), and let $g(z)$ be any function that is continuous on C. The series formed by multiplying each term of the power series by $g(z)$ can be integrated term by term over C; that is,

$$(2) \qquad\qquad \int_C g(z) S(z)\, dz = \sum_{n=0}^{\infty} a_n \int_C g(z)(z - z_0)^n\, dz.$$

To prove this theorem, we note that since both $g(z)$ and the sum $S(z)$ of the power series are continuous on C, the integral over C of the product

$$g(z) S(z) = \sum_{n=0}^{N-1} a_n g(z)(z - z_0)^n + g(z)\rho_N(z),$$

where $\rho_N(z)$ is the remainder of the given series after N terms, exists. The terms of the finite sum here are also continuous on the contour C, and so their integrals over C exist. Consequently, the integral of the quantity $g(z)\rho_N(z)$ must exist; and we may write

$$(3) \qquad \int_C g(z)S(z)\,dz = \sum_{n=0}^{N-1} a_n \int_C g(z)(z - z_0)^n\,dz + \int_C g(z)\rho_N(z)\,dz.$$

Let M be the maximum value of $|g(z)|$ on C, and let L denote the length of C. In view of the uniform convergence of the given power series (Sec. 48), we know that for each positive number ε there exists a positive integer N_ε such that, for all points z on C,

$$|\rho_N(z)| < \varepsilon \qquad \text{whenever} \qquad N > N_\varepsilon.$$

Since N_ε is independent of z, we find that

$$\left| \int_C g(z)\rho_N(z)\,dz \right| < M\varepsilon L \qquad \text{whenever} \qquad N > N_\varepsilon;$$

that is,

$$\lim_{N\to\infty} \int_C g(z)\rho_N(z)\,dz = 0.$$

It follows, therefore, from equation (3) that

$$\int_C g(z)S(z)\,dz = \lim_{N\to\infty} \sum_{n=0}^{N-1} a_n \int_C g(z)(z - z_0)^n\,dz.$$

This is the same as equation (2), and Theorem 1 is proved.

If $g(z) = 1$ for each value of z in the open disk bounded by the circle of convergence of power series (1), then

$$\int_C g(z)(z - z_0)^n\,dz = \int_C (z - z_0)^n\,dz = 0 \qquad (n = 0, 1, 2, \ldots)$$

for every *closed* contour C lying in that domain. According to equation (2), then,

$$\int_C S(z)\,dz = 0$$

for every such contour; and, by Morera's theorem (Sec. 40), the function $S(z)$ is analytic throughout the domain. We state this result as a corollary.

Corollary. *The sum $S(z)$ of power series (1) is analytic at each point z interior to the circle of convergence of that series.*

This corollary is often helpful in establishing the analyticity of functions and in evaluating limits.

EXAMPLE 1. To illustrate, let us show that the function defined by the equations

$$f(z) = \begin{cases} (\sin z)/z & \text{when } z \neq 0, \\ 1 & \text{when } z = 0 \end{cases}$$

is entire. Since the Maclaurin series expansion

$$\sin z = \sum_{n=0}^{\infty}(-1)^n \frac{z^{2n+1}}{(2n+1)!}$$

represents $\sin z$ for every value of z, the series

(4) $$\sum_{n=0}^{\infty}(-1)^n \frac{z^{2n}}{(2n+1)!} = 1 - \frac{z^2}{3!} + \frac{z^4}{5!} - \cdots,$$

obtained by dividing each term of that Maclaurin series by z, converges to $f(z)$ when $z \neq 0$. But series (4) clearly converges to $f(0)$ when $z = 0$. Hence $f(z)$ is represented by the convergent power series (4) for all z, and f is therefore an entire function. Note that, since f is continuous at $z = 0$ and since $(\sin z)/z = f(z)$ when $z \neq 0$,

(5) $$\lim_{z\to 0}\frac{\sin z}{z} = \lim_{z\to 0} f(z) = f(0) = 1.$$

This is a result known beforehand because the limit here is the definition of the derivative of $\sin z$ at $z = 0$.

We observed at the beginning of Sec. 45 that the Taylor series for a function f about a point z_0 converges to $f(z)$ at each point z interior to the circle centered at z_0 and passing through the nearest point z_1 where f fails to be analytic. In view of the above corollary, we now know that *there is no larger circle about z_0 such that at each point z interior to it the Taylor series converges to $f(z)$*. For if there were such a circle, f would be analytic at z_1; but f is not analytic at z_1.

We now present a companion to Theorem 1.

Theorem 2. *The power series (1) can be differentiated term by term. That is, at each point z interior to the circle of convergence of that series,*

(6) $$S'(z) = \sum_{n=1}^{\infty} na_n(z - z_0)^{n-1}.$$

To prove this, let z denote any point interior to the circle of convergence of series (1); and let C be some positively oriented simple closed contour surrounding z and interior to that circle. Also, define the function

(7) $$g(s) = \frac{1}{2\pi i} \cdot \frac{1}{(s-z)^2}$$

at each point s on C. Since $g(s)$ is continuous on C, Theorem 1 tells us that

(8) $$\int_C g(s)S(s)\, ds = \sum_{n=0}^{\infty} a_n \int_C g(s)(s - z_0)^n\, ds.$$

Now $S(s)$ is analytic inside and on C, and this enables us to write

$$\int_C g(s)S(s)\,ds = \frac{1}{2\pi i}\int_C \frac{S(s)\,ds}{(s-z)^2} = S'(z)$$

with the aid of the integral representation for derivatives in Sec. 40. Furthermore,

$$\int_C g(s)(s-z_0)^n\,ds = \frac{1}{2\pi i}\int_C \frac{(s-z_0)^n}{(s-z)^2}\,ds = \frac{d}{dz}(z-z_0)^n \qquad (n=0,1,2,\ldots).$$

Thus equation (8) reduces to

$$S'(z) = \sum_{n=0}^{\infty} a_n \frac{d}{dz}(z-z_0)^n,$$

which is the same as equation (6). This completes the proof.

EXAMPLE 2. In Example 4, Sec. 45, we saw that

$$\frac{1}{z} = \sum_{n=0}^{\infty} (-1)^n (z-1)^n \qquad (|z-1|<1).$$

Differentiation of each side of this equation reveals that

$$-\frac{1}{z^2} = \sum_{n=1}^{\infty} (-1)^n n (z-1)^{n-1} \qquad (|z-1|<1),$$

or

$$\frac{1}{z^2} = \sum_{n=0}^{\infty} (-1)^n (n+1)(z-1)^n \qquad (|z-1|<1).$$

50. UNIQUENESS OF SERIES REPRESENTATIONS

The uniqueness of Taylor and Laurent series representations, anticipated in Secs. 45 and 47, respectively, follows readily from Theorem 1 in Sec. 49. We consider first the uniqueness of Taylor series representations.

Theorem 1. *If a series*

(1)
$$\sum_{n=0}^{\infty} a_n(z-z_0)^n$$

converges to $f(z)$ at all points interior to some circle $|z-z_0|=R$, then it is the Taylor series expansion for f in powers of $z-z_0$.

To prove this, we write the series representation

(2)
$$f(z) = \sum_{n=0}^{\infty} a_n(z-z_0)^n \qquad (|z-z_0|<R)$$

in the hypothesis of the theorem using the index of summation m:

$$f(z) = \sum_{m=0}^{\infty} a_m(z - z_0)^m \qquad (|z - z_0| < R).$$

Then, by appealing to Theorem 1 in Sec. 49, we may write

$$(3) \qquad \int_C g(z)f(z)\, dz = \sum_{m=0}^{\infty} a_m \int_C g(z)(z - z_0)^m\, dz,$$

where $g(z)$ is any one of the functions

$$(4) \qquad g(z) = \frac{1}{2\pi i} \cdot \frac{1}{(z - z_0)^{n+1}} \qquad (n = 0, 1, 2, \ldots)$$

and C is some circle centered at z_0 and with radius less than R.

In view of the generalized form (5), Sec. 40, of the Cauchy integral formula (see also the corollary in Sec. 49), we find that

$$(5) \qquad \int_C g(z)f(z)\, dz = \frac{1}{2\pi i} \int_C \frac{f(z)\, dz}{(z - z_0)^{n+1}} = \frac{f^{(n)}(z_0)}{n!};$$

and, since (see Exercise 13, Sec. 33)

$$(6) \qquad \int_C g(z)(z - z_0)^m\, dz = \frac{1}{2\pi i} \int_C \frac{dz}{(z - z_0)^{n-m+1}} = \begin{cases} 0 & \text{when } m \neq n, \\ 1 & \text{when } m = n, \end{cases}$$

it is clear that

$$(7) \qquad \sum_{m=0}^{\infty} a_m \int_C g(z)(z - z_0)^m\, dz = a_n.$$

Because of equations (5) and (7), equation (3) now reduces to

$$\frac{f^{(n)}(z_0)}{n!} = a_n,$$

and this shows that series (2) is, in fact, the Taylor series for f about the point z_0.

Note how it follows from Theorem 1 that if series (1) converges to zero throughout some neighborhood of z_0, then the coefficients a_n must all be zero.

Our second theorem here concerns the uniqueness of Laurent series representations.

Theorem 2. *If a series*

$$(8) \qquad \sum_{n=-\infty}^{\infty} c_n(z - z_0)^n = \sum_{n=0}^{\infty} a_n(z - z_0)^n + \sum_{n=1}^{\infty} \frac{b_n}{(z - z_0)^n}$$

converges to $f(z)$ at all points in some annular domain about z_0, then it is the Laurent series expansion for f in powers of $z - z_0$ for that domain.

The method of proof here is similar to the one used in proving Theorem 1. The hypothesis of this theorem tells us that there is an annular domain about z_0 such that

$$f(z) = \sum_{n=-\infty}^{\infty} c_n(z - z_0)^n$$

for each point z in it. Let $g(z)$ be as defined by equation (4), but now allow n to be a negative integer too. Also, let C be any circle around the annulus, centered at z_0 and taken in the positive sense. Then, using the index of summation m and adapting Theorem 1 in Sec. 49 to series involving both nonnegative *and* negative powers of $z - z_0$ (Exercise 18, Sec. 51), write

$$\int_C g(z)f(z)\, dz = \sum_{m=-\infty}^{\infty} c_m \int_C g(z)(z - z_0)^m\, dz,$$

or

(9)
$$\frac{1}{2\pi i}\int_C \frac{f(z)\, dz}{(z - z_0)^{n+1}} = \sum_{m=-\infty}^{\infty} c_m \int_C g(z)(z - z_0)^m\, dz.$$

Since equations (6) are also valid when the integers m and n are allowed to be negative, equation (9) reduces to

$$\frac{1}{2\pi i}\int_C \frac{f(z)\, dz}{(z - z_0)^{n+1}} = c_n,$$

which is an expression in Sec. 46 for the coefficients in the Laurent series for f in the annulus.

51. MULTIPLICATION AND DIVISION OF POWER SERIES

Suppose that each of the power series

(1)
$$\sum_{n=0}^{\infty} a_n(z - z_0)^n \qquad \text{and} \qquad \sum_{n=0}^{\infty} b_n(z - z_0)^n$$

converges within some circle $|z - z_0| = R$. Their sums $f(z)$ and $g(z)$, respectively, are then analytic functions in the disk $|z - z_0| < R$ (Sec. 49), and the product of those sums has a Taylor series expansion which is valid there:

(2)
$$f(z)g(z) = \sum_{n=0}^{\infty} c_n(z - z_0)^n \qquad (|z - z_0| < R).$$

According to Theorem 1 in Sec. 50, the series (1) are themselves Taylor series. Hence the first three coefficients in series (2) are given by the equations

$$c_0 = f(z_0)g(z_0) = a_0 b_0,$$

$$c_1 = \frac{f(z_0)g'(z_0) + f'(z_0)g(z_0)}{1!} = a_0 b_1 + a_1 b_0,$$

and

$$c_2 = \frac{f(z_0)g''(z_0) + 2f'(z_0)g'(z_0) + f''(z_0)g(z_0)}{2!} = a_0 b_2 + a_1 b_1 + a_2 b_0.$$

The general expression for any coefficient c_n is easily obtained by referring to *Leibniz's rule* (Exercise 14)

$$(3) \qquad [f(z)g(z)]^{(n)} = \sum_{k=0}^{n} \binom{n}{k} f^{(k)}(z) g^{(n-k)}(z),$$

where

$$\binom{n}{k} = \frac{n!}{k!(n-k)!} \qquad (k = 0, 1, 2, \ldots, n),$$

for the nth derivative of the product of two differentiable functions. As usual, $f^{(0)}(z) = f(z)$ and $0! = 1$. Evidently,

$$c_n = \sum_{k=0}^{n} \frac{f^{(k)}(z_0)}{k!} \cdot \frac{g^{(n-k)}(z_0)}{(n-k)!} = \sum_{k=0}^{n} a_k b_{n-k};$$

and so expansion (2) can be written

$$(4) \qquad f(z)g(z) = a_0 b_0 + (a_0 b_1 + a_1 b_0)(z - z_0)$$

$$+ (a_0 b_2 + a_1 b_1 + a_2 b_0)(z - z_0)^2 + \cdots$$

$$+ \left(\sum_{k=0}^{n} a_k b_{n-k} \right) (z - z_0)^n + \cdots \qquad (|z - z_0| < R).$$

Series (4) is the same as the series obtained by formally multiplying the two series (1) term by term and collecting the resulting terms in like powers of $z - z_0$; it is called the *Cauchy product* of the two given series.

EXAMPLE 1.   The Maclaurin series for $e^z/(1 + z)$ is valid in the disk $|z| < 1$. The first three nonzero terms are easily found by writing

$$\frac{e^z}{1 + z} = \left(1 + z + \frac{1}{2}z^2 + \frac{1}{6}z^3 + \cdots \right) (1 - z + z^2 - z^3 + \cdots)$$

and multiplying these two series term by term. To be precise, we may multiply each term in the first series by 1, then each term in that series by $-z$, etc. The following systematic approach is suggested, where like powers of z are assembled vertically so that their coefficients can be readily added:

$$1 + z + \frac{1}{2}z^2 + \frac{1}{6}z^3 + \cdots$$

$$- z - z^2 - \frac{1}{2}z^3 - \frac{1}{6}z^4 - \cdots$$

$$z^2 + z^3 + \frac{1}{2}z^4 + \frac{1}{6}z^5 + \cdots$$

$$- z^3 - z^4 - \frac{1}{2}z^5 - \frac{1}{6}z^6 - \cdots$$

$$\vdots$$

The desired result is

$$\frac{e^z}{1+z} = 1 + \frac{1}{2}z^2 - \frac{1}{3}z^3 + \cdots \qquad (|z| < 1).$$

Continuing to let $f(z)$ and $g(z)$ denote the sums of series (1), suppose that $g(z) \neq 0$ when $|z - z_0| < R$. Since the quotient $f(z)/g(z)$ is analytic throughout the disk $|z - z_0| < R$, it has a Taylor series representation

$$(5) \qquad \frac{f(z)}{g(z)} = \sum_{n=0}^{\infty} d_n(z - z_0)^n \qquad (|z - z_0| < R),$$

where the coefficients d_n can be found by differentiating $f(z)/g(z)$ successively and evaluating the derivatives at $z = z_0$. The results are the same as those found by formally carrying out the division of the first of series (1) by the second. Since it is usually only the first few terms that are needed in practice, this method is not difficult.

EXAMPLE 2.   As pointed out in Sec. 25, the zeros of the entire function sinh z are the numbers $z = n\pi i$ ($n = 0, \pm 1, \pm 2, \ldots$). So the quotient

$$\frac{1}{z^2 \sinh z} = \frac{1}{z^3} \left(\frac{1}{1 + z^2/3! + z^4/5! + \cdots} \right)$$

has a Laurent series representation in the punctured disk $0 < |z| < \pi$. The denominator of the fraction in parentheses here is a power series that converges to $(\sinh z)/z$ when $z \neq 0$ and to 1 when $z = 0$. Thus the sum of that series is not zero anywhere in the disk $|z| < \pi$; and a power series representation of the fraction can be found by dividing the series into unity as follows:

$$1 + \frac{1}{3!}z^2 + \frac{1}{5!}z^4 + \cdots \overline{\Big)1}$$

$$\begin{array}{r} 1 - \frac{1}{3!}z^2 + \left[\frac{1}{(3!)^2} - \frac{1}{5!}\right]z^4 + \cdots \\ \hline 1 + \frac{1}{3!}z^2 + \qquad \frac{1}{5!}z^4 + \cdots \\ \hline -\frac{1}{3!}z^2 - \qquad \frac{1}{5!}z^4 + \cdots \\ -\frac{1}{3!}z^2 - \qquad \frac{1}{(3!)^2}z^4 - \cdots \\ \hline \left[\frac{1}{(3!)^2} - \frac{1}{5!}\right]z^4 + \cdots \\ \left[\frac{1}{(3!)^2} - \frac{1}{5!}\right]z^4 + \cdots \\ \hline \vdots \end{array}$$

That is,

$$\frac{1}{1 + z^2/3! + z^4/5! + \cdots} = 1 - \frac{1}{3!}z^2 + \left[\frac{1}{(3!)^2} - \frac{1}{5!}\right]z^4 + \cdots,$$

or

$$(6) \qquad \frac{1}{1 + z^2/3! + z^4/5! + \cdots} = 1 - \frac{1}{6}z^2 + \frac{7}{360}z^4 + \cdots \qquad (|z| < \pi).$$

Hence

$$(7) \qquad \frac{1}{z^2 \sinh z} = \frac{1}{z^3} - \frac{1}{6}\cdot\frac{1}{z} + \frac{7}{360}z + \cdots \qquad (0 < |z| < \pi).$$

Although we have given only the first three nonzero terms of this Laurent series, any number of terms can, of course, be found by continuing the division.

EXERCISES

1. By differentiating the Maclaurin series representation

$$\frac{1}{1-z} = \sum_{n=0}^{\infty} z^n \qquad (|z| < 1),$$

obtain the expansions

$$\frac{1}{(1-z)^2} = \sum_{n=0}^{\infty}(n+1)z^n, \qquad \frac{2}{(1-z)^3} = \sum_{n=0}^{\infty}(n+1)(n+2)z^n \qquad (|z| < 1).$$

2. Find the Taylor series for the function

$$\frac{1}{z} = \frac{1}{2 + (z - 2)} = \frac{1}{2} \cdot \frac{1}{1 + (z - 2)/2}$$

about the point $z_0 = 2$. Then, by differentiating that series term by term, show that

$$\frac{1}{z^2} = \frac{1}{4} \sum_{n=0}^{\infty} (-1)^n (n + 1) \left(\frac{z - 2}{2} \right)^n \qquad (|z - 2| < 2).$$

3. By substituting $1/(1 - z)$ for z in the expansion

$$\frac{1}{(1 - z)^2} = \sum_{n=0}^{\infty} (n + 1) z^n \qquad (|z| < 1),$$

found in Exercise 1, derive the Laurent series representation

$$\frac{1}{z^2} = \sum_{n=2}^{\infty} \frac{(-1)^n (n - 1)}{(z - 1)^n} \qquad (1 < |z - 1| < \infty).$$

(Compare Example 2, Sec. 49.)

4. Show how the expansion

$$\cos z = \sum_{n=0}^{\infty} (-1)^n \frac{z^{2n}}{(2n)!} \qquad (|z| < \infty)$$

can be derived by differentiating the series

$$\sin z = \sum_{n=0}^{\infty} (-1)^n \frac{z^{2n+1}}{(2n + 1)!} \qquad (|z| < \infty)$$

term by term.

5. With the aid of series, prove that the function f defined by means of the equations

$$f(z) = \begin{cases} (e^z - 1)/z & \text{when } z \neq 0 \\ 1 & \text{when } z = 0 \end{cases}$$

is entire.

6. Prove that if

$$f(z) = \begin{cases} \dfrac{\cos z}{z^2 - (\pi/2)^2} & \text{when } z \neq \pm \pi/2, \\ -\dfrac{1}{\pi} & \text{when } z = \pm \pi/2, \end{cases}$$

then f is an entire function.

7. In the w plane, integrate the Taylor series expansion (see Example 4, Sec. 45)

$$\frac{1}{w} = \sum_{n=0}^{\infty} (-1)^n (w - 1)^n \qquad (|w - 1| < 1)$$

along a contour interior to the circle of convergence from $w = 1$ to $w = z$ to obtain the representation

$$\operatorname{Log} z = \sum_{n=1}^{\infty} \frac{(-1)^{n+1}}{n}(z-1)^n \qquad (|z-1| < 1).$$

8. Use the result in Exercise 7 to show that if

$$f(z) = \begin{cases} \dfrac{\operatorname{Log} z}{z-1} & \text{when } z \neq 1, \\ 1 & \text{when } z = 1, \end{cases}$$

then f is analytic throughout the domain $0 < |z| < \infty, -\pi < \operatorname{Arg} z < \pi$.

9. Prove that if f is analytic at z_0 and $f(z_0) = f'(z_0) = \cdots = f^{(m)}(z_0) = 0$, then the function g defined by the equations

$$g(z) = \begin{cases} \dfrac{f(z)}{(z-z_0)^{m+1}} & \text{when } z \neq z_0, \\ \dfrac{f^{(m+1)}(z_0)}{(m+1)!} & \text{when } z = z_0 \end{cases}$$

is analytic at z_0.

10. Use multiplication of series to show that

$$\frac{e^z}{z(z^2+1)} = \frac{1}{z} + 1 - \frac{1}{2}z - \frac{5}{6}z^2 + \cdots \qquad (0 < |z| < 1).$$

11. In each case, use division to show that

(a) $\csc z = \dfrac{1}{z} + \dfrac{1}{3!}z + \left[\dfrac{1}{(3!)^2} - \dfrac{1}{5!}\right]z^3 + \cdots \qquad (0 < |z| < \pi);$

(b) $\dfrac{1}{e^z - 1} = \dfrac{1}{z} - \dfrac{1}{2} + \dfrac{1}{12}z - \dfrac{1}{720}z^3 + \cdots \qquad (0 < |z| < 2\pi).$

12. Use the expansion

$$\frac{1}{z^2 \sinh z} = \frac{1}{z^3} - \frac{1}{6}\cdot\frac{1}{z} + \frac{7}{360}z + \cdots \qquad (0 < |z| < \pi)$$

in Example 2, Sec. 51, to show that if C is the circle $|z| = 1$, taken counterclockwise, then (see Example 1, Sec. 47)

$$\int_C \frac{dz}{z^2 \sinh z} = -\frac{\pi i}{3}.$$

13. Follow the steps below, which illustrate an alternative to straightforward division of series, to obtain representation (6) in Example 2, Sec. 51.

(a) Write

$$\frac{1}{1 + z^2/3! + z^4/5! + \cdots} = d_0 + d_1 z + d_2 z^2 + d_3 z^3 + d_4 z^4 + \cdots,$$

where the coefficients in the power series on the right are to be determined by multiplying the two series in the equation

$$1 = \left(1 + \frac{1}{3!}z^2 + \frac{1}{5!}z^4 + \cdots\right)(d_0 + d_1 z + d_2 z^2 + d_3 z^3 + d_4 z^4 + \cdots).$$

Perform this multiplication to show that

$$(d_0 - 1) + d_1 z + \left(d_2 + \frac{1}{3!}d_0\right)z^2 + \left(d_3 + \frac{1}{3!}d_1\right)z^3$$

$$+ \left(d_4 + \frac{1}{3!}d_2 + \frac{1}{5!}d_0\right)z^4 + \cdots = 0$$

when $|z| < \pi$.

(b) By setting the coefficients in the last series in part (a) equal to zero, find the values of d_0, d_1, d_2, d_3, and d_4. With these values, the first equation in part (a) becomes equation (6), Sec. 51.

14. Use mathematical induction to verify formula (3), Sec. 51, for the nth derivative of the product of two differentiable functions.

15. Let $f(z)$ be an entire function that is represented by a series of the form

$$f(z) = z + a_2 z^2 + a_3 z^3 + \cdots \qquad (|z| < \infty).$$

(a) By differentiating the composite function $g(z) = f[f(z)]$ successively, find the first three nonzero terms in the Maclaurin series for $g(z)$ and thus show that

$$f[f(z)] = z + 2a_2 z^2 + 2(a_2^2 + a_3)z^3 + \cdots \qquad (|z| < \infty).$$

(b) Obtain the result in part (a) in a *formal* manner by writing

$$f[f(z)] = f(z) + a_2[f(z)]^2 + a_3[f(z)]^3 + \cdots,$$

replacing $f(z)$ on the right-hand side here by its series representation, and then collecting terms in like powers of z.

(c) By applying the result in part (a) to the function $f(z) = \sin z$, show that

$$\sin(\sin z) = z - \frac{1}{3}z^3 + \cdots \qquad (|z| < \infty).$$

16. The *Euler numbers* are the numbers E_n $(n = 0, 1, 2, \ldots)$ in the Maclaurin series representation

$$\frac{1}{\cosh z} = \sum_{n=0}^{\infty} \frac{E_n}{n!}z^n \qquad (|z| < \pi/2).$$

Point out why this representation is valid in the indicated disk and why $E_{2n+1} = 0$ $(n = 0, 1, 2, \ldots)$. Then show that

$$E_0 = 1, \qquad E_2 = -1, \qquad E_4 = 5, \qquad \text{and} \qquad E_6 = -61.$$

17. Suppose that a function $f(z)$ has a power series representation

$$f(z) = \sum_{n=0}^{\infty} a_n (z - z_0)^n$$

inside some circle $|z - z_0| = R$. Use Theorem 2 in Sec. 49, regarding term by term differentiation of such a series, and mathematical induction to show that

$$f^{(n)}(z) = \sum_{k=0}^{\infty} \frac{(n+k)!}{k!} a_{n+k}(z - z_0)^k \qquad (n = 0, 1, 2, \ldots)$$

when $|z - z_0| < R$. Then, by setting $z = z_0$, show that the coefficients a_n $(n = 0, 1, 2, \ldots)$ are the coefficients in the Taylor series for f about z_0. Thus give an alternative proof of Theorem 1 in Sec. 50.

18. Consider two series

$$S_1(z) = \sum_{n=0}^{\infty} a_n(z - z_0)^n, \qquad S_2(z) = \sum_{n=1}^{\infty} \frac{b_n}{(z - z_0)^n},$$

which converge in some annular domain centered at z_0. Let C denote any contour lying in that annulus, and let $g(z)$ be a function which is continuous on C. Modify the proof of Theorem 1, Sec. 49, which tells us that

$$\int_C g(z)S_1(z)\, dz = \sum_{n=0}^{\infty} a_n \int_C g(z)(z - z_0)^n\, dz,$$

to prove that

$$\int_C g(z)S_2(z)\, dz = \sum_{n=1}^{\infty} b_n \int_C \frac{g(z)}{(z - z_0)^n}\, dz.$$

Conclude from these results that if

$$S(z) = \sum_{n=-\infty}^{\infty} c_n(z - z_0)^n = \sum_{n=0}^{\infty} a_n(z - z_0)^n + \sum_{n=1}^{\infty} \frac{b_n}{(z - z_0)^n},$$

then

$$\int_C g(z)S(z)\, dz = \sum_{n=-\infty}^{\infty} c_n \int_C g(z)(z - z_0)^n\, dz.$$

52. ANALYTIC CONTINUATION

The reader may at this time pass directly to Chap. 6 without disruption since the material in this section is not used elsewhere in the book. The material might, however, be included in an introductory course, and we place it here since power series are useful in illustrating it.

The *intersection* of two domains D_1 and D_2 is the domain $D_1 \cap D_2$ consisting of all points that lie in both D_1 and D_2. If we have two domains D_1 and D_2 with points in common (Fig. 56) and a function f_1 that is analytic in D_1, there *may* exist a function f_2, which is analytic in D_2, such that $f_2(z) = f_1(z)$ for each z in the intersection $D_1 \cap D_2$. If so, we call f_2 an *analytic continuation* of f_1 into the second domain D_2.

Whenever that analytic continuation exists, it is unique, according to a corollary to be proved in Sec. 58 (Chap. 6). More precisely, not more than one function

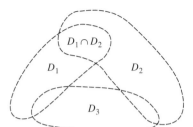

FIGURE 56

can be analytic in D_2 and also assume the value $f_1(z)$ at each point z of the domain $D_1 \cap D_2$ interior to D_2. However, if there is an analytic continuation f_3 of f_2 from D_2 into a domain D_3 which intersects D_1, as indicated in Fig. 56, it is not necessarily true that $f_3(z) = f_1(z)$ for each z in $D_1 \cap D_3$. In Example 3 below, we illustrate the fact that such a chain of continuations of a given function from a domain D_1 may lead to a different function defined on D_1.

If f_2 is the analytic continuation of f_1 from a domain D_1 into a domain D_2, then the function F defined by the equations

$$F(z) = \begin{cases} f_1(z) & \text{when } z \text{ is in } D_1, \\ f_2(z) & \text{when } z \text{ is in } D_2 \end{cases}$$

is analytic in the *union* $D_1 \cup D_2$, which is the domain consisting of all points that lie in either D_1 or D_2. The function F is the analytic continuation into $D_1 \cup D_2$ of either f_1 or f_2; and f_1 and f_2 are called *elements* of F.

EXAMPLE 1.   Consider first the function f_1 defined by the equation

(1) $$f_1(z) = \sum_{n=0}^{\infty} z^n.$$

The power series here converges to $1/(1-z)$ when $|z| < 1$ (Sec. 45). It diverges when $|z| \geq 1$ since then the nth term z^n does not approach zero as n tends to infinity. Hence

$$f_1(z) = \frac{1}{1-z} \qquad \text{when} \qquad |z| < 1,$$

and f_1 is not defined when $|z| \geq 1$.
    Now the function

(2) $$f_2(z) = \frac{1}{1-z} \qquad (z \neq 1)$$

is defined and analytic everywhere except at the point $z = 1$. Since $f_2(z) = f_1(z)$ inside the circle $|z| = 1$, the function f_2 is the analytic continuation of f_1 into the domain consisting of all points in the z plane except for $z = 1$. It is the only possible analytic continuation of f_1 into that domain, according to remarks at the beginning of the section. In this example, f_1 is also an element of f_2.

EXAMPLE 2. Consider the function

$$(3) \qquad f_1(z) = \int_0^\infty e^{-zt}\, dt.$$

As shown in Exercise 1(c), Sec. 31, this integral exists when $\operatorname{Re} z > 0$, its value being $1/z$. Hence we can write

$$(4) \qquad f_1(z) = \frac{1}{z} \qquad (\operatorname{Re} z > 0).$$

The domain of definition $\operatorname{Re} z > 0$ is denoted by D_1 in Fig. 57, and f_1 is analytic there. Let f_2 be defined in terms of a geometric series:

$$(5) \qquad f_2(z) = i \sum_{n=0}^{\infty} \left(\frac{z+i}{i} \right)^n \qquad (|z+i| < 1).$$

Within its circle of convergence, which is the unit circle centered at the point $z = -i$, the series is convergent. To be specific,

$$(6) \qquad f_2(z) = \frac{i}{1 - (z+i)/i} = \frac{1}{z}$$

when z is in the domain $|z + i| < 1$, denoted by D_2. Evidently, then, $f_2(z) = f_1(z)$ for each z in the intersection $D_1 \cap D_2$; and f_2 is the analytic continuation of f_1 into D_2.

The function $F(z) = 1/z$ $(z \neq 0)$ is the analytic continuation of both f_1 and f_2 into the domain D_3 consisting of all points in the z plane except the origin. The functions f_1 and f_2 are elements of F.

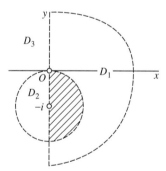

FIGURE 57

EXAMPLE 3. Finally, we consider this branch of $z^{1/2}$:

$$f_1(z) = \sqrt{r}\, e^{i\theta/2} \qquad (r > 0, 0 < \theta < \pi).$$

An analytic continuation of f_1 across the negative real axis into the lower half plane is

$$f_2(z) = \sqrt{r}\, e^{i\theta/2} \qquad \left(r > 0, \frac{\pi}{2} < \theta < 2\pi \right).$$

An analytic continuation of f_2 across the positive real axis into the first quadrant is, then,

$$f_3(z) = \sqrt{r}e^{i\theta/2} \qquad \left(r > 0, \pi < \theta < \frac{5\pi}{2}\right).$$

Note that $f_3(z) \neq f_1(z)$ in the first quadrant; in fact, $f_3(z) = -f_1(z)$ there.

EXERCISES

1. Show that the function

$$f_2(z) = \frac{1}{z^2 + 1} \qquad (z \neq \pm i)$$

is the analytic continuation of the function

$$f_1(z) = \sum_{n=0}^{\infty}(-1)^n z^{2n} \qquad (|z| < 1)$$

into the domain consisting of all points in the z plane except $z = \pm i$.

2. Show that the function $f_2(z) = 1/z^2$ $(z \neq 0)$ is the analytic continuation of the function

$$f_1(z) = \sum_{n=0}^{\infty}(n + 1)(z + 1)^n \qquad (|z + 1| < 1)$$

into the domain consisting of all points in the z plane except $z = 0$.

3. State why the function

$$f_4(z) = \sqrt{r}e^{i\theta/2} \qquad (r > 0, -\pi < \theta < \pi)$$

is the analytic continuation of the function (Example 3, Sec. 52)

$$f_1(z) = \sqrt{r}e^{i\theta/2} \qquad (r > 0, 0 < \theta < \pi)$$

across the positive real axis into the lower half plane.

4. Find the analytic continuation of Log z from the upper half plane Im $z > 0$ into the lower half plane across the negative real axis. Note that this analytic continuation is different from Log z in the lower half plane.

$\qquad$ *Ans.* $\ln r + i\theta$ $(r > 0, 0 < \theta < 2\pi)$.

5. Find the analytic continuation of the function

$$f(z) = \int_0^{\infty} te^{-zt}\, dt \qquad (\text{Re } z > 0)$$

into the domain consisting of all points in the z plane except the origin.

$\qquad$ *Ans.* $1/z^2$.

6. Show that the function $1/(z^2 + 1)$ is the analytic continuation of the function

$$f(z) = \int_0^{\infty} e^{-zt} \sin t\, dt \qquad (\text{Re } z > 0)$$

into the domain consisting of all points in the z plane except $z = \pm i$.

CHAPTER
6

RESIDUES
AND POLES

The Cauchy-Goursat theorem (Sec. 36) states that if a function is analytic at all points interior to and on a simple closed contour C, then the value of the integral of the function around that contour is zero. If, however, the function fails to be analytic at a finite number of points interior to C, there is, as we shall see in this chapter, a specific number, called a residue, which each of those points contributes to the value of the integral. We develop here the theory of residues; and, in Chap. 7, we shall illustrate their use in certain areas of applied mathematics.

53. RESIDUES

Recall (Sec. 20) that a point z_0 is called a singular point of a function f if f fails to be analytic at z_0 but is analytic at some point in every neighborhood of z_0. A singular point z_0 is said to be *isolated* if, in addition, there is a deleted neighborhood $0 < |z - z_0| < \varepsilon$ of z_0 throughout which f is analytic.

EXAMPLE 1. The function

$$\frac{z + 1}{z^3(z^2 + 1)}$$

has the three isolated singular points $z = 0$ and $z = \pm i$.

EXAMPLE 2. The origin is a singular point of Log z, but it is *not* an isolated singular point since every deleted neighborhood of the origin contains points on the negative real axis and Log z fails to be analytic at each of those points (see Sec. 26).

EXAMPLE 3.   The function

$$\frac{1}{\sin(\pi/z)}$$

has the singular points $z = 0$ and $z = 1/n$ $(n = \pm 1, \pm 2, \ldots)$, all lying on the segment of the real axis from $z = -1$ to $z = 1$. Each singular point except $z = 0$ is isolated. The singular point $z = 0$ is not isolated because every deleted neighborhood of the origin contains other singular points of the function.

When z_0 is an isolated singular point of a function f, there is a positive number R_2 such that f is analytic at each point z for which $0 < |z - z_0| < R_2$. Consequently, $f(z)$ is represented by a Laurent series

$$(1) \quad f(z) = \sum_{n=0}^{\infty} a_n(z - z_0)^n + \frac{b_1}{z - z_0} + \frac{b_2}{(z - z_0)^2} + \cdots + \frac{b_n}{(z - z_0)^n} + \cdots$$

$$(0 < |z - z_0| < R_2),$$

where the coefficients a_n and b_n have certain integral representations (Sec. 46). In particular,

$$b_n = \frac{1}{2\pi i} \int_C \frac{f(z)\, dz}{(z - z_0)^{-n+1}} \quad (n = 1, 2, \ldots)$$

where C is any positively oriented simple closed contour around z_0 and lying in the punctured disk $0 < |z - z_0| < R_2$. When $n = 1$, this expression for b_n can be written

$$(2) \qquad\qquad \int_C f(z)\, dz = 2\pi i b_1.$$

The complex number b_1, which is the coefficient of $1/(z - z_0)$ in expansion (1), is called the *residue* of f at the isolated singular point z_0. We shall often use the notation

$$\operatorname*{Res}_{z=z_0} f(z),$$

or simply B when the point z_0 and the function f are clearly indicated, to denote the residue b_1.

Equation (2) provides a powerful method for evaluating certain integrals around simple closed contours.

EXAMPLE 4.   Consider the integral

$$(3) \qquad\qquad \int_C \frac{dz}{z(z - 2)^4},$$

where C is the positively oriented circle $|z - 2| = 1$ (Fig. 58). Since the integrand is analytic everywhere in the finite plane except at the points $z = 0$ and $z = 2$, it has a Laurent series representation that is valid in the punctured disk $0 < |z - 2| < 2$,

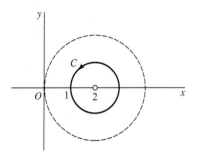

FIGURE 58

also shown in Fig. 58. Thus, according to equation (2), the value of integral (3) is $2\pi i$ times the residue of its integrand at $z = 2$. To determine that residue, we recall (Sec. 45) the Maclaurin series expansion

$$\frac{1}{1-z} = \sum_{n=0}^{\infty} z^n \qquad (|z| < 1)$$

and use it to write

$$\frac{1}{z(z-2)^4} = \frac{1}{(z-2)^4} \cdot \frac{1}{2 + (z-2)}$$

$$= \frac{1}{2(z-2)^4} \cdot \frac{1}{1 - \left(-\dfrac{z-2}{2}\right)}$$

$$= \sum_{n=0}^{\infty} \frac{(-1)^n}{2^{n+1}}(z-2)^{n-4} \qquad (0 < |z-2| < 2).$$

In this Laurent series, which could be written in the form (1), the coefficient of $1/(z-2)$ is the desired residue, namely $-1/16$. Consequently,

(4) $$\int_C \frac{dz}{z(z-2)^4} = 2\pi i\left(-\frac{1}{16}\right) = -\frac{\pi i}{8}.$$

EXAMPLE 5.   Let us show that

(5) $$\int_C \exp\left(\frac{1}{z^2}\right) dz = 0,$$

where C is the unit circle $|z| = 1$. Since $1/z^2$ is analytic everywhere except at the origin, so is the integrand. The isolated singular point $z = 0$ is interior to C; and, with the aid of the Maclaurin series (Sec. 45)

$$e^z = 1 + \frac{z}{1!} + \frac{z^2}{2!} + \frac{z^3}{3!} + \cdots \qquad (|z| < \infty),$$

one can write the Laurent series expansion

$$\exp\left(\frac{1}{z^2}\right) = 1 + \frac{1}{1!} \cdot \frac{1}{z^2} + \frac{1}{2!} \cdot \frac{1}{z^4} + \frac{1}{3!} \cdot \frac{1}{z^6} + \cdots \qquad (0 < |z| < \infty).$$

The residue of the integrand at its isolated singular point $z = 0$ is, therefore, zero $(b_1 = 0)$, and the value of integral (5) is established.

We are reminded in this example that, although the analyticity of a function within and on a simple closed contour C is a sufficient condition for the value of the integral around C to be zero, it is not a necessary condition.

54. RESIDUE THEOREMS

If a function f has only a finite number of singular points interior to a given simple closed contour C, then they must be isolated. The following theorem, which is known as *Cauchy's residue theorem*, is a precise statement of the fact that if f is also analytic on C and C is described in the positive sense, then the value of the integral of f around C is $2\pi i$ times the *sum* of the residues of f at those singular points.

Theorem 1. *Let C be a positively oriented simple closed contour. If a function f is analytic inside and on C except for a finite number of singular points z_k ($k = 1, 2, \ldots, n$) inside C, then*

$$(1) \qquad \int_C f(z)\, dz = 2\pi i \sum_{k=1}^{n} \operatorname*{Res}_{z=z_k} f(z).$$

To prove the theorem, let the points z_k ($k = 1, 2, \ldots, n$) be centers of positively oriented circles C_k which are interior to C and are so small that no two of them have points in common (Fig. 59). The circles C_k, together with the simple closed contour C, form the boundary of a closed region throughout which f is analytic and whose interior is a multiply connected domain. Hence, according to the extension of the Cauchy-Goursat theorem to such regions (Theorem 2, Sec. 38),

$$\int_C f(z)\, dz - \sum_{k=1}^{n} \int_{C_k} f(z)\, dz = 0.$$

This reduces to equation (1) because (Sec. 53)

$$\int_{C_k} f(z)\, dz = 2\pi i \operatorname*{Res}_{z=z_k} f(z) \qquad (k = 1, 2, \ldots, n),$$

and the proof is complete.

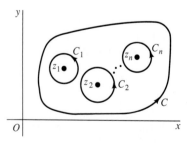

FIGURE 59

EXAMPLE 1.   Let us use the theorem to evaluate the integral

$$\int_C \frac{5z-2}{z(z-1)}\,dz$$

when C is the circle $|z| = 2$, described counterclockwise. The integrand has the two isolated singularities $z = 0$ and $z = 1$, both of which are interior to C. We can find the residues B_1 at $z = 0$ and B_2 at $z = 1$ with the aid of the Maclaurin series

$$\frac{1}{1-z} = 1 + z + z^2 + \cdots \qquad (|z| < 1).$$

We observe first that when $0 < |z| < 1$,

$$\frac{5z-2}{z(z-1)} = \frac{5z-2}{z} \cdot \frac{-1}{1-z} = \left(5 - \frac{2}{z}\right)(-1 - z - z^2 - \cdots);$$

and, by identifying the coefficient of $1/z$ in the product on the right here, we find that $B_1 = 2$. Also, since

$$\frac{5z-2}{z(z-1)} = \frac{5(z-1)+3}{z-1} \cdot \frac{1}{1+(z-1)}$$

$$= \left(5 + \frac{3}{z-1}\right)[1 - (z-1) + (z-1)^2 - \cdots]$$

when $0 < |z-1| < 1$, it is clear that $B_2 = 3$. Thus

$$\int_C \frac{5z-2}{z(z-1)}\,dz = 2\pi i(B_1 + B_2) = 10\pi i.$$

In this example, it is actually simpler to write the integrand as the sum of its partial fractions:

$$\frac{5z-2}{z(z-1)} = \frac{2}{z} + \frac{3}{z-1}.$$

Then, since $2/z$ is already a Laurent series when $0 < |z| < 1$ and since $3/(z-1)$ is one when $0 < |z-1| < 1$, it follows that

$$\int_C \frac{5z-2}{z(z-1)}\,dz = 2\pi i(2) + 2\pi i(3) = 10\pi i.$$

(See also Example 2 below.)

If the function f in Theorem 1 is, in addition, *analytic at each point in the finite plane exterior to C*, it is sometimes more efficient to evaluate the integral of f around C by finding a *single* residue of a certain related function. We state the method as a theorem.*

---

*This result arises in the theory of residues at infinity, which we shall not develop. For some details of that theory, see, for instance, R. P. Boas, "Invitation to Complex Analysis," pp. 76–77, 1987.

Theorem 2. *If a function f is analytic everywhere in the finite plane except for a finite number of singular points interior to a positively oriented simple closed contour C, then*

(2)
$$\int_C f(z)\, dz = 2\pi i \operatorname*{Res}_{z=0}\left[\frac{1}{z^2} f\left(\frac{1}{z}\right)\right].$$

We begin the derivation of expression (2) by constructing a circle $|z| = R_1$ which is large enough so that the contour C is interior to it (Fig. 60). Then if C_0 denotes a positively oriented circle $|z| = R_0$, where $R_0 > R_1$, we know from Laurent's theorem (Sec. 46) that

(3)
$$f(z) = \sum_{n=-\infty}^{\infty} c_n z^n \qquad (R_1 < |z| < \infty),$$

where

(4)
$$c_n = \frac{1}{2\pi i} \int_{C_0} \frac{f(z)\, dz}{z^{n+1}} \qquad (n = 0, \pm 1, \pm 2, \dots).$$

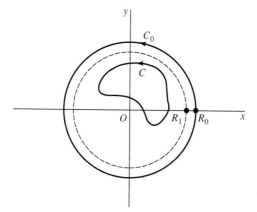

FIGURE 60

By writing $n = -1$ in expression (4), we find that

(5)
$$\int_{C_0} f(z)\, dz = 2\pi i c_{-1}.$$

Observe that, since the condition of validity with representation (3) is not of the type $0 < |z| < R_2$, the coefficient c_{-1} is *not* the residue of f at the point $z = 0$, which may not even be a singular point of f. But, if we replace z by $1/z$ in representation (3) and its condition of validity, we see that

$$\frac{1}{z^2} f\left(\frac{1}{z}\right) = \sum_{n=-\infty}^{\infty} \frac{c_n}{z^{n+2}} = \sum_{n=-\infty}^{\infty} \frac{c_{n-2}}{z^n} \qquad \left(0 < |z| < \frac{1}{R_1}\right)$$

and hence that

(6) $$c_{-1} = \operatorname*{Res}_{z=0}\left[\frac{1}{z^2}f\left(\frac{1}{z}\right)\right].$$

Then, in view of equations (5) and (6),

$$\int_{C_0} f(z)\,dz = 2\pi i \operatorname*{Res}_{z=0}\left[\frac{1}{z^2}f\left(\frac{1}{z}\right)\right].$$

Finally, since f is analytic throughout the closed region bounded by C and C_0, the principle of deformation of paths (Corollary 2, Sec. 38) yields the desired result (2).

EXAMPLE 2.   In Example 1, the integrand

$$f(z) = \frac{5z - 2}{z(z - 1)}$$

is analytic at each point z that is exterior to the circle C there. Since

$$\frac{1}{z^2}f\left(\frac{1}{z}\right) = \frac{5 - 2z}{z(1 - z)} = \frac{5 - 2z}{z} \cdot \frac{1}{1 - z}$$

$$= \left(\frac{5}{z} - 2\right)(1 + z + z^2 + \cdots)$$

$$= \frac{5}{z} + 3 + 3z + \cdots \qquad (0 < |z| < 1),$$

we see that the residue to be used in equation (2) is 5. Hence

$$\int_C \frac{5z - 2}{z(z - 1)}\,dz = 2\pi i(5) = 10\pi i,$$

as already demonstrated in Example 1.

55.   THE THREE TYPES OF ISOLATED SINGULAR POINTS

We saw in Sec. 53 that the theory of residues is based on the fact that if f has an isolated singular point z_0, then $f(z)$ can be represented by a Laurent series

(1) $$f(z) = \sum_{n=0}^{\infty} a_n(z - z_0)^n + \frac{b_1}{z - z_0} + \frac{b_2}{(z - z_0)^2} + \cdots + \frac{b_n}{(z - z_0)^n} + \cdots$$

in a punctured disk $0 < |z - z_0| < R_2$. The portion

$$\frac{b_1}{z - z_0} + \frac{b_2}{(z - z_0)^2} + \cdots + \frac{b_n}{(z - z_0)^n} + \cdots$$

of the series, involving negative powers of $z - z_0$, is called the *principal part* of f at z_0. We now use the principal part to identify the isolated singular point z_0 as one of three special types. This classification will aid us in the development of residue theory that appears in following sections.

If the principal part of f at z_0 contains at least one nonzero term but the number of such terms is finite, then there exists a positive integer m such that

$$b_m \neq 0 \qquad \text{and} \qquad b_{m+1} = b_{m+2} = \cdots = 0.$$

That is, expansion (1) takes the form

$$(2) \qquad f(z) = \sum_{n=0}^{\infty} a_n(z - z_0)^n + \frac{b_1}{z - z_0} + \frac{b_2}{(z - z_0)^2} + \cdots + \frac{b_m}{(z - z_0)^m}$$

$$(0 < |z - z_0| < R_2),$$

where $b_m \neq 0$. In this case, the isolated singular point z_0 is called a *pole of order m*.* A pole of order $m = 1$ is usually referred to as a *simple pole*.

EXAMPLE 1.   Observe that the function

$$\frac{z^2 - 2z + 3}{z - 2} = \frac{z(z - 2) + 3}{z - 2} = z + \frac{3}{z - 2} = 2 + (z-2) + \frac{3}{z - 2} \quad (0 < |z-2| < \infty)$$

has a simple pole ($m = 1$) at $z_0 = 2$. Its residue b_1 there is 3.

EXAMPLE 2.   The function

$$\frac{\sinh z}{z^4} = \frac{1}{z^4}\left(z + \frac{z^3}{3!} + \frac{z^5}{5!} + \frac{z^7}{7!} + \cdots\right) = \frac{1}{z^3} + \frac{1}{3!} \cdot \frac{1}{z} + \frac{z}{5!} + \frac{z^3}{7!} + \cdots \quad (0 < |z| < \infty)$$

has a pole of order $m = 3$ at $z_0 = 0$, with residue $b_1 = 1/6$.

There remain two extremes, the case in which all of the coefficients b_n's in expansion (1) are zero and the one in which an infinite number of them are nonzero. When all of the b_n's in expansion (1) are zero, so that

$$(3) \qquad f(z) = \sum_{n=0}^{\infty} a_n(z - z_0)^n = a_0 + a_1(z - z_0) + a_2(z - z_0)^2 + \cdots$$

$$(0 < |z - z_0| < R_2),$$

the point z_0 is known as a *removable singular point*. Note that the residue at a removable singular point is always zero. If we define, or possibly redefine, f at z_0 so that $f(z_0) = a_0$, expansion (3) becomes valid throughout the entire disk $|z - z_0| < R_2$. Since a power series always represents an analytic function interior to its circle of convergence (Sec. 49), it follows that f is analytic at z_0 when it is assigned the value a_0 there. The singularity at z_0 is, therefore, *removed*.

---

*Reasons for the terminology *pole* have been suggested on p. 70 of the book mentioned in the footnote in Sec. 54.

EXAMPLE 3.   The point $z_0 = 0$ is a removable singular point of the function

$$f(z) = \frac{1 - \cos z}{z^2} = \frac{1}{z^2}\left[1 - \left(1 - \frac{z^2}{2!} + \frac{z^4}{4!} - \frac{z^6}{6!} + \cdots\right)\right]$$

$$= \frac{1}{2!} - \frac{z^2}{4!} + \frac{z^4}{6!} - \cdots \qquad (0 < |z| < \infty).$$

When the value $f(0) = 1/2$ is assigned, f becomes entire.

When an infinite number of the coefficients b_n in series (1) are nonzero, z_0 is said to be an *essential singular point* of f. An important result concerning the behavior of a function near an essential singular point is due to Picard. It states that *in each neighborhood of an essential singular point, a function assumes every finite value, with one possible exception, an infinite number of times.*[*]

EXAMPLE 4.   The function

$$\exp\left(\frac{1}{z}\right) = \sum_{n=0}^{\infty} \frac{1}{n!} \cdot \frac{1}{z^n} = 1 + \frac{1}{1!} \cdot \frac{1}{z} + \frac{1}{2!} \cdot \frac{1}{z^2} + \cdots \qquad (0 < |z| < \infty)$$

has an essential singular point at $z_0 = 0$, where the residue b_1 is unity. For an illustration of Picard's theorem, let us show that $\exp(1/z)$ assumes the value -1 an infinite number of times in each neighborhood of the origin. To do this, we recall from the example in Sec. 23 that $\exp z = -1$ when $z = (2n + 1)\pi i$ $(n = 0, \pm 1, \pm 2, \ldots)$. This means that $\exp(1/z) = -1$ when

$$z = \frac{1}{(2n + 1)\pi i} \cdot \frac{i}{i} = -\frac{i}{(2n + 1)\pi} \qquad (n = 0, \pm 1, \pm 2, \ldots),$$

and an infinite number of these points clearly lie in any given neighborhood of the origin. Since $\exp(1/z) \neq 0$ for any value of z, zero is the exceptional value in Picard's theorem.

In the remaining sections of this chapter, we shall develop in greater depth the theory of the three types of isolated singular points just described. The emphasis will be on useful and efficient methods for identifying poles and finding the corresponding residues. Two famous theorems regarding removable and essential singularities will, however, be proved in the final section of the chapter.

EXERCISES

1. Find the residue at $z = 0$ of the function

(a) $\dfrac{1}{z + z^2}$;   (b) $z\cos\left(\dfrac{1}{z}\right)$;   (c) $\dfrac{z - \sin z}{z}$;   (d) $\dfrac{\cot z}{z^4}$;   (e) $\dfrac{\sinh z}{z^4(1 - z^2)}$.

   Ans. (a) 1;   (b) $-1/2$;   (c) 0;   (d) $-1/45$;   (e) 7/6.

---

[*]For a proof of Picard's theorem, see Sec. 51 in Vol. III of the book by Markushevich, cited in Appendix 1.

*2. Use Theorem 1, Sec. 54, to evaluate the integral of each of these functions around the circle $|z| = 3$ in the positive sense:

(a) $\dfrac{\exp(-z)}{z^2}$;   (b) $\dfrac{\exp(-z)}{(z-1)^2}$;   (c) $z^2 \exp\left(\dfrac{1}{z}\right)$;   (d) $\dfrac{z+1}{z^2 - 2z}$.

Ans. (a) $-2\pi i$;   (b) $-2\pi i/e$;   (c) $\pi i/3$;   (d) $2\pi i$.

*3. Use Theorem 2, Sec. 54, to evaluate the integral of each of these functions around the circle $|z| = 2$ in the positive sense:

(a) $\dfrac{z^5}{1 - z^3}$;   (b) $\dfrac{1}{1 + z^2}$;   (c) $\dfrac{1}{z}$.

Ans. (a) $-2\pi i$;   (b) 0;   (c) $2\pi i$.

4. In each case, write the principal part of the function at its isolated singular point and determine whether that point is a pole, a removable singular point, or an essential singular point:

(a) $z \exp\left(\dfrac{1}{z}\right)$;   (b) $\dfrac{z^2}{1 + z}$;   (c) $\dfrac{\sin z}{z}$;   (d) $\dfrac{\cos z}{z}$;   (e) $\dfrac{1}{(2 - z)^3}$.

5. Show that the singular point of each of the following functions is a pole. Determine the order m of that pole and the corresponding residue B.

(a) $\dfrac{1 - \cosh z}{z^3}$;   (b) $\dfrac{1 - \exp(2z)}{z^4}$;   (c) $\dfrac{\exp(2z)}{(z-1)^2}$.

Ans. (a) $m = 1, B = -1/2$;   (b) $m = 3, B = -4/3$;   (c) $m = 2, B = 2e^2$.

6. Suppose that a function f is analytic at z_0, and consider the quotient

$$g(z) = \frac{f(z)}{z - z_0}.$$

Show that

(a) if $f(z_0) \neq 0$, then z_0 is a simple pole of g, with residue $f(z_0)$;

(b) if $f(z_0) = 0$, then z_0 is a removable singular point of g.

      Suggestion: As pointed out in Sec. 44, there is a Taylor series for $f(z)$ about z_0 since f is analytic there. Start each part of this exercise by writing out a few terms of that series.

7. Let the degrees of the polynomials

$$P(z) = a_0 + a_1 z + a_2 z^2 + \cdots + a_n z^n \qquad (a_n \neq 0)$$

and

$$Q(z) = b_0 + b_1 z + b_2 z^2 + \cdots + b_m z^m \qquad (b_m \neq 0)$$

be such that $m \geq n + 2$.

(a) Write

$$\frac{1}{z^2} \cdot \frac{P(1/z)}{Q(1/z)} \qquad (z \neq 0)$$

as the quotient of two polynomials, and point out why $z = 0$ is a removable singular point of that quotient.

(b) Use the final result in part (a) and Theorem 2 in Sec. 54 to show that if all of the zeros of $Q(z)$ are interior to a given simple closed contour C, then

$$\int_C \frac{P(z)}{Q(z)} \, dz = 0.$$

[Compare Exercise 3(b)].

8. Let C denote the circle $|z| = 1$, taken counterclockwise, and follow the steps below to show that

$$\int_C \exp\left(z + \frac{1}{z}\right) dz = 2\pi i \sum_{n=0}^{\infty} \frac{1}{n!(n+1)!}.$$

(a) By using the Maclaurin series for e^z and referring to Theorem 1 in Sec. 49, which justifies the term by term integration that is to be used, write the above integral as

$$\sum_{n=0}^{\infty} \frac{1}{n!} \int_C z^n \exp\left(\frac{1}{z}\right) dz.$$

(b) Apply Theorem 1 in Sec. 54 to evaluate the integrals appearing in part (a) to arrive at the desired result.

9. Write the function

$$f(z) = \frac{8a^3 z^2}{(z^2 + a^2)^3} \qquad (a > 0)$$

as

$$f(z) = \frac{\phi(z)}{(z - ai)^3} \qquad \text{where} \qquad \phi(z) = \frac{8a^3 z^2}{(z + ai)^3}.$$

Point out why $\phi(z)$ has a Taylor series representation about $z = ai$, and then use it to show that the principal part of f at that point is

$$\frac{\phi''(ai)/2}{z - ai} + \frac{\phi'(ai)}{(z - ai)^2} + \frac{\phi(ai)}{(z - ai)^3} = -\frac{i/2}{z - ai} - \frac{a/2}{(z - ai)^2} - \frac{a^2 i}{(z - ai)^3}.$$

56. RESIDUES AT POLES

When a function f has an isolated singularity at a point z_0, the basic method for identifying z_0 as a pole and finding the residue there is to write the appropriate Laurent series and to note the coefficient of $1/(z - z_0)$. The following theorem provides an alternative characterization of poles and another way of finding the corresponding residues.

Theorem. *An isolated singular point z_0 of a function f is a pole of order m if and only if $f(z)$ can be written in the form*

$$(1) \qquad\qquad f(z) = \frac{\phi(z)}{(z - z_0)^m},$$

where $\phi(z)$ is analytic and nonzero at z_0. Moreover,

(2)
$$\underset{z=z_0}{\text{Res}}\, f(z) = \phi(z_0) \qquad \text{if } m = 1$$

and

(3)
$$\underset{z=z_0}{\text{Res}}\, f(z) = \frac{\phi^{(m-1)}(z_0)}{(m-1)!} \qquad \text{if } m \geq 2.$$

Observe that expression (2) need not have been written separately since, with the convention that $\phi^{(0)}(z_0) = \phi(z_0)$ and $0! = 1$, expression (3) reduces to it when $m = 1$.

To prove the theorem, we first assume that $f(z)$ has the form (1) and recall (Sec. 44) that since $\phi(z)$ is analytic at z_0, it has a Taylor series representation

$$\phi(z) = \phi(z_0) + \frac{\phi'(z_0)}{1!}(z - z_0) + \frac{\phi''(z_0)}{2!}(z - z_0)^2 + \cdots$$

$$+ \frac{\phi^{(m-1)}(z_0)}{(m-1)!}(z - z_0)^{m-1} + \sum_{n=m}^{\infty} \frac{\phi^{(n)}(z_0)}{n!}(z - z_0)^n$$

in some neighborhood $|z - z_0| < \varepsilon$ of z_0; and from expression (1) it follows that

(4)
$$f(z) = \frac{\phi(z_0)}{(z - z_0)^m} + \frac{\phi'(z_0)/1!}{(z - z_0)^{m-1}} + \frac{\phi''(z_0)/2!}{(z - z_0)^{m-2}} + \cdots$$

$$+ \frac{\phi^{(m-1)}(z_0)/(m-1)!}{z - z_0} + \sum_{n=m}^{\infty} \frac{\phi^{(n)}(z_0)}{n!}(z - z_0)^{n-m}$$

when $0 < |z - z_0| < \varepsilon$. This Laurent series representation, together with the fact that $\phi(z_0) \neq 0$, reveals that z_0 is, indeed, a pole of order m of $f(z)$. The coefficient of $1/(z - z_0)$ tells us, of course, that the residue of $f(z)$ at z_0 is as in the statement of the theorem.

Suppose, on the other hand, that we know only that z_0 is a pole of order m of f, or that $f(z)$ has a Laurent series representation

$$f(z) = \sum_{n=0}^{\infty} a_n(z - z_0)^n + \frac{b_1}{z - z_0} + \frac{b_2}{(z - z_0)^2} + \cdots + \frac{b_{m-1}}{(z - z_0)^{m-1}} + \frac{b_m}{(z - z_0)^m}$$

$$(b_m \neq 0)$$

which is valid in a punctured disk $0 < |z - z_0| < R_2$. The function $\phi(z)$ defined by means of the equations

$$\phi(z) = \begin{cases} (z - z_0)^m f(z) & \text{when } z \neq z_0, \\ b_m & \text{when } z = z_0 \end{cases}$$

evidently has the power series representation

$$\phi(z) = b_m + b_{m-1}(z - z_0) + \cdots + b_2(z - z_0)^{m-2} + b_1(z - z_0)^{m-1} + \sum_{n=0}^{\infty} a_n(z - z_0)^{m+n}$$

throughout the entire disk $|z - z_0| < R_2$. Consequently, $\phi(z)$ is analytic in that disk (Sec. 49) and, in particular, at z_0. Inasmuch as $\phi(z_0) = b_m \neq 0$, expression (1) is established; and the proof of the theorem is complete.

Note how it follows from expression (1) that

$$\lim_{z \to z_0} \frac{1}{f(z)} = \lim_{z \to z_0} \frac{(z - z_0)^m}{\phi(z)} = \frac{0}{\phi(z_0)} = 0.$$

Thus, if z_0 is a pole of a function f, it is always true that

(5) $$\lim_{z \to z_0} f(z) = \infty,$$

according to the definition (Sec. 13) of such a limit involving the point at infinity.

EXAMPLE 1. The function $f(z) = (z + 1)/(z^2 + 9)$ has an isolated singular point at $z = 3i$ and can be written as

$$f(z) = \frac{\phi(z)}{z - 3i} \qquad \text{where} \qquad \phi(z) = \frac{z + 1}{z + 3i}.$$

Since $\phi(z)$ is analytic at $z = 3i$ and $\phi(3i) = (3 - i)/6 \neq 0$, that point is a simple pole of the function f; the residue there is $(3 - i)/6$. The point $z = -3i$ is also a simple pole of f, with residue $(3 + i)/6$.

EXAMPLE 2. If $f(z) = (z^3 + 2z)/(z - i)^3$, then

$$f(z) = \frac{\phi(z)}{(z - i)^3} \qquad \text{where} \qquad \phi(z) = z^3 + 2z.$$

The function $\phi(z)$ is entire, and $\phi(i) = i \neq 0$. Hence f has a pole of order 3 at $z = i$. The residue is

$$b_1 = \frac{\phi''(i)}{2!} = 3i.$$

While the above theorem can be extremely useful, the identification of an isolated singular point as a pole of a certain order is sometimes done most efficiently by appealing directly to a Laurent series.

EXAMPLE 3. If, for instance, the residue of the function $f(z) = (\sinh z)/z^4$ is needed at the singularity $z = 0$, it would be incorrect to write $f(z) = \phi(z)/z^4$, where $\phi(z) = \sinh z$, and to attempt an application of formula (3) with $m = 4$. For it is necessary that $\phi(z_0) \neq 0$ if formula (3) is to be used. The simplest way to find the residue at $z = 0$ is to write out a few terms of the Laurent series for $f(z)$, as was done in Example 2 of Sec. 55. There it was shown that $z = 0$ is a pole of the *third* order and that the residue is $1/6$.

In some cases, the series approach can be effectively combined with the theorem.

EXAMPLE 4. Since $z(e^z - 1)$ is entire and its zeros are $z = 2n\pi i$ ($n = 0, \pm 1, \pm 2, \ldots$), the point $z = 0$ is clearly an isolated singular point of the function

$$f(z) = \frac{1}{z(e^z - 1)}.$$

From the Maclaurin series

$$e^z = 1 + \frac{z}{1!} + \frac{z^2}{2!} + \frac{z^3}{3!} + \cdots \qquad (|z| < \infty),$$

we see that

$$z(e^z - 1) = z\left(\frac{z}{1!} + \frac{z^2}{2!} + \frac{z^3}{3!} + \cdots\right) = z^2\left(1 + \frac{z}{2!} + \frac{z^2}{3!} + \cdots\right) \qquad (|z| < \infty).$$

Thus

(6) $$f(z) = \frac{\phi(z)}{z^2} \qquad \text{where} \qquad \phi(z) = \frac{1}{1 + z/2! + z^2/3! + \cdots}.$$

Since $\phi(z)$ is analytic at $z = 0$ and $\phi(0) = 1 \neq 0$, the point $z = 0$ is a pole of the *second* order; and, according to formula (3), the residue is $b_1 = \phi'(0)$. Because

$$\phi'(z) = \frac{-(1/2! + 2z/3! + \cdots)}{(1 + z/2! + z^2/3! + \cdots)^2}$$

in a neighborhood of the origin, then, $b_1 = -1/2$.

The residue can also be found by dividing the series representation for $z(e^z - 1)$ into 1, or by multiplying the Laurent series for $1/(e^z - 1)$ in Exercise 11(b), Sec. 51, by $1/z$.

57. ZEROS AND POLES OF ORDER m

Consider a function f that is analytic at a point z_0. We know from Sec. 40 that all of the derivatives $f^{(n)}(z)$ ($n = 1, 2, \ldots$) exist at z_0. If $f(z_0) = 0$ and there is a positive integer m such that $f^{(m)}(z_0) \neq 0$ and each derivative of lower order at z_0 vanishes, then f is said to have a *zero of order m* at z_0.

Lemma. *A function f that is analytic at a point z_0 has a zero of order m there if and only if there is a function g, which is analytic and nonzero at z_0, such that*

(1) $$f(z) = (z - z_0)^m g(z).$$

Both parts of the proof that follows use the fact (Sec. 44) that if a function is analytic at a point z_0, then it must have a valid Taylor series representation in powers of $z - z_0$ which is valid throughout a neighborhood $|z - z_0| < \varepsilon$ of that point.

We start the first part of the proof by assuming that expression (1) holds and noting that, since $g(z)$ is analytic at z_0, it has a Taylor series representation

$$g(z) = g(z_0) + \frac{g'(z_0)}{1!}(z - z_0) + \frac{g''(z_0)}{2!}(z - z_0)^2 + \cdots$$

in some neighborhood $|z - z_0| < \varepsilon$ of z_0. Expression (1) thus takes the form

$$f(z) = g(z_0)(z-z_0)^m + \frac{g'(z_0)}{1!}(z-z_0)^{m+1} + \frac{g''(z_0)}{2!}(z-z_0)^{m+2} + \cdots \qquad (|z-z_0| < \varepsilon).$$

Since this is actually a Taylor series expansion for $f(z)$, according to Theorem 1 in Sec. 50, it follows that

(2) $$f(z_0) = f'(z_0) = f''(z_0) = \cdots = f^{(m-1)}(z_0) = 0$$

and $f^{(m)}(z_0) = m!g(z_0) \neq 0$. Hence z_0 is a zero of order m of f.

Conversely, if we assume that f has a zero of order m at z_0, its analyticity at z_0 and the fact that conditions (2) hold tell us that, in some neighborhood $|z - z_0| < \varepsilon$, there is a Taylor series

$$f(z) = \sum_{n=m}^{\infty} \frac{f^{(n)}(z_0)}{n!}(z - z_0)^n$$

$$= (z - z_0)^m \left[\frac{f^{(m)}(z_0)}{m!} + \frac{f^{(m+1)}(z_0)}{(m+1)!}(z - z_0) + \frac{f^{(m+2)}(z_0)}{(m+2)!}(z - z_0)^2 + \cdots \right].$$

Consequently, $f(z)$ has the form (1), where

$$g(z) = \frac{f^{(m)}(z_0)}{m!} + \frac{f^{(m+1)}(z_0)}{(m+1)!}(z - z_0) + \frac{f^{(m+2)}(z_0)}{(m+2)!}(z - z_0)^2 + \cdots \qquad (|z - z_0| < \varepsilon).$$

The convergence of this last series when $|z - z_0| < \varepsilon$ ensures that g is analytic in that neighborhood and, in particular, at z_0 (Sec. 49). Moreover,

$$g(z_0) = \frac{f^{(m)}(z_0)}{m!} \neq 0.$$

This completes the proof of the lemma.

EXAMPLE 1.   The entire function $f(z) = z(e^z - 1)$ has a zero of order $m = 2$ at the point $z_0 = 0$ since

$$f(0) = f'(0) = 0 \qquad \text{and} \qquad f''(0) = 2 \neq 0.$$

The function g in the statement of the lemma is, in this case, defined by means of the equations

$$g(z) = \begin{cases} (e^z - 1)/z & \text{when } z \neq 0, \\ 1 & \text{when } z = 0. \end{cases}$$

It is analytic at $z = 0$ and, in fact, entire (see Exercise 5, Sec. 51).

Zeros of order m are a source of poles of order m, as the following theorem points out.

Theorem. *Let two functions p and q be analytic at a point z_0, and suppose that $p(z_0) \neq 0$. If q has a zero of order m at z_0, then the quotient $p(z)/q(z)$ has a pole of order m there.*

The proof is easy and depends on the above lemma as well as the theorem in Sec. 56. Assuming that z_0 is a zero of order m of q, we know that there is a deleted neighborhood of z_0 in which $q(z) \neq 0$ (see Exercise 13), and this means that z_0 is an *isolated* singular point of the quotient $p(z)/q(z)$. Also,

$$q(z) = (z - z_0)^m g(z),$$

where g is analytic and nonzero at z_0; and this enables us to write

(3) $$\frac{p(z)}{q(z)} = \frac{p(z)/g(z)}{(z - z_0)^m}.$$

Since $p(z)/g(z)$ is analytic and nonzero at z_0, it now follows that z_0 is a pole of order m of $p(z)/q(z)$.

EXAMPLE 2. In Example 1, we saw that the function $f(z) = z(e^z - 1)$ has a zero of order $m = 2$ at the point $z_0 = 0$. Consequently, the reciprocal

$$\frac{1}{f(z)} = \frac{1}{z(e^z - 1)}$$

has a pole of order 2 at that point. This was demonstrated in another way in Example 4, Sec. 56.

The above theorem leads us to the following useful method for identifying *simple* poles and finding the corresponding residues.

Corollary. *Let two functions p and q be analytic at a point z_0. If*

$$p(z_0) \neq 0, \qquad q(z_0) = 0, \qquad and \qquad q'(z_0) \neq 0,$$

then z_0 is a simple pole of the quotient $p(z)/q(z)$ and

(4) $$\operatorname*{Res}_{z=z_0} \frac{p(z)}{q(z)} = \frac{p(z_0)}{q'(z_0)}.$$

To show this, we assume that p and q are as stated and observe that, because of the conditions on q, the point z_0 is a zero of order $m = 1$ of that function. According to the lemma, then,

(5) $$q(z) = (z - z_0)g(z),$$

where $g(z)$ is analytic and nonzero at z_0. Furthermore, the theorem tells us that z_0 is a simple pole of $p(z)/q(z)$; and equation (3) in the proof of the theorem becomes

$$\frac{p(z)}{q(z)} = \frac{p(z)/g(z)}{z - z_0}.$$

Now $p(z)/g(z)$ is analytic and nonzero at z_0, and it follows from the theorem in Sec. 56 that

$$(6) \qquad\qquad \underset{z=z_0}{\text{Res}} \frac{p(z)}{q(z)} = \frac{p(z_0)}{g(z_0)}.$$

But $g(z_0) = q'(z_0)$, as is seen by differentiating each side of equation (5) and setting $z = z_0$. Expression (6) thus takes the form (4).

EXAMPLE 3. Consider the function

$$f(z) = \cot z = \frac{\cos z}{\sin z},$$

which is a quotient of the entire functions $p(z) = \cos z$ and $q(z) = \sin z$. The singularities of that quotient occur at the zeros of q, or at the points

$$z = n\pi \qquad (n = 0, \pm 1, \pm 2, \ldots).$$

Since

$$p(n\pi) = (-1)^n \neq 0, \qquad q(n\pi) = 0, \qquad \text{and} \qquad q'(n\pi) = (-1)^n \neq 0,$$

each singular point $z = n\pi$ of f is a simple pole, with residue

$$B_n = \frac{p(n\pi)}{q'(n\pi)} = \frac{(-1)^n}{(-1)^n} = 1.$$

EXAMPLE 4. Let us find the residue of the function

$$f(z) = \frac{z}{z^4 + 4}$$

at the isolated singular point

$$z_0 = \sqrt{2}e^{i\pi/4} = 1 + i$$

by writing $p(z) = z$ and $q(z) = z^4 + 4$. Since

$$p(z_0) = z_0 \neq 0, \qquad q(z_0) = 0, \qquad \text{and} \qquad q'(z_0) = 4z_0^3 \neq 0,$$

f has a simple pole at z_0. The corresponding residue is the number

$$B_0 = \frac{p(z_0)}{q'(z_0)} = \frac{z_0}{4z_0^3} = \frac{1}{4z_0^2} = \frac{1}{8i} = -\frac{i}{8}.$$

Although this residue could also be found by the method of Sec. 56, the computation would be somewhat more involved.

There are formulas similar to formula (4) for residues at poles of higher order, but they are lengthier and, in general, not practical.

EXERCISES

1. In each case, show that the singular points of the function are poles. Determine the order m of each pole, and find the corresponding residue B.

(a) $\dfrac{z^2 + 2}{z - 1}$;   (b) $\left(\dfrac{z}{2z + 1}\right)^3$;   (c) $\tanh z$;   (d) $\dfrac{\exp z}{z^2 + \pi^2}$.

Ans. (a) $m = 1, B = 3$;   (b) $m = 3, B = -3/16$;
(c) $m = 1, B = 1$;   (d) $m = 1, B = \pm i/2\pi$.

2. Show that the point $z = 0$ is a simple pole of the function $f(z) = \csc z$ and that the residue there is 1 by appealing to

(a) the Laurent series for $\csc z$ in Exercise 11(a), Sec. 51;

(b) the corollary in Sec. 57.

3. Show that

(a) $\operatorname*{Res}\limits_{z=-1} \dfrac{z^{1/4}}{z + 1} = \dfrac{1 + i}{\sqrt{2}}$ $(|z| > 0, 0 < \arg z < 2\pi)$;

(b) $\operatorname*{Res}\limits_{z=i} \dfrac{\operatorname{Log} z}{(z^2 + 1)^2} = \dfrac{\pi + 2i}{8}$;

(c) $\operatorname*{Res}\limits_{z=z_n} (z \sec z) = (-1)^{n+1} z_n$, where $z_n = \dfrac{\pi}{2} + n\pi$ $(n = 0, \pm 1, \pm 2, \ldots)$;

(d) $\operatorname*{Res}\limits_{z=\pi i} \dfrac{\exp(zt)}{\sinh z} + \operatorname*{Res}\limits_{z=-\pi i} \dfrac{\exp(zt)}{\sinh z} = -2 \cos \pi t$.

*4. Find the value of the integral

$$\int_C \frac{3z^3 + 2}{(z - 1)(z^2 + 9)}\, dz,$$

taken counterclockwise around the circle   (a) $|z - 2| = 2$;   (b) $|z| = 4$.
Ans. (a) πi;   (b) $6\pi i$.

5. Find the value of the integral

$$\int_C \frac{dz}{z^3(z + 4)},$$

taken counterclockwise around the circle   (a) $|z| = 2$;   (b) $|z + 2| = 3$.
Ans. (a) $\pi i/32$;   (b) 0.

6. Let C be the circle $|z| = 2$, described in the positive sense, and evaluate the integral

(a) $\displaystyle\int_C \tan z\, dz$;   (b) $\displaystyle\int_C \frac{dz}{\sinh 2z}$;   (c) $\displaystyle\int_C \frac{\cosh \pi z\, dz}{z(z^2 + 1)}$.

Ans. (a) $-4\pi i$;   (b) $-\pi i$;   (c) $4\pi i$.

7. Use Theorem 2 in Sec. 54, involving a single residue, to evaluate the integral of $f(z)$ around the positively oriented circle $|z| = 3$ when

(a) $f(z) = \dfrac{(3z + 2)^2}{z(z - 1)(2z + 5)}$;   (b) $f(z) = \dfrac{z^3(1 - 3z)}{(1 + z)(1 + 2z^4)}$;   (c) $f(z) = \dfrac{z^3 e^{1/z}}{1 + z^3}$.

Ans. (a) $9\pi i$;   (b) $-3\pi i$;   (c) $2\pi i$.

8. Let C_N denote the positively oriented boundary of the square whose edges lie along the lines

$$x = \pm\left(N + \frac{1}{2}\right)\pi \quad \text{and} \quad y = \pm\left(N + \frac{1}{2}\right)\pi,$$

where N is a positive integer. Show that

$$\int_{C_N} \frac{dz}{z^2 \sin z} = 2\pi i \left[\frac{1}{6} + 2\sum_{n=1}^{N} \frac{(-1)^n}{n^2 \pi^2} \right].$$

Then, using the fact that the value of this integral tends to zero as N tends to infinity (Exercise 17, Sec. 33), point out how it follows that

$$\sum_{n=1}^{\infty} \frac{(-1)^{n+1}}{n^2} = \frac{\pi^2}{12}.$$

9. Show that

$$\int_C \frac{dz}{(z^2 - 1)^2 + 3} = \frac{\pi}{2\sqrt{2}},$$

where C is the positively oriented boundary of the rectangle whose sides lie along the lines $x = \pm 2$, $y = 0$, and $y = 1$.

   Suggestion: By observing that the four zeros of the polynomial $q(z) = (z^2 - 1)^2 + 3$ are the square roots of the numbers $1 \pm \sqrt{3}i$, show that the reciprocal $1/q(z)$ is analytic inside and on C except at the points

$$z_0 = \frac{\sqrt{3} + i}{\sqrt{2}} \quad \text{and} \quad -\bar{z}_0 = \frac{-\sqrt{3} + i}{\sqrt{2}}.$$

Then apply the corollary in Sec. 57.

10. Consider the function

$$f(z) = \frac{1}{[q(z)]^2},$$

where q is analytic at z_0, $q(z_0) = 0$, and $q'(z_0) \neq 0$. Show that z_0 is a pole of order $m = 2$ of the function f, with residue

$$B_0 = -\frac{q''(z_0)}{[q'(z_0)]^3}.$$

   Suggestion: Note that z_0 is a zero of order $m = 1$ of the function q, so that equation (5), Sec. 57, holds. Then write

$$f(z) = \frac{\phi(z)}{(z - z_0)^2} \quad \text{where} \quad \phi(z) = \frac{1}{[g(z)]^2}.$$

The desired form of the residue $B_0 = \phi'(z_0)$ can be obtained by showing that $q'(z_0) = g(z_0)$ and $q''(z_0) = 2g'(z_0)$.

11. Apply the result in Exercise 10 to find the residue at $z = 0$ of the function

   (a) $f(z) = \csc^2 z$; (b) $f(z) = \dfrac{1}{(z + z^2)^2}$.

   Ans. (a) 0; (b) -2.

12. Let p and q denote functions that are analytic at a point z_0, where $p(z_0) \neq 0$ and $q(z_0) = 0$. Show that if the quotient $p(z)/q(z)$ has a pole of order m at z_0, then z_0 is a zero of order m of q. (Compare the theorem in Sec. 57.)

Suggestion: Note that the theorem in Sec. 56 enables one to write

$$\frac{p(z)}{q(z)} = \frac{\phi(z)}{(z - z_0)^m},$$

where $\phi(z)$ is analytic and nonzero at z_0. Then solve for $q(z)$.

13. Suppose that a function $f(z)$ is analytic and has a zero at a point z_0 but is not identically equal to zero in any neighborhood of z_0. Point out why z_0 must be a zero of some finite order m. Then use the lemma in Sec. 57 to show how it follows that $f(z) \neq 0$ throughout some deleted neighborhood $0 < |z - z_0| < \varepsilon$ of z_0. Thus show that z_0 is *isolated* (see Sec. 53) from other zeros of f.

    Suggestion: Recall from Sec. 14 that if a function is continuous and nonzero at a point z_0, then it is nonzero throughout some neighborhood of z_0.

14. Recall (Sec. 8) that a point z_0 is an accumulation point of a set S if each deleted neighborhood of z_0 contains at least one point of S. One form of the *Bolzano-Weierstrass theorem* can be stated as follows: *an infinite set of points lying in a closed bounded region R has at least one accumulation point in R.*[*] Use that theorem and the result in Exercise 13 to show that if a function f is analytic in the region R consisting of all points inside and on a simple closed contour C, except possibly for poles inside C, and if all the zeros of f in R are interior to C and of finite order, then those zeros must be finite in number.

15. Let R denote the region consisting of all points inside and on a simple closed contour C. Use the Bolzano-Weierstrass theorem (see Exercise 14) and the fact that poles are isolated singular points to show that if f is analytic in the region R except for poles interior to C, then those poles must be finite in number.

58. CONDITIONS UNDER WHICH $f(z) \equiv 0$

While this section is not concerned with poles or residues, it is a continuation of the discussion of zeros of analytic functions that was begun in Sec. 57. We present here some important results that are often used in other areas of the theory of functions of a complex variable. This section has, in fact, been referred to earlier in the book.

Lemma. *If $f(z) = 0$ at each point z of a domain or arc containing a point z_0, then $f(z) \equiv 0$ in any neighborhood N_0 of z_0 throughout which f is analytic. That is, $f(z) = 0$ at each point z in N_0.*

We begin the proof of this lemma with the observation that, under the stated conditions, $f(z)$ is identically equal to zero in some neighborhood N of z_0. For, otherwise, there would be a deleted neighborhood of z_0 throughout which $f(z) \neq 0$ (see Exercise 13, Sec. 57) and that would be inconsistent with the condition that $f(z) \equiv 0$ in a domain or on an arc containing z_0. Since $f(z) \equiv 0$ in the neighborhood N, then, it follows that all of the coefficients $a_n = f^{(n)}(z_0)/n!$ $(n = 0, 1, 2, \ldots)$ in the Taylor series for $f(z)$ about z_0 must be zero. Thus $f(z) \equiv 0$ in the neighborhood N_0, since that Taylor series also represents $f(z)$ in N_0.

---

[*] See, for example, A. E. Taylor and W. R. Mann, "Advanced Calculus," 3d ed., pp. 517 and 521, 1983.

Theorem. *If a function f is analytic throughout a domain D and* $f(z) = 0$ *at each point z of a domain or arc contained in D, then* $f(z) \equiv 0$ *in D.*

The method of proof here is similar to that used in proving the theorem in Sec. 42. We let z_0 be any point in the domain or arc at each point of which $f(z) = 0$; and we construct a polygonal line L extending from z_0 to any other point P in D. Also, d represents the shortest distance from points on L to the boundary of D. When D is the entire plane, d may have any positive value.

Now there exists a finite sequence of points $z_0, z_1, z_2, \ldots, z_n$ along L such that z_n coincides with the point P and

$$|z_k - z_{k-1}| < d \qquad (k = 1, 2, \ldots, n).$$

On forming a finite sequence of neighborhoods $N_0, N_1, N_2, \ldots, N_n$, where each N_k has center z_k and radius d, we note that f is analytic in each of those neighborhoods and that the center of each N_k $(k = 1, 2, \ldots, n)$ lies in the neighborhood N_{k-1} (see Fig. 46 in Sec. 42).

Since f is analytic in the neighborhood N_0 and because of the choice of the point z_0, our lemma tells us that $f(z) \equiv 0$ in N_0. But the point z_1 lies in the neighborhood, or domain, N_0. Hence a second application of the lemma reveals that $f(z) \equiv 0$ in N_1; and, by continuing in this manner, we arrive at the fact that $f(z) \equiv 0$ in N_n. Since N_n is centered at the point P and since P was arbitrarily selected in D, we may conclude that $f(z) \equiv 0$ in D. This completes the proof of the theorem.

Suppose now that two functions f and g are analytic in the same domain and that $f(z) = g(z)$ at each point z of some domain or arc contained in D. The function h defined by the equation $h(z) = f(z) - g(z)$ is also analytic in D, and $h(z) = 0$ throughout the subdomain or along the arc. According to our theorem, then, $h(z) = 0$ throughout D; that is, $f(z) = g(z)$ when z is in D. We thus arrive at the following corollary.

Corollary. *A function that is analytic in a domain D is uniquely determined over D by its values over a domain, or along an arc, contained in D.*

Note that the theorem and corollary here apply when the arcs in their statements are line segments. Often such a segment is the real axis or a portion of it (see Sec. 21).

EXAMPLE.   Since $\sin^2 x + \cos^2 x = 1$, the entire function

$$f(z) = \sin^2 z + \cos^2 z - 1$$

has zero values along the real axis. Consequently, according to the theorem above, $f(z) = 0$ throughout the complex plane; and this means that

$$\sin^2 z + \cos^2 z = 1$$

for all z. Note, too, how the corollary tells us that $\sin z$ and $\cos z$ are the only entire functions that can assume the values $\sin x$ and $\cos x$, respectively, along the real axis or any segment of it.

59.   BEHAVIOR OF f NEAR REMOVABLE AND ESSENTIAL SINGULAR POINTS

Although our interest in a function f at an isolated singular point mainly concerns poles, we conclude this chapter with proofs of two theorems regarding the behavior of f near removable and essential singular points. Since these theorems will not be used elsewhere in the book, the reader can skip them without loss of continuity.

We preface the first theorem with the observation that *a function f is always analytic and bounded in some deleted neighborhood* $0 < |z - z_0| < \varepsilon$ *of a removable singularity* z_0. For, as noted in Sec. 55, f is analytic in a disk $|z - z_0| < R_2$ when $f(z_0)$ is properly defined at such a point; and f is then continuous in any closed disk $|z - z_0| \leq \varepsilon$ when $\varepsilon < R_2$. Consequently, f is bounded in that disk, according to Sec. 14, and this means that f must be bounded in the deleted neighborhood $0 < |z - z_0| < \varepsilon$. The following theorem, which is known as *Riemann's theorem*, is related to the above observation.

Theorem 1. *Suppose that a function f is analytic and bounded in some deleted neighborhood* $0 < |z - z_0| < \varepsilon$ *of a point* z_0. *If f is not analytic at z_0, then it has a removable singularity there.*

To prove this, we assume that f is not analytic at z_0. The point z_0 is, then, an isolated singularity of f; and $f(z)$ is represented by a Laurent series

$$(1) \qquad f(z) = \sum_{n=0}^{\infty} a_n(z - z_0)^n + \sum_{n=1}^{\infty} \frac{b_n}{(z - z_0)^n}$$

throughout the deleted neighborhood $0 < |z - z_0| < \varepsilon$. If C denotes a positively oriented circle $|z - z_0| = \rho$, where $\rho < \varepsilon$ (Fig. 61), we know from Sec. 46 that the coefficients b_n in expansion (1) can be written

$$(2) \qquad b_n = \frac{1}{2\pi i} \int_C \frac{f(z)\, dz}{(z - z_0)^{-n+1}} \qquad (n = 1, 2, \ldots).$$

Now the boundedness condition on f tells us that there is a positive constant M such that $|f(z)| \leq M$ whenever $0 < |z - z_0| < \varepsilon$. Hence it follows from expression (2)

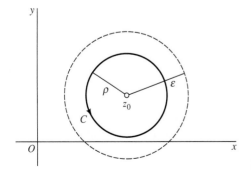

FIGURE 61

that

$$|b_n| \leq \frac{1}{2\pi} \cdot \frac{M}{\rho^{-n+1}} 2\pi\rho = M\rho^n \qquad (n = 1, 2, \ldots).$$

Since the coefficients b_n are constants and since ρ can be chosen arbitrarily small, we may conclude that $b_n = 0$ $(n = 1, 2, \ldots)$ in the Laurent series (1). This tells us that z_0 is a removable singularity of f, and the proof of the theorem is complete.

As already pointed out in Sec. 55, the behavior of a function near an essential singular point is quite irregular. The second theorem to be proved here is related to Picard's theorem in Sec. 55 and is usually referred to as the *Casorati-Weierstrass theorem*. It states that, in each deleted neighborhood of an essential singular point, a function assumes values arbitrarily close to any given number.

Theorem 2. *Suppose that z_0 is an essential singularity of a function f, and let w_0 be any complex number. Then, for any positive number ε, the inequality*

$$(3) \qquad\qquad |f(z) - w_0| < \varepsilon$$

is satisfied at some point z in each deleted neighborhood $0 < |z - z_0| < \delta$ of z_0 (Fig. 62).

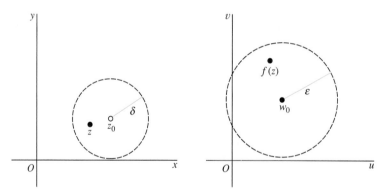

FIGURE 62

The proof is by contradiction. Since z_0 is an isolated singularity of f, there is a deleted neighborhood $0 < |z - z_0| < \delta$ throughout which f is analytic; and we assume that condition (3) is *not* satisfied for any point z there. Thus $|f(z) - w_0| \geq \varepsilon$ when $0 < |z - z_0| < \delta$, and the function

$$(4) \qquad\qquad g(z) = \frac{1}{f(z) - w_0} \qquad (0 < |z - z_0| < \delta)$$

is bounded and analytic in its domain of definition. According to Theorem 1, then, z_0 is a removable singularity of g; and we let g be defined at z_0 so that it is analytic there.

If $g(z_0) \neq 0$, the function $f(z)$, which can be written

(5)
$$f(z) = \frac{1}{g(z)} + w_0$$

when $0 < |z - z_0| < \delta$, becomes analytic at z_0 if it is defined there as

$$f(z_0) = \frac{1}{g(z_0)} + w_0.$$

But this means that z_0 is a removable singularity of f, not an essential one, and we have a contradiction.

If $g(z_0) = 0$, the function g must have a zero of some finite order m (Sec. 57) at z_0 because g is not identically equal to zero in the neighborhood $|z - z_0| < \delta$. In view of equation (5), then, f has a pole of order m at z_0 (see the theorem in Sec. 57). So, once again, we have a contradiction; and Theorem 2 here is proved.

EXERCISES

1. Recall that the hyperbolic sine and hyperbolic cosine functions, the exponential function, and the sine and cosine functions are all entire, and then use the theorem in Sec. 58 to obtain each of these identities for all complex z from the corresponding identities when z is real:

   (a) $\sinh z + \cosh z = e^z$;   (b) $\sin 2z = 2 \sin z \cos z$;

   (c) $\cosh^2 z - \sinh^2 z = 1$;   (d) $\sin\left(\frac{\pi}{2} - z\right) = \cos z$.

2. Use the corollary in Sec. 58 to show that if a function $f(z)$ is analytic and not constant throughout a domain D, then it is not constant throughout any neighborhood in D.

   Suggestion: Suppose that $f(z)$ does have a constant value w_0 throughout some neighborhood in D.

3. Show that if a function f is analytic in a deleted neighborhood of a point z_0 and if z_0 is an accumulation point (Sec. 8) of zeros of f, then either z_0 is an essential singular point of f or else $f(z)$ is identically equal to zero.

   Suggestion: Recall Exercise 13, Sec. 57.

4. Examine the set of zeros of the function $z^2 \sin(1/z)$, and apply the result obtained in Exercise 3 to show that the origin is an essential singular point of the function. Note that this conclusion also follows from the nature of the Laurent series that represents the function in the domain $0 < |z| < \infty$.

CHAPTER

7

APPLICATIONS
OF RESIDUES

We turn now to some important applications of the theory of residues, which was developed in the preceding chapter. The applications include evaluation of certain types of definite and improper integrals occurring in *real* analysis and applied mathematics. Considerable attention is also given to finding inverse Laplace transforms by the method of summing residues.

60. EVALUATION OF IMPROPER INTEGRALS

In calculus, the improper integral of a continuous function $f(x)$ over the semi-infinite interval $x \geq 0$ is defined by means of the equation

$$(1) \qquad \int_0^\infty f(x)\, dx = \lim_{R \to \infty} \int_0^R f(x)\, dx.$$

When the limit on the right exists, the improper integral is said to *converge* to that limit. If $f(x)$ is continuous for *all* x, its improper integral over the infinite interval $-\infty < x < \infty$ is defined by writing

$$(2) \qquad \int_{-\infty}^\infty f(x)\, dx = \lim_{R_1 \to \infty} \int_{-R_1}^0 f(x)\, dx + \lim_{R_2 \to \infty} \int_0^{R_2} f(x)\, dx;$$

and when both of the limits here exist, integral (2) converges to their sum. Another value that is assigned to integral (2) is often useful. Namely, the *Cauchy principal value* (P.V.) of integral (2) is the number

(3) $$\text{P.V.} \int_{-\infty}^{\infty} f(x)\, dx = \lim_{R \to \infty} \int_{-R}^{R} f(x)\, dx,$$

provided this single limit exists.

If integral (2) converges, its Cauchy principal value (3) exists; and that value is the number to which integral (2) converges. This is because

$$\int_{-R}^{R} f(x)\, dx = \int_{-R}^{0} f(x)\, dx + \int_{0}^{R} f(x)\, dx$$

and the limit as $R \to \infty$ of each of the integrals on the right exists when integral (2) converges. It is *not,* however, always true that integral (2) converges when its Cauchy principal value exists (see Exercise 8).

But suppose that $f(x)\,(-\infty < x < \infty)$ is an *even* function, one where $f(-x) = f(x)$ for each x. The symmetry of the graph of $y = f(x)$ with respect to the y axis enables us to write

$$\int_{0}^{R} f(x)\, dx = \frac{1}{2} \int_{-R}^{R} f(x)\, dx,$$

and we see that integral (1) converges to one half the Cauchy principal value (3) when that value exists. Moreover, since integral (1) converges and since

$$\int_{-R_1}^{0} f(x)\, dx = \int_{0}^{R_1} f(x)\, dx,$$

integral (2) converges to twice the value of integral (1). We have thus shown that *when $f(x)\,(-\infty < x < \infty)$ is even and the Cauchy principal value (3) exists, both of the integrals (1) and (2) converge and*

(4) $$2 \int_{0}^{\infty} f(x)\, dx = \int_{-\infty}^{\infty} f(x)\, dx = \text{P.V.} \int_{-\infty}^{\infty} f(x)\, dx.$$

The following example illustrates a method, involving residues, that is often used to evaluate improper integrals of rational functions

$$f(x) = \frac{p(x)}{q(x)},$$

where $p(x)$ and $q(x)$ are polynomials with real coefficients and no factors in common and where $q(x)$ has no real zeros.

EXAMPLE. Let us show that the integral

$$\int_{0}^{\infty} \frac{x^2}{x^6 + 1}\, dx$$

converges and find its value. We start with the observation that the function

$$f(z) = \frac{z^2}{z^6 + 1}$$

has isolated singularities at the sixth roots of -1 and is analytic everywhere else. Those roots are

$$c_k = \exp\left[i\left(\frac{\pi}{6} + \frac{2k\pi}{6}\right)\right] \qquad (k = 0, 1, 2, \ldots, 5),$$

and none of them lies on the real axis. The first three roots,

$$c_0 = e^{i\pi/6}, \qquad c_1 = i, \qquad \text{and} \qquad c_2 = e^{i5\pi/6},$$

lie in the upper half plane (Fig. 63), and the other three lie in the lower one. When $R > 1$, the singular points c_k $(k = 0, 1, 2)$ of f lie in the interior of the semicircular region bounded by the segment $z = x$ $(-R \le x \le R)$ of the real axis and the upper half C_R of the circle $|z| = R$ from $z = R$ to $z = -R$. Integrating $f(z)$ counterclockwise around the boundary of this semicircular region, we see that

(5)
$$\int_{-R}^{R} f(x)\,dx + \int_{C_R} f(z)\,dz = 2\pi i(B_0 + B_1 + B_2),$$

where B_k is the residue of $f(z)$ at c_k $(k = 0, 1, 2)$.

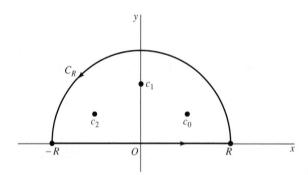

FIGURE 63

With the aid of the corollary in Sec. 57, we find that the points c_k are simple poles of f and that

$$B_k = \operatorname*{Res}_{z=c_k} \frac{z^2}{z^6 + 1} = \frac{c_k^2}{6c_k^5} = \frac{1}{6c_k^3} \qquad (k = 0, 1, 2).$$

Thus

$$2\pi i(B_0 + B_1 + B_2) = 2\pi i\left(\frac{1}{6i} - \frac{1}{6i} + \frac{1}{6i}\right) = \frac{\pi}{3};$$

and equation (5) can be put in the form

(6)
$$\int_{-R}^{R} f(x)\,dx = \frac{\pi}{3} - \int_{C_R} f(z)\,dz,$$

which is valid for all values of R greater than 1.

Next, we show that the value of the integral on the right in equation (6) tends to 0 as R tends to ∞. To do this, we observe that when $|z| = R$,

$$|z^2| = |z|^2 = R^2$$

and

$$|z^6 + 1| \geq ||z|^6 - 1| = R^6 - 1.$$

So, if z is any point on C_R,

$$|f(z)| = \frac{|z^2|}{|z^6 + 1|} \leq M_R \qquad \text{where} \qquad M_R = \frac{R^2}{R^6 - 1};$$

and this means that

(7)
$$\left| \int_{C_R} f(z)\,dz \right| \leq M_R \pi R,$$

πR being the length of the semicircle C_R. Since the number

$$M_R \pi R = \frac{\pi R^3}{R^6 - 1}$$

is a quotient of polynomials in R and since the degree of the numerator is less than the degree of the denominator, that quotient must tend to zero as R tends to ∞. More precisely, if we divide both numerator and denominator by R^6 and write

$$M_R \pi R = \frac{\dfrac{\pi}{R^3}}{1 - \dfrac{1}{R^6}},$$

it is evident that $M_R \pi R$ tends to zero. Consequently, in view of inequality (7),

$$\lim_{R \to \infty} \int_{C_R} f(z)\,dz = 0.$$

It now follows from equation (6) that

$$\lim_{R \to \infty} \int_{-R}^{R} \frac{x^2}{x^6 + 1}\,dx = \frac{\pi}{3},$$

or

$$\text{P.V.} \int_{-\infty}^{\infty} \frac{x^2}{x^6 + 1}\,dx = \frac{\pi}{3}.$$

Since the integrand here is even, we know from equations (4) and the statement in italics just prior to it that

(8)
$$\int_0^\infty \frac{x^2}{x^6 + 1}\, dx = \frac{\pi}{6}.$$

EXERCISES

Use residues to evaluate the improper integrals in Exercises 1 through 5.

1. $\displaystyle\int_0^\infty \frac{dx}{x^2 + 1}.$

   Ans. $\pi/2$.

2. $\displaystyle\int_0^\infty \frac{dx}{(x^2 + 1)^2}.$

   Ans. $\pi/4$.

3. $\displaystyle\int_0^\infty \frac{dx}{x^4 + 1}.$

   Ans. $\pi/(2\sqrt{2})$.

4. $\displaystyle\int_0^\infty \frac{x^2\, dx}{(x^2 + 1)(x^2 + 4)}.$

   Ans. $\pi/6$.

5. $\displaystyle\int_0^\infty \frac{x^2\, dx}{(x^2 + 9)(x^2 + 4)^2}.$

   Ans. $\pi/200$.

Use residues to find the Cauchy principal values of the integrals in Exercises 6 and 7.

6. $\displaystyle\int_{-\infty}^\infty \frac{dx}{x^2 + 2x + 2}.$

7. $\displaystyle\int_{-\infty}^\infty \frac{x\, dx}{(x^2 + 1)(x^2 + 2x + 2)}.$

   Ans. $-\pi/5$.

8. Show that the improper integral of $f(x) = x$ over the interval $-\infty < x < \infty$ does not converge but that the Cauchy principal value of the integral nonetheless exists.

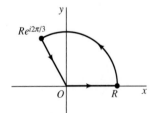

FIGURE 64

9. Use residues and the contour shown in Fig. 64, where $R > 1$, to establish the integration formula

$$\int_0^\infty \frac{dx}{x^3 + 1} = \frac{2\pi}{3\sqrt{3}}.$$

10. The integration formula

$$\int_0^\infty \frac{dx}{[(x^2 - a)^2 + 1]^2} = \frac{\pi}{8\sqrt{2A^3}}\left[(2a^2 + 3)\sqrt{A + a} + a\sqrt{A - a}\right],$$

where a is any real number and $A = \sqrt{a^2 + 1}$, arises in the theory of case-hardening of steel by means of radio-frequency heating.* Follow the steps below to derive it.

(a) Point out why the four zeros of the polynomial

$$q(z) = (z^2 - a)^2 + 1$$

are the square roots of the numbers $a \pm i$. Then, using the fact that the numbers

$$z_0 = \frac{1}{\sqrt{2}}\left(\sqrt{A + a} + i\sqrt{A - a}\right)$$

and $-z_0$ are the square roots of $a + i$ (Exercise 4, Sec. 7), verify that $\pm\overline{z}_0$ are the square roots of $a - i$ and hence that z_0 and $-\overline{z}_0$ are the only zeros of $q(z)$ in the upper half plane $\mathrm{Im}\, z \geq 0$.

(b) Using the method derived in Exercise 10, Sec. 57, and keeping in mind that $z_0^2 = a+i$ for purposes of simplification, show that the point z_0 in part (a) is a pole of order 2 of the function $f(z) = 1/[q(z)]^2$ and that the residue B_1 at z_0 can be written

$$B_1 = -\frac{q''(z_0)}{[q'(z_0)]^3} = \frac{a - i(2a^2 + 3)}{16A^2 z_0}.$$

After observing that $q'(-\overline{z}) = -\overline{q'(z)}$ and $q''(-\overline{z}) = \overline{q''(z)}$, use the same method to show that the point $-\overline{z}_0$ in part (a) is also a pole of order 2 of the function $f(z)$, with residue

$$B_2 = \overline{\left\{\frac{q''(z_0)}{[q'(z_0)]^3}\right\}} = -\overline{B}_1.$$

Then obtain the expression

$$B_1 + B_2 = \frac{1}{8A^2 i}\,\mathrm{Im}\left[\frac{-a + i(2a^2 + 3)}{z_0}\right]$$

for the sum of these residues.

(c) Refer to part (a) and show that $|q(z)| \geq (R - |z_0|)^4$ if $|z| = R$, where $R > |z_0|$. Then, with the aid of the final result in part (b), complete the derivation of the integration formula.

---

*See pp. 359–364 of the book by Brown, Hoyler, and Bierwirth that is listed in Appendix 1.

11. Let m and n be integers, where $0 \leq m < n$. Follow the steps below to derive the integration formula

$$\int_0^\infty \frac{x^{2m}}{x^{2n} + 1}\, dx = \frac{\pi}{2n} \csc\left(\frac{2m + 1}{2n}\pi\right).$$

(a) Show that the zeros of the polynomial $z^{2n} + 1$ lying above the real axis are

$$c_k = \exp\left[i\frac{(2k + 1)\pi}{2n}\right] \qquad (k = 0, 1, 2, \ldots, n - 1)$$

and that there are none on that axis.

(b) With the aid of the corollary in Sec. 57, show that

$$\operatorname*{Res}_{z=c_k} \frac{z^{2m}}{z^{2n} + 1} = -\frac{1}{2n} e^{i(2k+1)\alpha} \qquad (k = 0, 1, 2, \ldots, n - 1),$$

where c_k are the zeros found in part (a) and

$$\alpha = \frac{2m + 1}{2n}\pi.$$

Then use the identity (see Exercise 13, Sec. 6)

$$\sum_{k=0}^{n-1} z^k = \frac{1 - z^n}{1 - z} \qquad (z \neq 1)$$

to obtain the expression

$$2\pi i \sum_{k=0}^{n-1} \operatorname*{Res}_{z=c_k} \frac{z^{2m}}{z^{2n} + 1} = \frac{\pi}{n \sin \alpha}.$$

(c) Use the final result in part (b) to complete the derivation of the integration formula.

61. IMPROPER INTEGRALS INVOLVING SINES AND COSINES

Residue theory can be useful in evaluating convergent improper integrals of the form

(1) $$\int_{-\infty}^{\infty} f(x) \sin ax\, dx \qquad \text{or} \qquad \int_{-\infty}^{\infty} f(x) \cos ax\, dx,$$

where a denotes a positive constant. As in Sec. 60, we assume that $f(x) = p(x)/q(x)$, where $p(x)$ and $q(x)$ are polynomials with real coefficients and no factors in common. Also, $q(x)$ has no real zeros. The method described in Sec. 60 cannot be applied directly here since $|\sin az|$ and $|\cos az|$ increase like $\sinh ay$, or e^{ay}, as y tends to infinity (Sec. 24). The modification illustrated below is suggested by the fact that

$$\int_{-R}^{R} f(x) \cos ax\, dx + i \int_{-R}^{R} f(x) \sin ax\, dx = \int_{-R}^{R} f(x) e^{iax}\, dx,$$

together with the fact that the modulus e^{-ay} of e^{iaz} is bounded in the upper half plane $y \geq 0$.

EXAMPLE 1.   Let us show that

(2)
$$\int_{-\infty}^{\infty} \frac{\cos 3x}{(x^2 + 1)^2} \, dx = \frac{2\pi}{e^3}.$$

Because the integrand is even, it is sufficient to show that the Cauchy principal value of the integral exists and to find that value.

We introduce the function

(3)
$$f(z) = \frac{1}{(z^2 + 1)^2}$$

and observe that the product $f(z)e^{i3z}$ is analytic everywhere on and above the real axis except at the point $z = i$. The singularity $z = i$ lies in the interior of the semi-circular region whose boundary consists of the segment $-R \le x \le R$ of the real axis and the upper half C_R of the circle $|z| = R$ $(R > 1)$ from $z = R$ to $z = -R$. Integration of $f(z)e^{i3z}$ around that boundary yields the equation

(4)
$$\int_{-R}^{R} \frac{e^{i3x}}{(x^2 + 1)^2} \, dx = 2\pi i B_1 - \int_{C_R} f(z)e^{i3z} \, dz,$$

where

$$B_1 = \operatorname*{Res}_{z=i} [f(z)e^{i3z}].$$

Since

$$f(z)e^{i3z} = \frac{\phi(z)}{(z - i)^2} \qquad \text{where} \qquad \phi(z) = \frac{e^{i3z}}{(z + i)^2},$$

the point $z = i$ is evidently a pole of order 2 of $f(z)e^{i3z}$; and

$$B_1 = \phi'(i) = \frac{1}{ie^3}.$$

By equating the real parts on each side of equation (4), then, we find that

(5)
$$\int_{-R}^{R} \frac{\cos 3x}{(x^2 + 1)^2} \, dx = \frac{2\pi}{e^3} - \operatorname{Re} \int_{C_R} f(z)e^{i3z} \, dz.$$

Finally, we observe that when z is a point on C_R,

$$|f(z)| \le M_R \qquad \text{where} \qquad M_R = \frac{1}{(R^2 - 1)^2}$$

and that $|e^{i3z}| = e^{-3y} \le 1$ for such a point. Consequently,

(6)
$$\left| \operatorname{Re} \int_{C_R} f(z)e^{i3z} \, dz \right| \le \left| \int_{C_R} f(z)e^{i3z} \, dz \right| \le M_R \pi R.$$

Since the quantity

$$M_R \pi R = \frac{\pi R}{(R^2 - 1)^2}$$

tends to 0 as R tends to ∞ and because of inequalities (6), we need only let R tend to ∞ in equation (5) to arrive at the desired result (2).

In evaluating integrals of type (1), it is sometimes necessary to use a result based on *Jordan's inequality*:

$$(7) \qquad\qquad \int_0^{\pi} e^{-R\sin\theta}\,d\theta < \frac{\pi}{R} \qquad (R > 0).$$

To verify this inequality, we first note from the graphs of the functions $y = \sin\theta$ and $y = 2\theta/\pi$ when $0 \le \theta \le \pi/2$ (Fig. 65) that $\sin\theta \ge 2\theta/\pi$ for all values of θ in that interval. Consequently, if $R > 0$,

$$e^{-R\sin\theta} \le e^{-2R\theta/\pi} \qquad \text{when} \qquad 0 \le \theta \le \frac{\pi}{2};$$

and so

$$\int_0^{\pi/2} e^{-R\sin\theta}\,d\theta \le \int_0^{\pi/2} e^{-2R\theta/\pi}\,d\theta = \frac{\pi}{2R}(1 - e^{-R}).$$

Hence

$$(8) \qquad\qquad \int_0^{\pi/2} e^{-R\sin\theta}\,d\theta < \frac{\pi}{2R} \qquad (R > 0).$$

But this is just another form of inequality (7), since the graph of $y = \sin\theta$ is symmetric with respect to the vertical line $\theta = \pi/2$ on the interval $0 \le \theta \le \pi$.

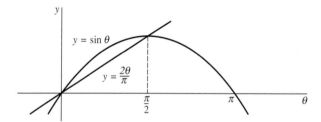

FIGURE 65

Suppose now that a function $f(z)$ is analytic at all points in the upper half plane $y \ge 0$ that lie above a semicircle $z = R_0 e^{i\theta} (0 \le \theta \le \pi)$; and let C_R denote any semicircle $z = R e^{i\theta} (0 \le \theta \le \pi)$, where $R > R_0$. Jordan's inequality (7) enables us to show that *if, for all points z on C_R, there is a positive constant M_R such that $|f(z)| \le M_R$, where M_R tends to zero as R tends to infinity, then*

$$(9) \qquad\qquad \lim_{R\to\infty} \int_{C_R} f(z)e^{iaz}\,dz = 0 \qquad (a > 0).$$

Our verification of limit (9), which is known as *Jordan's lemma,* begins with the fact that

$$\int_{C_R} f(z)e^{iaz}\,dz = \int_0^\pi f(Re^{i\theta})\exp(iaRe^{i\theta})iRe^{i\theta}\,d\theta.$$

Since

$$|f(Re^{i\theta})| \le M_R \qquad \text{and} \qquad |\exp(iaRe^{i\theta})| \le e^{-aR\sin\theta}$$

and in view of Jordan's inequality (7), it follows that

$$\left|\int_{C_R} f(z)e^{iaz}\,dz\right| \le M_R R \int_0^\pi e^{-aR\sin\theta}\,d\theta < \frac{M_R\pi}{a}.$$

Limit (9) is then evident, since $M_R \to 0$ as $R \to \infty$.

EXAMPLE 2.   Let us find the Cauchy principal value of the integral

$$\int_{-\infty}^\infty \frac{x\sin x\,dx}{x^2 + 2x + 2}.$$

As in Example 1, the existence of the value in question will be established by our actually finding it.

We write

$$f(z) = \frac{z}{z^2 + 2z + 2} = \frac{z}{(z - z_1)(z - \bar{z}_1)},$$

where $z_1 = -1 + i$. The point z_1, which lies above the x axis, is a simple pole of the function $f(z)e^{iz}$, with residue

$$(10) \qquad\qquad B_1 = \frac{z_1 e^{iz_1}}{z_1 - \bar{z}_1}.$$

Hence, when $R > \sqrt{2}$ and C_R denotes the upper half of the positively oriented circle $|z| = R$,

$$\int_{-R}^R \frac{xe^{ix}\,dx}{x^2 + 2x + 2} = 2\pi i B_1 - \int_{C_R} f(z)e^{iz}\,dz;$$

and this means that

$$(11) \qquad \int_{-R}^R \frac{x\sin x\,dx}{x^2 + 2x + 2} = \text{Im}(2\pi i B_1) - \text{Im}\int_{C_R} f(z)e^{iz}\,dz.$$

Now

$$(12) \qquad\qquad \left|\text{Im}\int_{C_R} f(z)e^{iz}\,dz\right| \le \left|\int_{C_R} f(z)e^{iz}\,dz\right|;$$

and we note that, when z is a point on C_R,

$$|f(z)| \le M_R \qquad \text{where} \qquad M_R = \frac{R}{\left(R - \sqrt{2}\right)^2}$$

and that $|e^{iz}| = e^{-y} \le 1$ for such a point. By proceeding as we did in Example 1, we *cannot* conclude that the right-hand side of inequality (12), and hence the left-hand side, tends to zero as R tends to infinity. For the quantity

$$M_R \pi R = \frac{\pi R^2}{\left(R - \sqrt{2}\right)^2}$$

does not tend to zero. Limit (9) does, however, provide the desired result.

So it does, indeed, follow from inequality (12) that the left-hand side there tends to zero as R tends to infinity. Consequently, equation (11), together with expression (10) for the residue B_1, tells us that

(13) $$\text{P.V.} \int_{-\infty}^{\infty} \frac{x \sin x \, dx}{x^2 + 2x + 2} = \text{Im}(2\pi i B_1) = \frac{\pi}{e}(\sin 1 + \cos 1).$$

EXERCISES

Use residues to evaluate the improper integrals in Exercises 1 through 8.

1. $\displaystyle\int_{-\infty}^{\infty} \frac{\cos x \, dx}{(x^2 + a^2)(x^2 + b^2)} \ (a > b > 0).$

   Ans. $\dfrac{\pi}{a^2 - b^2} \left(\dfrac{e^{-b}}{b} - \dfrac{e^{-a}}{a} \right).$

2. $\displaystyle\int_{0}^{\infty} \frac{\cos ax}{x^2 + 1} \, dx \ (a > 0).$

   Ans. $\dfrac{\pi}{2} e^{-a}.$

3. $\displaystyle\int_{0}^{\infty} \frac{\cos ax}{(x^2 + b^2)^2} \, dx \ (a > 0, b > 0).$

   Ans. $\dfrac{\pi}{4b^3}(1 + ab)e^{-ab}.$

4. $\displaystyle\int_{0}^{\infty} \frac{x \sin 2x}{x^2 + 3} \, dx.$

   Ans. $\dfrac{\pi}{2} \exp(-2\sqrt{3}).$

5. $\displaystyle\int_{-\infty}^{\infty} \frac{x \sin ax}{x^4 + 4} \, dx \ (a > 0).$

   Ans. $\dfrac{\pi}{2} e^{-a} \sin a.$

6. $\displaystyle\int_{-\infty}^{\infty} \frac{x^3 \sin ax}{x^4 + 4}\, dx\ (a > 0).$

Ans. $\pi e^{-a} \cos a.$

7. $\displaystyle\int_{-\infty}^{\infty} \frac{x \sin x\, dx}{(x^2 + 1)(x^2 + 4)}.$

8. $\displaystyle\int_{0}^{\infty} \frac{x^3 \sin x\, dx}{(x^2 + 1)(x^2 + 9)}.$

Use residues to find the Cauchy principal values of the integrals in Exercises 9 through 11.

9. $\displaystyle\int_{-\infty}^{\infty} \frac{\sin x\, dx}{x^2 + 4x + 5}.$

Ans. $-\dfrac{\pi}{e} \sin 2.$

10. $\displaystyle\int_{-\infty}^{\infty} \frac{(x + 1)\cos x}{x^2 + 4x + 5}\, dx.$

Ans. $\dfrac{\pi}{e}(\sin 2 - \cos 2).$

11. $\displaystyle\int_{-\infty}^{\infty} \frac{\cos x\, dx}{(x + a)^2 + b^2}\ (b > 0).$

12. Follow the steps below to evaluate the *Fresnel integrals,* which are important in diffraction theory:

$$\int_0^{\infty} \cos(x^2)\, dx = \int_0^{\infty} \sin(x^2)\, dx = \frac{1}{2}\sqrt{\frac{\pi}{2}}.$$

(a) By integrating the function $\exp(iz^2)$ around the positively oriented boundary of the sector $0 \le r \le R, 0 \le \theta \le \pi/4$ (Fig. 66) and appealing to the Cauchy-Goursat theorem, show that

$$\int_0^{R} \cos(x^2)\, dx = \frac{1}{\sqrt{2}} \int_0^{R} e^{-r^2}\, dr - \mathrm{Re} \int_{C_R} e^{iz^2}\, dz$$

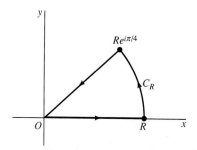

FIGURE 66

and

$$\int_0^R \sin(x^2)\,dx = \frac{1}{\sqrt{2}}\int_0^R e^{-r^2}\,dr - \operatorname{Im}\int_{C_R} e^{iz^2}\,dz,$$

where C_R is the arc $z = Re^{i\theta}\,(0 \le \theta \le \pi/4)$.

(b) Show that the value of the integral along the arc C_R in part (a) tends to zero as R tends to infinity by obtaining the inequality

$$\left|\int_{C_R} e^{iz^2}\,dz\right| \le \frac{R}{2}\int_0^{\pi/2} e^{-R^2 \sin\phi}\,d\phi$$

and then referring to the form (8), Sec. 61, of Jordan's inequality.

(c) Use the results in parts (a) and (b), together with the known integration formula*

$$\int_0^\infty e^{-x^2}\,dx = \frac{\sqrt{\pi}}{2},$$

to complete the exercise.

13. Use the method described below to derive the integration formula

$$\int_0^\infty e^{-x^2}\cos(2bx)\,dx = \frac{\sqrt{\pi}}{2}e^{-b^2} \qquad (b > 0).$$

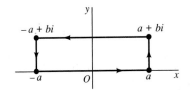

FIGURE 67

(a) Show that the sum of the integrals of $\exp(-z^2)$ along the lower and upper horizontal legs of the rectangular path in Fig. 67 can be written

$$2\int_0^a e^{-x^2}\,dx - 2e^{b^2}\int_0^a e^{-x^2}\cos(2bx)\,dx$$

and that the sum of the integrals along the vertical legs on the right and left can be written

$$ie^{-a^2}\int_0^b e^{y^2}e^{-i2ay}\,dy - ie^{-a^2}\int_0^b e^{y^2}e^{i2ay}\,dy.$$

---

*The usual way to evaluate this integral is by writing its square as

$$\int_0^\infty e^{-x^2}\,dx \int_0^\infty e^{-y^2}\,dy = \int_0^\infty \int_0^\infty e^{-(x^2+y^2)}\,dx\,dy$$

and then evaluating the iterated integral by changing to polar coordinates. Details are given in, for example, A. E. Taylor and W. R. Mann, "Advanced Calculus," 3d ed., pp. 680–681, 1983.

Thus, with the aid of the Cauchy-Goursat theorem, show that

$$\int_0^a e^{-x^2} \cos(2bx)\, dx = e^{-b^2} \int_0^a e^{-x^2}\, dx + e^{-(a^2+b^2)} \int_0^b e^{y^2} \sin(2ay)\, dy.$$

(b) Given that (see the footnote to Exercise 12)

$$\int_0^\infty e^{-x^2}\, dx = \frac{\sqrt{\pi}}{2},$$

show that the desired integration formula follows by letting a tend to infinity in the equation obtained at the end of part (a).

62. DEFINITE INTEGRALS INVOLVING SINES AND COSINES

The method of residues is also useful in evaluating certain definite integrals of the type

(1)
$$\int_0^{2\pi} F(\sin\theta, \cos\theta)d\theta.$$

The fact that θ varies from 0 to 2π suggests that we consider θ as an argument of a point z on the unit circle C centered at the origin; hence we write $z = e^{i\theta}\,(0 \le \theta \le 2\pi)$. When we make this substitution using the equations

(2)
$$\sin\theta = \frac{z - z^{-1}}{2i}, \qquad \cos\theta = \frac{z + z^{-1}}{2}, \qquad d\theta = \frac{dz}{iz},$$

integral (1) becomes the contour integral

(3)
$$\int_C F\left(\frac{z - z^{-1}}{2i}, \frac{z + z^{-1}}{2}\right)\frac{dz}{iz}$$

of a function of z around the circle C in the positive direction. The original integral (1) is, of course, simply a parametric form of integral (3), in accordance with expression (2), Sec. 32. When the integrand of integral (3) is a rational function of z, we can evaluate that integral by means of Cauchy's residue theorem once the zeros of the polynomial in the denominator have been located and provided none lie on C.

EXAMPLE. Let us show that

(4)
$$\int_0^{2\pi} \frac{d\theta}{1 + a\sin\theta} = \frac{2\pi}{\sqrt{1 - a^2}} \qquad (-1 < a < 1).$$

This integration formula is clearly valid when $a = 0$, and we exclude that case in our derivation. With substitutions (2), the integral takes the form

(5)
$$\int_C \frac{2/a}{z^2 + (2i/a)z - 1}\, dz,$$

where C is the positively oriented circle $|z| = 1$. The denominator of the integrand here has the pure imaginary zeros

$$z_1 = \left(\frac{-1 + \sqrt{1 - a^2}}{a}\right)i, \qquad z_2 = \left(\frac{-1 - \sqrt{1 - a^2}}{a}\right)i.$$

So if $f(z)$ denotes the integrand, then

$$f(z) = \frac{2/a}{(z - z_1)(z - z_2)}.$$

Note that, because $|a| < 1$,

$$|z_2| = \frac{1 + \sqrt{1 - a^2}}{|a|} > 1.$$

Also, since $|z_1 z_2| = 1$, it follows that $|z_1| < 1$. Hence there are no singular points on C, and the only one interior to it is the point z_1. The corresponding residue B_1 is found by writing

$$f(z) = \frac{\phi(z)}{z - z_1} \qquad \text{where} \qquad \phi(z) = \frac{2/a}{z - z_2}.$$

This shows that z_1 is a simple pole and that

$$B_1 = \phi(z_1) = \frac{2/a}{z_1 - z_2} = \frac{1}{i\sqrt{1 - a^2}}.$$

Consequently,

$$\int_C \frac{2/a}{z^2 + (2i/a)z - 1} \, dz = 2\pi i B_1 = \frac{2\pi}{\sqrt{1 - a^2}};$$

and integration formula (4) follows.

The method just illustrated applies equally well when the arguments of the sine and cosine are integral multiples of θ. One can write, for example, $\cos 2\theta = (z^2 + z^{-2})/2$ when $z = e^{i\theta}$.

EXERCISES

Use residues to evaluate the definite integrals in Exercises 1 through 7.

1. $\displaystyle\int_0^{2\pi} \frac{d\theta}{5 + 4\sin\theta}.$

   Ans. $\dfrac{2\pi}{3}.$

2. $\displaystyle\int_{-\pi}^{\pi} \frac{d\theta}{1 + \sin^2\theta}.$

Ans. $\sqrt{2}\pi.$

3. $\displaystyle\int_{0}^{2\pi} \frac{\cos^2 3\theta d\theta}{5 - 4\cos 2\theta}.$

Ans. $\dfrac{3\pi}{8}.$

4. $\displaystyle\int_{0}^{2\pi} \frac{d\theta}{1 + a\cos\theta} \quad (-1 < a < 1).$

Ans. $\dfrac{2\pi}{\sqrt{1 - a^2}}.$

5. $\displaystyle\int_{0}^{\pi} \frac{\cos 2\theta d\theta}{1 - 2a\cos\theta + a^2} \quad (-1 < a < 1).$

Ans. $\dfrac{a^2\pi}{1 - a^2}.$

6. $\displaystyle\int_{0}^{\pi} \frac{d\theta}{(a + \cos\theta)^2} \quad (a > 1).$

Ans. $\dfrac{a\pi}{\left(\sqrt{a^2 - 1}\right)^3}.$

7. $\displaystyle\int_{0}^{\pi} \sin^{2n}\theta d\theta \quad (n = 1, 2, \ldots).$

Ans. $\dfrac{(2n)!}{2^{2n}(n!)^2}\pi.$

63. INDENTED PATHS

In the last three sections, two fundamentally different paths of integration have been used to evaluate integrals from real analysis. There are still other paths that can be taken, depending on the nature of the integral to be evaluated. In this section, we illustrate the use of *indented* paths.

We begin with the statement of an important limit that will be used in our first example. The statement, which is to be proved in Exercise 8, Sec. 64, can be applied to any function $f(z)$ that has a simple pole at a point $z = x_0$ on the real axis, with a Laurent series representation in a punctured disk $0 < |z - x_0| < R_2$ and with residue B_0. Namely, if C_ρ denotes the upper half of a circle $|z - x_0| = \rho$, where $\rho < R_2$ and the clockwise direction is taken (Fig. 68), then

(1) $$\lim_{\rho \to 0} \int_{C_\rho} f(z)\, dz = -B_0\pi i.$$

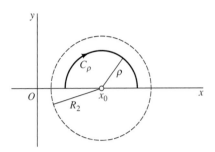

FIGURE 68

EXAMPLE 1. Modifying the method used in Sec. 60, we derive here the integration formula*

$$(2) \qquad \int_0^\infty \frac{\sin x}{x} \, dx = \frac{\pi}{2}$$

by integrating e^{iz}/z around the simple closed contour shown in Fig. 69. In that figure, ρ and R denote positive real numbers, where $\rho < R$; and L_1 and L_2 represent the intervals $\rho \le x \le R$ and $-R \le x \le -\rho$, respectively, on the real axis. While the semicircle C_R is as in Sec. 60, the semicircle C_ρ is introduced here in order to avoid integrating through the singularity $z = 0$ of e^{iz}/z.

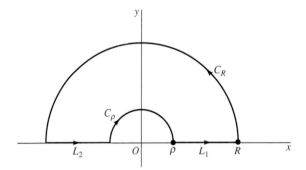

FIGURE 69

The Cauchy-Goursat theorem tells us that

$$\int_{L_1} \frac{e^{iz}}{z} \, dz + \int_{C_R} \frac{e^{iz}}{z} \, dz + \int_{L_2} \frac{e^{iz}}{z} \, dz + \int_{C_\rho} \frac{e^{iz}}{z} \, dz = 0,$$

or

$$(3) \qquad \int_{L_1} \frac{e^{iz}}{z} \, dz + \int_{L_2} \frac{e^{iz}}{z} \, dz = -\int_{C_\rho} \frac{e^{iz}}{z} \, dz - \int_{C_R} \frac{e^{iz}}{z} \, dz.$$

---

*This formula arises in the theory of the *Fourier integral*. See the authors' "Fourier Series and Boundary Value Problems," 5th ed., pp. 219–220, 1993, where it is derived in a completely different way.

Moreover, since the legs L_1 and $-L_2$ have parametric representations

(4) $z = re^{i0} = r \ (\rho \leq r \leq R)$ and $z = re^{i\pi} = -r \ (\rho \leq r \leq R),$

respectively, the left-hand side of equation (3) can be written

$$\int_{L_1} \frac{e^{iz}}{z} \, dz - \int_{-L_2} \frac{e^{iz}}{z} \, dz = \int_{\rho}^{R} \frac{e^{ir}}{r} \, dr - \int_{\rho}^{R} \frac{e^{-ir}}{r} \, dr = 2i \int_{\rho}^{R} \frac{\sin r}{r} \, dr.$$

Consequently,

(5) $$2i \int_{\rho}^{R} \frac{\sin r}{r} \, dr = -\int_{C_\rho} \frac{e^{iz}}{z} \, dz - \int_{C_R} \frac{e^{iz}}{z} \, dz.$$

Now, from the Laurent series representation

$$\frac{e^{iz}}{z} = \frac{1}{z} \left[1 + \frac{(iz)}{1!} + \frac{(iz)^2}{2!} + \frac{(iz)^3}{3!} + \cdots \right]$$

$$= \frac{1}{z} + \frac{i}{1!} + \frac{i^2}{2!} z + \frac{i^3}{3!} z^2 + \cdots \qquad (0 < |z| < \infty),$$

it is clear that e^{iz}/z has a simple pole at the origin, with residue unity. So, according to the statement in italics just prior to this example,

$$\lim_{\rho \to 0} \int_{C_\rho} \frac{e^{iz}}{z} \, dz = -\pi i.$$

Also, since

$$\left| \frac{1}{z} \right| = \frac{1}{|z|} = \frac{1}{R}$$

when z is a point on C_R, we know from limit (9) in Sec. 61 that

$$\lim_{R \to \infty} \int_{C_R} \frac{e^{iz}}{z} \, dz = 0.$$

Thus, by letting ρ tend to 0 in equation (5) and then letting R tend to ∞, we arrive at the result

$$2i \int_{0}^{\infty} \frac{\sin r}{r} \, dr = \pi i,$$

which is, in fact, formula (2).

EXAMPLE 2.   We can use the same closed path (Fig. 69) as in Example 1 to show that

(6) $$\int_{0}^{\infty} \frac{\ln x}{(x^2 + 4)^2} \, dx = \frac{\pi}{32} (\ln 2 - 1).$$

To do this, we consider the branch

$$f(z) = \frac{\log z}{(z^2 + 4)^2} \qquad \left(|z| > 0, -\frac{\pi}{2} < \arg z < \frac{3\pi}{2} \right)$$

of the multiple-valued function $(\log z)/(z^2 + 4)^2$. This branch, whose branch cut consists of the origin and the negative imaginary axis, is analytic everywhere in the indicated domain except at the point $z = 2i$. In order that the singularity $2i$ always be inside the closed path, we require that $\rho < 2 < R$. Then, according to the residue theorem,

$$\int_{L_1} f(z)\, dz + \int_{C_R} f(z)\, dz + \int_{L_2} f(z)\, dz + \int_{C_\rho} f(z)\, dz = 2\pi i \operatorname*{Res}_{z=2i} f(z).$$

That is,

(7) $\quad \displaystyle\int_{L_1} f(z)\, dz + \int_{L_2} f(z)\, dz = 2\pi i \operatorname*{Res}_{z=2i} f(z) - \int_{C_\rho} f(z)\, dz - \int_{C_R} f(z)\, dz.$

Since

$$f(z) = \frac{\ln r + i\theta}{(r^2 e^{i2\theta} + 4)^2} \qquad (z = re^{i\theta}),$$

the parametric representations (4) for the legs L_1 and $-L_2$ can be used to write the left-hand side of equation (7) as

$$\int_{L_1} f(z)\, dz - \int_{-L_2} f(z)\, dz = \int_\rho^R \frac{\ln r}{(r^2 + 4)^2}\, dr + \int_\rho^R \frac{\ln r + i\pi}{(r^2 + 4)^2}\, dr.$$

Also, since

$$f(z) = \frac{\phi(z)}{(z - 2i)^2} \qquad \text{where} \qquad \phi(z) = \frac{\log z}{(z + 2i)^2},$$

the singularity $z = 2i$ of $f(z)$ is a pole of order 2, with residue

$$\phi'(2i) = \frac{\pi}{64} + i\frac{1 - \ln 2}{32}.$$

Equation (7) thus becomes

(8) $\quad \displaystyle 2\int_\rho^R \frac{\ln r}{(r^2 + 4)^2}\, dr + i\pi \int_\rho^R \frac{dr}{(r^2 + 4)^2} = \frac{\pi}{16}(\ln 2 - 1) + i\frac{\pi^2}{32}$

$$- \int_{C_\rho} f(z)\, dz - \int_{C_R} f(z)\, dz;$$

and, by equating the real parts on each side here, we find that

(9) $\quad \displaystyle 2\int_\rho^R \frac{\ln r}{(r^2 + 4)^2}\, dr = \frac{\pi}{16}(\ln 2 - 1) - \operatorname{Re} \int_{C_\rho} f(z)\, dz - \operatorname{Re} \int_{C_R} f(z)\, dz.$

It remains only to show that

(10) $$\lim_{\rho \to 0} \operatorname{Re} \int_{C_\rho} f(z)\, dz = 0 \qquad \text{and} \qquad \lim_{R \to \infty} \operatorname{Re} \int_{C_R} f(z)\, dz = 0.$$

For, by letting ρ and R tend to 0 and ∞, respectively, in equation (9), we then arrive at

$$2 \int_0^\infty \frac{\ln r}{(r^2 + 4)^2}\, dr = \frac{\pi}{16}(\ln 2 - 1),$$

which is the same as equation (6).

Limits (10) are established as follows. First, we note that if $\rho < 1$ and $z = \rho e^{i\theta}$ is a point on C_ρ, then

$$|\log z| = |\ln \rho + i\theta| \le |\ln \rho| + |i\theta| \le -\ln \rho + \pi$$

and

$$|z^2 + 4| \ge ||z|^2 - 4| = 4 - \rho^2.$$

As a consequence,

$$\left| \operatorname{Re} \int_{C_\rho} f(z)\, dz \right| \le \left| \int_{C_\rho} f(z)\, dz \right| \le \frac{-\ln \rho + \pi}{(4 - \rho^2)^2}\, \pi \rho = \pi \frac{\pi \rho - \rho \ln \rho}{(4 - \rho^2)^2};$$

and, by l'Hospital's rule, the product $\rho \ln \rho$ in the numerator on the far right here tends to 0 as ρ tends to 0. So the first of limits (10) clearly holds. Likewise, by writing

$$\left| \operatorname{Re} \int_{C_R} f(z)\, dz \right| \le \left| \int_{C_R} f(z)\, dz \right| \le \frac{\ln R + \pi}{(R^2 - 4)^2}\, \pi R = \pi \frac{\dfrac{\pi}{R} + \dfrac{\ln R}{R}}{\left(R - \dfrac{4}{R} \right)^2}$$

and using l'Hospital's rule to show that the quotient $(\ln R)/R$ tends to 0 as R tends to ∞, we obtain the second of limits (10).

Note how another integration formula, namely

(11) $$\int_0^\infty \frac{dx}{(x^2 + 4)^2} = \frac{\pi}{32},$$

follows by equating imaginary, rather than real, parts on each side of equation (8):

(12) $$\pi \int_\rho^R \frac{dr}{(r^2 + 4)^2} = \frac{\pi^2}{32} - \operatorname{Im} \int_{C_\rho} f(z)\, dz - \operatorname{Im} \int_{C_R} f(z)\, dz.$$

Formula (11) is then obtained by letting ρ and R tend to 0 and ∞, respectively, since

$$\left| \operatorname{Im} \int_{C_\rho} f(z)\, dz \right| \le \left| \int_{C_\rho} f(z)\, dz \right| \qquad \text{and} \qquad \left| \operatorname{Im} \int_{C_R} f(z)\, dz \right| \le \left| \int_{C_R} f(z)\, dz \right|.$$

64. INTEGRATION ALONG A BRANCH CUT

Cauchy's residue theorem can be useful in evaluating a real integral when part of the path of integration of the function $f(z)$ to which the theorem is applied lies along a branch cut of that function.

EXAMPLE. Let x^{-a}, where $x > 0$ and $0 < a < 1$, denote the principal value of the indicated power of x; that is, x^{-a} is the positive real number $\exp(-a \ln x)$. We shall evaluate here the improper real integral

$$(1) \qquad \int_0^\infty \frac{x^{-a}}{x+1}\, dx \qquad (0 < a < 1),$$

which is important in the study of the *gamma function*.* Note that integral (1) is improper not only because of its upper limit of integration but also because its integrand has an infinite discontinuity at $x = 0$. The integral converges when $0 < a < 1$ since the integrand behaves like x^{-a} near $x = 0$ and like x^{-a-1} as x tends to infinity. We do not, however, need to establish convergence separately; for that will be contained in our evaluation of the integral.

We begin by letting C_ρ and C_R denote the circles $|z| = \rho$ and $|z| = R$, respectively, where $\rho < 1 < R$; and we assign them the orientations shown in Fig. 70.

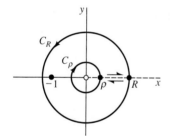

FIGURE 70

Since the branch

$$(2) \qquad f(z) = \frac{z^{-a}}{z+1} \qquad (|z| > 0, 0 < \arg z < 2\pi)$$

of the multiple-valued function $z^{-a}/(z+1)$, with branch cut $\arg z = 0$, is piecewise continuous on C_ρ and C_R, the integrals

$$(3) \qquad \int_{C_\rho} f(z)\, dz \qquad \text{and} \qquad \int_{C_R} f(z)\, dz$$

exist. Our technique for evaluating integral (1) is based on an expression for the sum of the contour integrals (3).

---

*See, for example, p. 4 of the book by Lebedev cited in Appendix 1.

To obtain that expression, we consider the closed contour, indicated in Fig. 70, that is traced out by a point moving from ρ to R along the branch cut for $f(z)$, next around C_R back to R, then along the cut to ρ, and finally around C_ρ back to ρ. If we write

$$f(z) = \frac{\exp(-a \log z)}{z + 1} = \frac{\exp[-a(\ln r + i\theta)]}{re^{i\theta} + 1} \qquad (z = re^{i\theta})$$

and use $\theta = 0$ and $\theta = 2\pi$ along the upper and lower "edges," respectively, of the cut annulus, we see that

$$f(z) = \frac{\exp[-a(\ln r + i0)]}{r + 1} = \frac{r^{-a}}{r + 1} \qquad (z = re^{i0})$$

on the upper edge and

$$f(z) = \frac{\exp[-a(\ln r + i2\pi)]}{r + 1} = \frac{r^{-a}e^{-i2a\pi}}{r + 1} \qquad (z = re^{i2\pi})$$

on the lower edge. The residue theorem thus suggests that

(4) $$\int_\rho^R \frac{r^{-a}}{r + 1}\, dr + \int_{C_R} f(z)\, dz - \int_\rho^R \frac{r^{-a}e^{-i2a\pi}}{r + 1}\, dr + \int_{C_\rho} f(z)\, dz$$
$$= 2\pi i \operatorname*{Res}_{z=-1} f(z).$$

Our derivation of equation (4) is, of course, only *formal* since $f(z)$ is not analytic, or even defined, on the branch cut involved. It is, nevertheless, valid and can be fully justified by an argument such as the one in Exercise 9.

The residue in equation (4) can be found by noting that the function

$$\phi(z) = z^{-a} = \exp(-a \log z) \qquad (|z| > 0, 0 < \arg z < 2\pi)$$

is analytic at $z = -1$ and that

$$\phi(-1) = \exp[-a(\ln 1 + i\pi)] = e^{-ia\pi}.$$

This shows that the point $z = -1$ is a simple pole of the function $f(z)$, defined by equation (2), and that

$$\operatorname*{Res}_{z=-1} f(z) = e^{-ia\pi}.$$

Equation (4) can, therefore, be written as the desired expression for the sum of integrals (3):

(5) $$\int_{C_\rho} f(z)\, dz + \int_{C_R} f(z)\, dz = 2\pi i e^{-ia\pi} + (e^{-i2a\pi} - 1)\int_\rho^R \frac{r^{-a}}{r + 1}\, dr.$$

Referring now to definition (2) of $f(z)$, we see that

$$\left| \int_{C_\rho} f(z)\, dz \right| \le \frac{\rho^{-a}}{1 - \rho} 2\pi\rho = \frac{2\pi}{1 - \rho}\rho^{1-a}$$

and

$$\left| \int_{C_R} f(z)\, dz \right| \le \frac{R^{-a}}{R-1} 2\pi R = \frac{2\pi R}{R-1} \cdot \frac{1}{R^a}.$$

Since $0 < a < 1$, the values of these two integrals tend to 0 as ρ and R tend to 0 and ∞, respectively. Rewriting equation (5) as an expression for the last integral there and then letting ρ tend to 0, we thus find that

$$\int_0^R \frac{r^{-a}}{r+1}\, dr = \frac{1}{e^{-i2a\pi}-1}\left[\int_{C_R} f(z)\, dz - 2\pi i e^{-ia\pi} \right].$$

Finally, by letting R tend to ∞ in this last equation, we arrive at the result

$$\int_0^\infty \frac{r^{-a}}{r+1}\, dr = 2\pi i \frac{e^{-ia\pi}}{1-e^{-i2a\pi}} \cdot \frac{e^{ia\pi}}{e^{ia\pi}} = \pi \frac{2i}{e^{ia\pi}-e^{-ia\pi}},$$

which is the same as

$$(6) \qquad \int_0^\infty \frac{x^{-a}}{x+1}\, dx = \frac{\pi}{\sin a\pi} \qquad (0 < a < 1).$$

EXERCISES

In Exercises 1 through 4, take the indented contour in Fig. 69 (Sec. 63).

1. Derive the integration formula

$$\int_0^\infty \frac{\cos(ax) - \cos(bx)}{x^2}\, dx = \frac{\pi}{2}(b-a) \qquad (a \ge 0, b \ge 0).$$

Then, with the aid of the trigonometric identity $1 - \cos(2x) = 2\sin^2 x$, point out how it follows that

$$\int_0^\infty \frac{\sin^2 x}{x^2}\, dx = \frac{\pi}{2}.$$

2. Evaluate the improper integral

$$\int_0^\infty \frac{x^a}{(x^2+1)^2}\, dx, \text{ where } -1 < a < 3 \text{ and } x^a = \exp(a \ln x).$$

Ans. $\dfrac{(1-a)\pi}{4\cos(a\pi/2)}.$

3. Use the function

$$f(z) = \frac{z^{1/3}\log z}{z^2+1} = \frac{e^{(1/3)\log z}\log z}{z^2+1} \qquad \left(|z| > 0, -\frac{\pi}{2} < \arg z < \frac{3\pi}{2} \right)$$

to derive this pair of integration formulas:

$$\int_0^\infty \frac{\sqrt[3]{x}\ln x}{x^2+1}\, dx = \frac{\pi^2}{6}, \qquad \int_0^\infty \frac{\sqrt[3]{x}}{x^2+1}\, dx = \frac{\pi}{\sqrt{3}}.$$

4. Use the function

$$f(z) = \frac{(\log z)^2}{z^2 + 1} \qquad \left(|z| > 0, -\frac{\pi}{2} < \arg z < \frac{3\pi}{2}\right)$$

to show that

$$\int_0^\infty \frac{(\ln x)^2}{x^2 + 1}\, dx = \frac{\pi^3}{8}, \qquad \int_0^\infty \frac{\ln x}{x^2 + 1}\, dx = 0.$$

Suggestion: The integration formula obtained in Exercise 1, Sec. 60, is needed here.

5. Use the function

$$f(z) = \frac{z^{1/3}}{(z + a)(z + b)} = \frac{e^{(1/3)\log z}}{(z + a)(z + b)} \qquad (|z| > 0, 0 < \arg z < 2\pi)$$

and the closed contour in Fig. 70 (Sec. 64) to show formally that

$$\int_0^\infty \frac{\sqrt[3]{x}}{(x + a)(x + b)}\, dx = \frac{2\pi}{\sqrt{3}} \cdot \frac{\sqrt[3]{a} - \sqrt[3]{b}}{a - b} \qquad (a > b > 0).$$

6. Show that

$$\int_0^\infty \frac{dx}{\sqrt{x}(x^2 + 1)} = \frac{\pi}{\sqrt{2}}$$

by integrating an appropriate branch of the multiple-valued function

$$f(z) = \frac{z^{-1/2}}{z^2 + 1} = \frac{e^{(-1/2)\log z}}{z^2 + 1}$$

over (a) the indented path in Fig. 69, Sec. 63; (b) the closed contour in Fig. 70, Sec. 64.

7. The *beta function* is this function of two real variables:

$$B(p, q) = \int_0^1 t^{p-1}(1 - t)^{q-1}\, dt \qquad (p > 0, q > 0).$$

Make the substitution $t = 1/(x + 1)$ and use the result obtained in the example in Sec. 64 to show that

$$B(p, 1 - p) = \frac{\pi}{\sin(p\pi)} \qquad (0 < p < 1).$$

8. Suppose that a function $f(z)$ has a simple pole at a point $z = x_0$ on the real axis, with a Laurent series representation in a punctured disk $0 < |z - x_0| < R_2$ and with residue B_0.

(a) Let C_ρ denote the upper half of a circle $|z - x_0| = \rho$, where $\rho < R_2$ and the direction is clockwise (see Fig. 68 in Sec. 63). After choosing a number ρ_0 such that $\rho < \rho_0 < R_2$, point out why

$$f(z) = g(z) + \frac{B_0}{z - x_0} \qquad (0 < |z - x_0| \le \rho_0),$$

where $g(z)$ is continuous and thus bounded in the closed disk $|z - x_0| \le \rho_0$.

(b) Use the result in part (a) and a parametric representation for $-C_\rho$ to show that

$$\lim_{\rho \to 0} \int_{C_\rho} f(z)\, dz = -B_0\pi i.$$

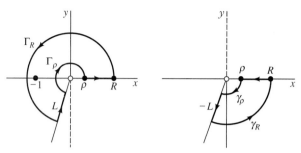

FIGURE 71

9. Consider the two simple closed contours shown in Fig. 71 and obtained by dividing into two pieces the annulus formed by the circles C_ρ and C_R in Fig. 70 (Sec. 64). The legs L and $-L$ of those contours are directed line segments along any ray $\arg z = \theta_0$, where $\pi < \theta_0 < 3\pi/2$. Also, Γ_ρ and γ_ρ are the indicated portions of C_ρ, while Γ_R and γ_R make up C_R.

(a) Show how it follows from Cauchy's residue theorem that when the branch

$$f_1(z) = \frac{z^{-a}}{z+1} \qquad \left(|z| > 0, -\frac{\pi}{2} < \arg z < \frac{3\pi}{2}\right)$$

of the multiple-valued function $z^{-a}/(z+1)$ is integrated around the closed contour on the left in Fig. 71,

$$\int_\rho^R \frac{r^{-a}}{r+1}\,dr + \int_{\Gamma_R} f_1(z)\,dz + \int_L f_1(z)\,dz + \int_{\Gamma_\rho} f_1(z)\,dz = 2\pi i \operatorname*{Res}_{z=-1} f_1(z).$$

(b) Apply the Cauchy-Goursat theorem to the branch

$$f_2(z) = \frac{z^{-a}}{z+1} \qquad \left(|z| > 0, \frac{\pi}{2} < \arg z < \frac{5\pi}{2}\right)$$

of $z^{-a}/(z+1)$, integrated around the closed contour on the right in Fig. 71, to show that

$$-\int_\rho^R \frac{r^{-a}e^{-i2a\pi}}{r+1}\,dr + \int_{\gamma_\rho} f_2(z)\,dz - \int_L f_2(z)\,dz + \int_{\gamma_R} f_2(z)\,dz = 0.$$

(c) Point out why, in the last three integrals in parts (a) and (b), the branches $f_1(z)$ and $f_2(z)$ of $z^{-a}/(z+1)$ can be replaced by the branch

$$f(z) = \frac{z^{-a}}{z+1} \qquad (|z| > 0, 0 < \arg z < 2\pi).$$

Then, by adding corresponding sides of those two equations, derive equation (4), Sec. 64, which was obtained only formally there.

65. ARGUMENT PRINCIPLE AND ROUCHÉ'S THEOREM

A function f is said to be *meromorphic* in a domain D if it is analytic throughout D except possibly for poles. Suppose now that f is meromorphic in the domain interior to a positively oriented simple closed contour C and that it is analytic and nonzero on C. The image Γ of C under the transformation $w = f(z)$ is a closed contour, not necessarily simple, in the w plane (Fig. 72). As a point z traverses C in the positive direction, its image w traverses Γ in a particular direction that determines the orientation of Γ. Note that, since f has no zeros on C, the contour Γ does not pass through the origin in the w plane.

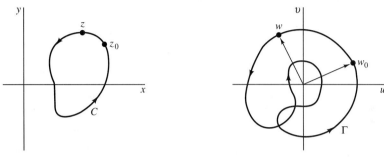

FIGURE 72
$w = f(z)$.

Let w and w_0 be points on Γ, where w_0 is fixed and ϕ_0 is a value of $\arg w_0$. Then let $\arg w$ vary continuously, starting with the value ϕ_0, as the point w begins at the point w_0 and traverses Γ once in the direction of orientation assigned to it by the mapping $w = f(z)$. When w returns to the starting point w_0, $\arg w$ assumes a particular value of $\arg w_0$, which we denote by ϕ_1. Thus the change in $\arg w$ as w describes Γ once in its direction of orientation is $\phi_1 - \phi_0$. This change is, of course, independent of the point w_0 chosen to determine it. Since $w = f(z)$, the number $\phi_1 - \phi_0$ is, in fact, the change in argument of $f(z)$ as z describes C once in the positive direction, starting with a point z_0; and we write

$$\Delta_C \arg f(z) = \phi_1 - \phi_0.$$

The value of $\Delta_C \arg f(z)$ is evidently an integral multiple of 2π, and the integer

$$\frac{1}{2\pi} \Delta_C \arg f(z)$$

represents the number of times the point w winds around the origin in the w plane. For that reason, this integer is sometimes called the *winding number* of Γ with respect to the origin $w = 0$. It is positive if Γ winds around the origin in the counterclockwise direction and negative if it winds clockwise around that point. The winding number is always zero when Γ does not enclose the origin. The verification of this fact for a special case is left to the exercises.

The winding number can be determined from the number of zeros and poles of f interior to C. The number of poles is necessarily finite, according to Exercise 15, Sec. 57. Likewise, with the understanding that $f(z)$ is not identically equal to zero everywhere else inside C, it is easily shown (Exercise 4) that the zeros of f are finite in number and are all of finite order. Suppose now that f has Z zeros and P poles in the domain interior to C. We agree that f has m_0 zeros at a point z_0 if it has a zero of order m_0 there; and if f has a pole of order m_p at z_0, that pole is to be counted m_p times. The following theorem, which is known as the *argument principle*, states that the winding number is simply the difference $Z - P$.

   Theorem 1. *Let a function f be meromorphic in the domain interior to a positively oriented simple closed contour C, and suppose that f is analytic and nonzero on C. If, counting multiplicities, Z is the number of zeros and P the number of poles inside C, then*

$$(1) \qquad\qquad \frac{1}{2\pi}\Delta_C \arg f(z) = Z - P.$$

   To prove this, we evaluate the integral of $f'(z)/f(z)$ around C in two different ways. First, we let $z = z(t)$ $(a \leq t \leq b)$ be a parametric representation for C, so that

$$(2) \qquad\qquad \int_C \frac{f'(z)}{f(z)}\, dz = \int_a^b \frac{f'[z(t)]z'(t)}{f[z(t)]}\, dt.$$

Since, under the transformation $w = f(z)$, the image Γ of C never passes through the origin in the w plane, the image of any point $z = z(t)$ on C can be expressed in exponential form as $w = \rho(t)e^{i\phi(t)}$. Thus

$$(3) \qquad\qquad f[z(t)] = \rho(t)e^{i\phi(t)} \qquad (a \leq t \leq b);$$

and, along each of the smooth arcs making up the contour Γ, it follows that (see Exercise 11, Sec. 31)

$$(4) \quad f'[z(t)]z'(t) = \frac{d}{dt}f[z(t)] = \frac{d}{dt}[\rho(t)e^{i\phi(t)}] = \rho'(t)e^{i\phi(t)} + i\rho(t)e^{i\phi(t)}\phi'(t).$$

Inasmuch as $\rho'(t)$ and $\phi'(t)$ are piecewise continuous on the interval $a \leq t \leq b$, we can now use expressions (3) and (4) to write integral (2) as follows:

$$\int_C \frac{f'(z)}{f(z)}\, dz = \int_a^b \frac{\rho'(t)}{\rho(t)}\, dt + i\int_a^b \phi'(t)\, dt = \ln\rho(t)\Big]_a^b + i\phi(t)\Big]_a^b.$$

But

$$\rho(b) = \rho(a) \qquad \text{and} \qquad \phi(b) - \phi(a) = \Delta_C \arg f(z).$$

Hence

$$(5) \qquad\qquad \int_C \frac{f'(z)}{f(z)}\, dz = i\Delta_C \arg f(z).$$

   Another way to evaluate integral (5) is to use Cauchy's residue theorem. To be specific, we observe that the integrand $f'(z)/f(z)$ is analytic inside and on C except at the points inside C at which the zeros and poles of f occur. If f has a zero of order

m_0 at z_0, then (Sec. 57)

$$(6) \qquad f(z) = (z - z_0)^{m_0} g(z),$$

where $g(z)$ is analytic and nonzero at z_0. Hence

$$f'(z_0) = m_0(z - z_0)^{m_0-1} g(z) + (z - z_0)^{m_0} g'(z),$$

or

$$(7) \qquad \frac{f'(z)}{f(z)} = \frac{m_0}{z - z_0} + \frac{g'(z)}{g(z)}.$$

Since $g'(z)/g(z)$ is analytic at z_0, it has a Taylor series representation about that point; and so equation (7) tells us that $f'(z)/f(z)$ has a simple pole at z_0, with residue m_0. If, on the other hand, f has a pole of order m_p at z_0, we know from the theorem in Sec. 56 that

$$(8) \qquad f(z) = (z - z_0)^{-m_p} \phi(z),$$

where $\phi(z)$ is analytic and nonzero at z_0. Because expression (8) has the same form as expression (6), with the positive integer m_0 in equation (6) replaced by $-m_p$, it is clear from equation (7) that $f'(z)/f(z)$ has a simple pole at z_0, with residue $-m_p$. Applying the residue theorem, then, we find that

$$(9) \qquad \int_C \frac{f'(z)}{f(z)} \, dz = 2\pi i(Z - P).$$

Expression (1) now follows by equating the right-hand sides of equations (5) and (9).

EXAMPLE 1. The only singularity of the function $1/z^2$ is a pole of order 2 at the origin, and there are no zeros in the finite plane. In particular, this function is analytic and nonzero on the unit circle $z = e^{i\theta}(0 \le \theta \le 2\pi)$. If we let C denote that positively oriented circle, Theorem 1 tells us that

$$\frac{1}{2\pi} \Delta_C \arg\left(\frac{1}{z^2}\right) = -2.$$

That is, the image Γ of C under the transformation $w = 1/z^2$ winds around the origin $w = 0$ twice in the clockwise direction. This can be verified directly by noting that Γ has the parametric representation $w = e^{-i2\theta}(0 \le \theta \le 2\pi)$.

The next theorem, which is known as *Rouché's theorem*, is an immediate consequence of Theorem 1 and is important in locating regions of the complex plane in which there may be zeros of a given analytic function.

Theorem 2. *Let two functions $f(z)$ and $g(z)$ be analytic inside and on a simple closed contour C, and suppose that $|f(z)| > |g(z)|$ at each point on C. Then $f(z)$ and $f(z) + g(z)$ have the same number of zeros, counting multiplicities, inside C.*

The orientation of C in the statement of the theorem is evidently immaterial. Thus, in the proof here, we may assume that the orientation is positive. We begin with the observation that neither the function $f(z)$ nor the sum $f(z) + g(z)$ has a zero on C, since

$$|f(z)| > |g(z)| \geq 0 \qquad \text{and} \qquad |f(z) + g(z)| \geq \|f(z)| - |g(z)\| > 0$$

when z is on C.

If Z_f and Z_{f+g} denote the number of zeros, counting multiplicities, of $f(z)$ and $f(z) + g(z)$, respectively, inside C, we know from Theorem 1 that

$$Z_f = \frac{1}{2\pi}\Delta_C \arg f(z) \qquad \text{and} \qquad Z_{f+g} = \frac{1}{2\pi}\Delta_C \arg[f(z) + g(z)].$$

Consequently, since

$$\Delta_C \arg[f(z) + g(z)] = \Delta_C \arg\left\{ f(z)\left[1 + \frac{g(z)}{f(z)}\right]\right\}$$

$$= \Delta_C \arg f(z) + \Delta_C \arg\left[1 + \frac{g(z)}{f(z)}\right],$$

it is clear that

(10) $$Z_{f+g} = Z_f + \frac{1}{2\pi}\Delta_C \arg F(z),$$

where

$$F(z) = 1 + \frac{g(z)}{f(z)}.$$

But

$$|F(z) - 1| = \frac{|g(z)|}{|f(z)|} < 1;$$

and this means that, under the transformation $w = F(z)$, the image of C lies in the open disk $|w - 1| < 1$. That image does not, then, enclose the origin $w = 0$. Hence $\Delta_C \arg F(z) = 0$ and, since equation (10) reduces to $Z_{f+g} = Z_f$, the theorem is proved.

EXAMPLE 2.   In order to determine the number of roots of the equation

(11) $$z^7 - 4z^3 + z - 1 = 0$$

inside the circle $|z| = 1$, write

$$f(z) = -4z^3 \qquad \text{and} \qquad g(z) = z^7 + z - 1.$$

Then observe that $|f(z)| = 4|z|^3 = 4$ and $|g(z)| \leq |z|^7 + |z| + 1 = 3$ when $|z| = 1$. The conditions in Rouché's theorem are thus satisfied. Consequently, since $f(z)$ has three zeros, counting multiplicities, inside the circle $|z| = 1$, so does $f(z) + g(z)$. That is, equation (11) has three roots there.

EXERCISES

*1. Let C denote the unit circle $|z| = 1$, described in the positive sense. Use Theorem 1 in Sec. 65 to determine the value of $\Delta_C \arg f(z)$ when

(a) $f(z) = z^2$;   (b) $f(z) = (z^3 + 2)/z$;   (c) $f(z) = (2z - 1)^7/z^3$.

    Ans. (a) 4π; (b) -2π; (c) 8π.

2. Let f be a function which is analytic inside and on a simple closed contour C, and suppose that $f(z)$ is never zero on C. Let the image of C under the transformation $w = f(z)$ be the closed contour Γ shown in Fig. 73. Determine the value of $\Delta_C \arg f(z)$ from that figure; and, with the aid of Theorem 1 in Sec. 65, determine the number of zeros, counting multiplicities, of f interior to C.

    Ans. 6π; 3.

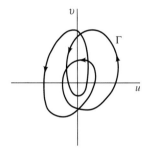

FIGURE 73

3. Using the notation in Sec. 65, suppose that Γ does not enclose the origin $w = 0$ and that there is a ray from that point which does not intersect Γ. By observing that the absolute value of $\Delta_C \arg f(z)$ must be less than 2π when a point z makes one cycle around C and recalling that $\Delta_C \arg f(z)$ is an integral multiple of 2π, point out why the winding number of Γ with respect to the origin $w = 0$ must be zero.

4. Suppose that a function f is meromorphic in the domain D interior to a simple closed contour C on which f is analytic and nonzero, and let D_0 denote the domain consisting of all points in D except for poles. Point out how it follows from the theorem in Sec. 58 and Exercise 14, Sec. 57, that if $f(z)$ is not identically equal to zero in D_0, then the zeros of f in D are all of finite order and that they are finite in number.

    Suggestion: Note that if a point z_0 in D is a zero of f that is not of finite order, then there must be a neighborhood of z_0 throughout which $f(z)$ is identically equal to zero.

5. Suppose that a function f is analytic inside and on a positively oriented simple closed contour C and that it has no zeros on C. Show that if f has n zeros z_k ($k = 1, 2, \ldots, n$) inside C, where each z_k is of multiplicity m_k, then

$$\int_C \frac{zf'(z)}{f(z)}\, dz = 2\pi i \sum_{k=1}^{n} m_k z_k.$$

[Compare equation (9), Sec. 65 when $P = 0$ there.]

6. Determine the number of zeros, counting multiplicities, of the polynomial

(a) $z^6 - 5z^4 + z^3 - 2z$;   (b) $2z^4 - 2z^3 + 2z^2 - 2z + 9$

inside the circle $|z| = 1$.

    Ans. (a) 4; (b) 0.

7. Determine the number of zeros, counting multiplicities, of the polynomial

   (a) $z^4 + 3z^3 + 6$;   (b) $z^4 - 2z^3 + 9z^2 + z - 1$;   (c) $z^5 + 3z^3 + z^2 + 1$

   inside the circle $|z| = 2$.
   Ans. (a) 3; (b) 2; (c) 5.

8. Determine the number of roots, counting multiplicities, of the equation

$$2z^5 - 6z^2 + z + 1 = 0$$

   in the annulus $1 \leq |z| < 2$.
   Ans. 3.

9. Show that if c is a complex number such that $|c| > e$, then the equation $cz^n = e^z$ has n roots, counting multiplicities, inside the circle $|z| = 1$.

10. Write $f(z) = z^n$ and $g(z) = a_0 + a_1 z + \cdots + a_{n-1} z^{n-1}$ and use Rouché's theorem to prove that any polynomial

$$P(z) = a_0 + a_1 z + \cdots + a_{n-1} z^{n-1} + a_n z^n \qquad (a_n \neq 0),$$

   where $n \geq 1$, has precisely n zeros, counting multiplicities. Thus give an alternative proof of the fundamental theorem of algebra (Theorem 2, Sec. 41).

   Suggestion: Note that one can let a_n be unity. Then show that $|g(z)| < |f(z)|$ on the circle $|z| = R$, where R is sufficiently large and, in particular, larger than

$$1 + |a_0| + |a_1| + \cdots + |a_{n-1}|.$$

11. Inequality (5), Sec. 41, ensures that the zeros of a polynomial

$$P(z) = a_0 + a_1 z + \cdots + a_{n-1} z^{n-1} + a_n z^n \qquad (a_n \neq 0)$$

   of degree $n \geq 1$ all lie inside some circle $|z| = R$ about the origin. Also, Exercise 4 above tells us that they are all of finite order and that there is a finite number N of them. Use expression (9), Sec. 65, and Theorem 2 in Sec. 54 to show that

$$N = \operatorname*{Res}_{z=0} \frac{P'(1/z)}{z^2 P(1/z)},$$

   where multiplicities of the zeros are to be counted. Then evaluate this residue to show that $N = n$. (Compare Exercise 10.)

12. Let the functions f and g be as in the statement of Rouché's theorem in Sec. 65, and let the orientation of the contour C there be positive. Then define the function

$$\Phi(t) = \frac{1}{2\pi i} \int_C \frac{f'(z) + tg'(z)}{f(z) + tg(z)} \, dz \qquad (0 \leq t \leq 1),$$

   and follow the steps below to give another proof of that theorem.

   (a) Point out why the denominator in the integrand of the integral defining $\Phi(t)$ is never zero on C. This ensures the existence of the integral.

   (b) Let t and t_0 be any two points in the interval $0 \leq t \leq 1$, and show that

$$|\Phi(t) - \Phi(t_0)| = \frac{|t - t_0|}{2\pi} \left| \int_C \frac{fg' - f'g}{(f + tg)(f + t_0 g)} \, dz \right|.$$

   Then, after pointing out why

$$\left| \frac{fg' - f'g}{(f + tg)(f + t_0 g)} \right| \leq \frac{|fg' - f'g|}{(|f| - |g|)^2}$$

at points on C, show that there is a positive constant A, which is independent of t and t_0, such that

$$|\Phi(t) - \Phi(t_0)| \le A|t - t_0|.$$

Conclude from this inequality that $\Phi(t)$ is continuous on the interval $0 \le t \le 1$.

(c) By referring to equation (9), Sec. 65, state why the value of the function Φ is, for each value of t in the interval $0 \le t \le 1$, an integer representing the number of zeros of $f(z) + tg(z)$ inside C. Then conclude from the fact that Φ is continuous, as shown in part (b), that $f(z)$ and $f(z) + g(z)$ have the same number of zeros, counting multiplicities, inside C.

66. INVERSE LAPLACE TRANSFORMS

Suppose that a function F of the complex variable s is analytic throughout the finite s plane except for a finite number of isolated singularities. Then let L_R denote a vertical line segment from $s = \gamma - iR$ to $s = \gamma + iR$, where the constant γ is positive and large enough that the singularities of F all lie to the left of that segment (Fig. 74). A new function f of the real variable t is defined for positive values of t by means of the equation

(1)
$$f(t) = \frac{1}{2\pi i} \lim_{R \to \infty} \int_{L_R} e^{st} F(s)\, ds \qquad (t > 0),$$

provided this limit exists. Expression (1) is usually written

(2)
$$f(t) = \frac{1}{2\pi i} \text{P.V.} \int_{\gamma - i\infty}^{\gamma + i\infty} e^{st} F(s)\, ds \qquad (t > 0)$$

[compare equation (3), Sec. 60)], and such an integral is called a *Bromwich integral*.

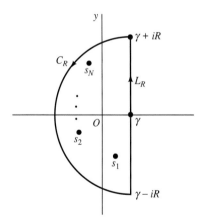

FIGURE 74

It can be shown that, when fairly general conditions are imposed on the functions involved, $f(t)$ is the *inverse* Laplace transform of $F(s)$. That is, if $F(s)$ is the *Laplace transform* of $f(t)$, defined by the equation

(3)
$$F(s) = \int_0^{\infty} e^{-st} f(t)\, dt,$$

then $f(t)$ is retrieved by means of equation (2), where the choice of the positive number γ is immaterial as long as the singularities of F all lie to the left of L_R.* Laplace transforms and their inverses are important in solving both ordinary and partial differential equations.

Residues can often be used to evaluate the limit in expression (1) when the function $F(s)$ is specified. To see how this is done, we let s_n $(n = 1, 2, \ldots, N)$ denote the singularities of $F(s)$. We then let R_0 denote the largest of their moduli and consider a semicircle C_R with parametric representation

$$(4) \qquad s = \gamma + Re^{i\theta} \qquad \left(\frac{\pi}{2} \leq \theta \leq \frac{3\pi}{2}\right),$$

where $R > R_0 + \gamma$. Note that, for each s_n,

$$|s_n - \gamma| \leq |s_n| + \gamma \leq R_0 + \gamma < R.$$

Hence the singularities all lie in the interior of the semicircular region bounded by C_R and L_R (see Fig. 74), and the residue theorem tells us that

$$(5) \qquad \int_{L_R} e^{st} F(s)\, ds = 2\pi i \sum_{n=1}^{N} \operatorname*{Res}_{s=s_n}[e^{st} F(s)] - \int_{C_R} e^{st} F(s)\, ds.$$

Suppose now that, *for all points s on C_R, there is a positive constant M_R such that $|F(s)| \leq M_R$, where M_R tends to zero as R tends to infinity.* We may use the parametric representation (4) for C_R to write

$$\int_{C_R} e^{st} F(s)\, ds = \int_{\pi/2}^{3\pi/2} \exp(\gamma t + Rte^{i\theta}) F(\gamma + Re^{i\theta}) Rie^{i\theta}\, d\theta.$$

Then, since

$$|\exp(\gamma t + Rte^{i\theta})| = e^{\gamma t} e^{Rt\cos\theta} \qquad \text{and} \qquad |F(\gamma + Re^{i\theta})| \leq M_R,$$

we find that

$$(6) \qquad \left| \int_{C_R} e^{st} F(s)\, ds \right| \leq e^{\gamma t} M_R R \int_{\pi/2}^{3\pi/2} e^{Rt\cos\theta}\, d\theta.$$

But the substitution $\phi = \theta - (\pi/2)$, together with Jordan's inequality (7), Sec. 61, reveals that

$$\int_{\pi/2}^{3\pi/2} e^{Rt\cos\theta}\, d\theta = \int_0^{\pi} e^{-Rt\sin\phi}\, d\phi < \frac{\pi}{Rt}.$$

---

*For an extensive treatment of such details regarding Laplace transforms, see R. V. Churchill, "Operational Mathematics," 3d ed., 1972, where transforms $F(s)$ with an infinite number of isolated singular points, or with branch cuts, are also discussed.

Inequality (6) thus becomes

$$(7) \qquad \left| \int_{C_R} e^{st} F(s)\, ds \right| \leq \frac{e^{\gamma t} M_R \pi}{t},$$

and this shows that

$$(8) \qquad \lim_{R \to \infty} \int_{C_R} e^{st} F(s)\, ds = 0.$$

Letting R tend to ∞ in equation (5), then, we see that the function $f(t)$, defined by equation (1), exists and that it can be written

$$(9) \qquad f(t) = \sum_{n=1}^{N} \operatorname*{Res}_{s=s_n}[e^{st} F(s)] \qquad (t > 0).$$

In many applications of Laplace transforms, such as the solution of partial differential equations arising in studies of heat conduction and mechanical vibrations, the function $F(s)$ is analytic for all values of s in the finite plane except for an *infinite* set of isolated singular points s_n $(n = 1, 2, \ldots)$ that lie to the left of some vertical line $\operatorname{Re} s = \gamma$. Often the method just described for finding $f(t)$ can then be modified in such a way that the finite sum (9) is replaced by an *infinite series* of residues:

$$(10) \qquad f(t) = \sum_{n=1}^{\infty} \operatorname*{Res}_{s=s_n}[e^{st} F(s)] \qquad (t > 0).$$

The basic modification is to replace the vertical line segments L_R by vertical line segments L_N $(N = 1, 2, \ldots)$ from $s = \gamma - ib_N$ to $s = \gamma + ib_N$. The circular arcs C_R are then replaced by contours C_N $(N = 1, 2, \ldots)$ from $\gamma + ib_N$ to $\gamma - ib_N$ such that, for each N, the sum $L_N + C_N$ is a simple closed contour enclosing the singular points $s_1, s_2, \ldots, s_N$. Once it is shown that

$$(11) \qquad \lim_{N \to \infty} \int_{C_N} e^{st} F(s)\, ds = 0,$$

expression (2) for $f(t)$ becomes expression (10).

The choice of the contours C_N depends on the nature of the function $F(s)$. Common choices include circular or parabolic arcs and rectangular paths. Also, the simple closed contour $L_N + C_N$ need not enclose precisely N singularities. When, for example, the region between $L_N + C_N$ and $L_{N+1} + C_{N+1}$ contains two singular points of $F(s)$, the pair of corresponding residues of $e^{st} F(s)$ are simply grouped together as a single term in series (10). Since it is often quite tedious to establish limit (11) in any case, we shall accept it in the examples and related exercises below that involve an infinite number of singularities.* Thus our use of expression (10) will be only formal.

---

*An extensive treatment of ways to obtain limit (11) appears in the book by R. V. Churchill that is cited in the footnote earlier in this section. In fact, the inverse transform to be found in Example 3 in the next section is fully verified on pp. 220–226 of that book.

67.   EXAMPLES

Calculation of the sums of the residues of $e^{st}F(s)$ in expressions (9) and (10), Sec. 66, is often facilitated by techniques developed in Exercises 12 and 13 of this section. We preface our examples here with a statement of those techniques.

Suppose that $F(s)$ has a pole of order m at a point s_0 and that its Laurent series representation in a punctured disk $0 < |s - s_0| < R_2$ has principal part

$$\frac{b_1}{s - s_0} + \frac{b_2}{(s - s_0)^2} + \cdots + \frac{b_m}{(s - s_0)^m} \qquad (b_m \neq 0).$$

Then

$$(1) \qquad \operatorname*{Res}_{s=s_0}[e^{st}F(s)] = e^{s_0 t}\left[b_1 + \frac{b_2}{1!}t + \cdots + \frac{b_m}{(m-1)!}t^{m-1}\right].$$

When the pole s_0 is of the form $s_0 = \alpha + i\beta$ ($\beta \neq 0$) and $\overline{F(s)} = F(\bar{s})$ at points of analyticity of $F(s)$ (see Sec. 21), the conjugate $\bar{s}_0 = \alpha - i\beta$ is also a pole of order m. Moreover,

$$(2) \qquad \operatorname*{Res}_{s=s_0}[e^{st}F(s)] + \operatorname*{Res}_{s=\bar{s}_0}[e^{st}F(s)]$$

$$= 2e^{\alpha t}\operatorname{Re}\left\{e^{i\beta t}\left[b_1 + \frac{b_2}{1!}t + \cdots + \frac{b_m}{(m-1)!}t^{m-1}\right]\right\}$$

when t is real. Note that if s_0 is a *simple* pole ($m = 1$), expressions (1) and (2) become

$$(3) \qquad \operatorname*{Res}_{s=s_0}[e^{st}F(s)] = e^{s_0 t}\operatorname*{Res}_{s=s_0}F(s)$$

and

$$(4) \qquad \operatorname*{Res}_{s=s_0}[e^{st}F(s)] + \operatorname*{Res}_{s=\bar{s}_0}[e^{st}F(s)] = 2e^{\alpha t}\operatorname{Re}\left[e^{i\beta t}\operatorname*{Res}_{s=s_0}F(s)\right],$$

respectively.

EXAMPLE 1.   Let us find the function $f(t)$ that corresponds to

$$(5) \qquad\qquad F(s) = \frac{s}{(s^2 + a^2)^2} \qquad (a > 0).$$

The singularities of $F(s)$ are the conjugate points

$$s_0 = ai \qquad \text{and} \qquad \bar{s}_0 = -ai.$$

Upon writing

$$F(s) = \frac{\phi(s)}{(s - ai)^2} \qquad \text{where} \qquad \phi(s) = \frac{s}{(s + ai)^2},$$

we see that $\phi(s)$ is analytic and nonzero at $s_0 = ai$. Hence s_0 is a pole of order $m = 2$ of $F(s)$. Furthermore, $\overline{F(s)} = F(\bar{s})$ at points where $F(s)$ is analytic. Consequently, $\bar{s}_0$ is also a pole of order 2 of $F(s)$; and we know from expression (2) that

$$(6) \qquad \operatorname*{Res}_{s=s_0}[e^{st}F(s)] + \operatorname*{Res}_{s=\bar{s}_0}[e^{st}F(s)] = 2\operatorname{Re}[e^{iat}(b_1 + b_2 t)],$$

where b_1 and b_2 are the coefficients in the principal part

$$\frac{b_1}{s - ai} + \frac{b_2}{(s - ai)^2}$$

of $F(s)$ at ai. These coefficients are readily found with the aid of the first two terms in the Taylor series for $\phi(s)$ about $s_0 = ai$:

$$F(s) = \frac{1}{(s-ai)^2}\phi(s) = \frac{1}{(s-ai)^2}\left[\phi(ai) + \frac{\phi'(ai)}{1!}(s-ai) + \cdots\right]$$

$$= \frac{\phi(ai)}{(s-ai)^2} + \frac{\phi'(ai)}{s-ai} + \cdots \qquad (0 < |s-ai| < 2a).$$

It is straightforward to show that $\phi(ai) = -i/4a$ and $\phi'(ai) = 0$, and we find that $b_1 = 0$ and $b_2 = -i/4a$. Hence expression (6) becomes

$$\operatorname*{Res}_{s=s_0}[e^{st}F(s)] + \operatorname*{Res}_{s=\bar{s}_0}[e^{st}F(s)] = 2\operatorname{Re}\left[e^{iat}\left(-\frac{i}{4a}t\right)\right] = \frac{1}{2a}t\sin at.$$

We can, then, conclude that

(7) $$f(t) = \frac{1}{2a}t\sin at \qquad (t > 0),$$

provided that $F(s)$ satisfies the boundedness condition stated in italics in Sec. 66.

To verify that boundedness condition, we let s be any point on the semicircle

$$s = \gamma + Re^{i\theta} \qquad \left(\frac{\pi}{2} \le \theta \le \frac{3\pi}{2}\right),$$

where $\gamma > 0$ and $R > a + \gamma$; and we note that

$$|s| = |\gamma + Re^{i\theta}| \le \gamma + R \qquad \text{and} \qquad |s| = |\gamma + Re^{i\theta}| \ge |\gamma - R| = R - \gamma > a.$$

Since

$$|s^2 + a^2| \ge ||s|^2 - a^2| \ge (R - \gamma)^2 - a^2 > 0,$$

it follows that

$$|F(s)| = \frac{|s|}{|s^2 + a^2|^2} \le M_R \qquad \text{where} \qquad M_R = \frac{\gamma + R}{[(R-\gamma)^2 - a^2]^2}.$$

The desired boundedness condition is now established, since $M_R \to 0$ as $R \to \infty$.

EXAMPLE 2. In order to find $f(t)$ when

$$F(s) = \frac{\tanh s}{s^2} = \frac{\sinh s}{s^2\cosh s},$$

we note that $F(s)$ has isolated singularities at $s = 0$ and at the zeros (Sec. 25)

$$s = \left(\frac{\pi}{2} + n\pi\right)i \qquad (n = 0, \pm1, \pm2, \ldots)$$

of cosh s. We list those singularities as

$$s_0 = 0 \quad \text{and} \quad s_n = \frac{(2n-1)\pi}{2}i, \quad \bar{s}_n = -\frac{(2n-1)\pi}{2}i \quad (n = 1, 2, \ldots).$$

Then, formally,

$$(8) \qquad f(t) = \underset{s=s_0}{\text{Res}}[e^{st}F(s)] + \sum_{n=1}^{\infty} \left\{ \underset{s=s_n}{\text{Res}}[e^{st}(F(s)] + \underset{s=\bar{s}_n}{\text{Res}}[e^{st}F(s)] \right\}.$$

Division of Maclaurin series yields the Laurent series representation

$$F(s) = \frac{1}{s^2} \cdot \frac{\sinh s}{\cosh s} = \frac{1}{s} - \frac{1}{3}s + \cdots \quad \left(0 < |s| < \frac{\pi}{2} \right),$$

which tells us that s_0 is a simple pole of $F(s)$, with residue unity. Thus

$$(9) \qquad \underset{s=s_0}{\text{Res}}[e^{st}F(s)] = \underset{s=s_0}{\text{Res}} F(s) = 1,$$

according to expression (3).

The residues of $F(s)$ at the points s_n ($n = 1, 2, \ldots$) are readily found by applying the method of the corollary in Sec. 57 for identifying simple poles and determining the residues at such points. To be specific, we write

$$F(s) = \frac{p(s)}{q(s)} \quad \text{where} \quad p(s) = \sinh s \quad \text{and} \quad q(s) = s^2 \cosh s$$

and observe that

$$\sinh s_n = \sinh\left[i\left(n\pi - \frac{\pi}{2} \right) \right] = i \sin\left(n\pi - \frac{\pi}{2} \right) = -i \cos n\pi = (-1)^{n+1}i \neq 0.$$

Then, since

$$p(s_n) = \sinh s_n \neq 0, \quad q(s_n) = 0, \quad \text{and} \quad q'(s_n) = s_n^2 \sinh s_n \neq 0,$$

we find that

$$\underset{s=s_n}{\text{Res}} F(s) = \frac{p(s_n)}{q'(s_n)} = \frac{1}{s_n^2} = -\frac{4}{\pi^2} \cdot \frac{1}{(2n-1)^2} \quad (n = 1, 2, \ldots).$$

The identities $\overline{\sinh s} = \sinh \bar{s}$ and $\overline{\cosh s} = \cosh \bar{s}$ (see Exercise 11, Sec. 25) ensure that $\overline{F(s)} = F(\bar{s})$ at points of analyticity of $F(s)$. Hence $\bar{s}_n$ is also a simple pole of $F(s)$, and expression (4) can be used to write

$$(10) \quad \underset{s=s_n}{\text{Res}}[e^{st}F(s)] + \underset{s=\bar{s}_n}{\text{Res}}[e^{st}F(s)] = 2 \,\text{Re} \left\{ -\frac{4}{\pi^2} \cdot \frac{1}{(2n-1)^2} \exp\left[i \frac{(2n-1)\pi t}{2} \right] \right\}$$

$$= -\frac{8}{\pi^2} \cdot \frac{1}{(2n-1)^2} \cos \frac{(2n-1)\pi t}{2} \quad (n = 1, 2, \ldots).$$

Finally, by substituting expressions (9) and (10) into equation (8), we arrive at the desired result:

$$(11) \qquad f(t) = 1 - \frac{8}{\pi^2} \sum_{n=1}^{\infty} \frac{1}{(2n-1)^2} \cos \frac{(2n-1)\pi t}{2} \quad (t > 0).$$

EXAMPLE 3.   We consider here the function

(12) $$F(s) = \frac{\sinh(xs^{1/2})}{s \sinh(s^{1/2})} \qquad (0 < x < 1),$$

where $s^{1/2}$ denotes any branch of this double-valued function. We agree, however, to use the *same* branch in the numerator and denominator, so that

(13) $$F(s) = \frac{xs^{1/2} + (xs^{1/2})^3/3! + \cdots}{s[s^{1/2} + (s^{1/2})^3/3! + \cdots]} = \frac{x + x^3 s/6 + \cdots}{s + s^2/6 + \cdots}$$

when s is not a singular point of $F(s)$. One such singular point is clearly $s = 0$. With the additional agreement that the branch cut of $s^{1/2}$ does not lie along the negative real axis, so that $\sinh(s^{1/2})$ is well defined along that axis, the other singular points occur if $s^{1/2} = \pm n\pi i$ $(n = 1, 2, \ldots)$. The points

$$s_0 = 0 \qquad \text{and} \qquad s_n = -n^2\pi^2 \quad (n = 1, 2, \ldots)$$

thus constitute the set of singular points of $F(s)$. The problem is now to evaluate the residues in the formal series representation

(14) $$f(t) = \operatorname*{Res}_{s=s_0}[e^{st}F(s)] + \sum_{n=1}^{\infty} \operatorname*{Res}_{s=s_n}[e^{st}F(s)].$$

Division of the power series on the far right in expression (13) reveals that s_0 is a simple pole of $F(s)$, with residue x. So expression (3) tells us that

(15) $$\operatorname*{Res}_{s=s_0}[e^{st}F(s)] = x.$$

As for the residues of $F(s)$ at the singular points $s_n = -n^2\pi^2$ $(n = 1, 2, \ldots)$, we write

$$F(s) = \frac{p(s)}{q(s)} \qquad \text{where} \qquad p(s) = \sinh(xs^{1/2}) \quad \text{and} \quad q(s) = s \sinh(s^{1/2}).$$

Appealing to the corollary in Sec. 57, as we did in Example 2, we note that

$$p(s_n) = \sinh(xs_n^{1/2}) \neq 0, \quad q(s_n) = 0, \quad \text{and} \quad q'(s_n) = \frac{1}{2}s_n^{1/2} \cosh(s^{1/2}) \neq 0;$$

and this tells us that each s_n is a simple pole of $F(s)$, with residue

$$\operatorname*{Res}_{s=s_n} F(s) = \frac{p(s_n)}{q'(s_n)} = \frac{2}{\pi} \cdot \frac{(-1)^n}{n} \sin n\pi x.$$

So, in view of expression (3),

(16) $$\operatorname*{Res}_{s=s_n}[e^{st}F(s)] = e^{s_n t} \operatorname*{Res}_{s=s_n} F(s) = \frac{2}{\pi} \cdot \frac{(-1)^n}{n} e^{-n^2\pi^2 t} \sin n\pi x.$$

Substituting expressions (15) and (16) into equation (14), we arrive at the function

(17) $$f(t) = x + \frac{2}{\pi} \sum_{n=1}^{\infty} \frac{(-1)^n}{n} e^{-n^2\pi^2 t} \sin n\pi x \qquad (t > 0).$$

EXERCISES

In Exercises 1 through 5, use the method described in Sec. 66 and illustrated in Example 1, Sec. 67, to find the function $f(t)$ corresponding to the given function $F(s)$.

1. $F(s) = \dfrac{2s^3}{s^4 - 4}$.

Ans. $f(t) = \cosh\sqrt{2}\,t + \cos\sqrt{2}\,t$.

2. $F(s) = \dfrac{2s - 2}{(s+1)(s^2 + 2s + 5)}$.

Ans. $f(t) = e^{-t}(\sin 2t + \cos 2t - 1)$.

3. $F(s) = \dfrac{12}{s^3 + 8}$.

Ans. $f(t) = e^{-2t} + e^{t}(\sqrt{3}\sin\sqrt{3}t - \cos\sqrt{3}t)$.

4. $F(s) = \dfrac{s^2 - a^2}{(s^2 + a^2)^2}$ $(a > 0)$.

Ans. $f(t) = t\cos at$.

5. $F(s) = \dfrac{8a^3 s^2}{(s^2 + a^2)^3}$ $(a > 0)$.

Suggestion: Refer to Exercise 9, Sec. 55, for the principal part of $F(s)$ at ai.
Ans. $f(t) = (1 + a^2 t^2)\sin at - at\cos at$.

In Exercises 6 through 11, use the formal method, involving an infinite series of residues and illustrated in Examples 2 and 3 in Sec. 67, to find the function $f(t)$ that corresponds to the given function $F(s)$.

6. $F(s) = \dfrac{\sinh(xs)}{s^2 \cosh s}$ $(0 < x < 1)$.

Ans. $f(t) = x + \dfrac{8}{\pi^2}\displaystyle\sum_{n=1}^{\infty}\dfrac{(-1)^n}{(2n-1)^2}\sin\dfrac{(2n-1)\pi x}{2}\cos\dfrac{(2n-1)\pi t}{2}$.

7. $F(s) = \dfrac{1}{s\cosh(s^{1/2})}$.

Ans. $f(t) = 1 + \dfrac{4}{\pi}\displaystyle\sum_{n=1}^{\infty}\dfrac{(-1)^n}{2n-1}\exp\left[-\dfrac{(2n-1)^2\pi^2 t}{4}\right]$.

8. $F(s) = \dfrac{\coth(\pi s/2)}{s^2 + 1}$.

Ans. $f(t) = \dfrac{2}{\pi} - \dfrac{4}{\pi}\displaystyle\sum_{n=1}^{\infty}\dfrac{\cos 2nt}{4n^2 - 1}$. *

---

*This is actually the rectified sine function $f(t) = |\sin t|$. See the authors' "Fourier Series and Boundary Value Problems," 5th ed., p. 51, 1993.

9. $F(s) = \dfrac{\sinh(xs^{1/2})}{s^2 \sinh(s^{1/2})}$ $(0 < x < 1)$.

Ans. $f(t) = \dfrac{1}{6}x(x^2 - 1) + xt + \dfrac{2}{\pi^3}\sum\limits_{n=1}^{\infty}\dfrac{(-1)^{n+1}}{n^3}e^{-n^2\pi^2 t}\sin n\pi x.$

10. $F(s) = \dfrac{1}{s^2} - \dfrac{1}{s\sinh s}.$

Ans. $f(t) = \dfrac{2}{\pi}\sum\limits_{n=1}^{\infty}\dfrac{(-1)^{n+1}}{n}\sin n\pi t.$

11. $F(s) = \dfrac{\sinh(xs)}{s(s^2 + \omega^2)\cosh s}$ $(0 < x < 1)$,

where $\omega > 0$ and $\omega \neq \omega_n = \dfrac{(2n-1)\pi}{2}$ $(n = 1, 2, \ldots)$.

Ans. $f(t) = \dfrac{\sin \omega x \sin \omega t}{\omega^2 \cos \omega} + 2\sum\limits_{n=1}^{\infty}\dfrac{(-1)^{n+1}}{\omega_n}\cdot\dfrac{\sin \omega_n x \sin \omega_n t}{\omega^2 - \omega_n^2}.$

12. Suppose that a function $F(s)$ has a pole of order m at $s = s_0$, with a Laurent series expansion

$$F(s) = \sum_{n=0}^{\infty} a_n(s-s_0)^n + \frac{b_1}{s-s_0} + \frac{b_2}{(s-s_0)^2} + \cdots + \frac{b_{m-1}}{(s-s_0)^{m-1}} + \frac{b_m}{(s-s_0)^m} \qquad (b_m \neq 0)$$

in the punctured disk $0 < |s - s_0| < R_2$, and note that $(s - s_0)^m F(s)$ is represented in that domain by the power series

$$b_m + b_{m-1}(s - s_0) + \cdots + b_2(s - s_0)^{m-2} + b_1(s - s_0)^{m-1} + \sum_{n=0}^{\infty} a_n(s - s_0)^{m+n}.$$

By collecting the terms that make up the coefficient of $(s - s_0)^{m-1}$ in the product (Sec. 51) of this power series and the Taylor series expansion

$$e^{st} = e^{s_0 t}\left[1 + \frac{t}{1!}(s - s_0) + \cdots + \frac{t^{m-2}}{(m-2)!}(s - s_0)^{m-2} + \frac{t^{m-1}}{(m-1)!}(s - s_0)^{m-1} + \cdots\right]$$

of the entire function $e^{st} = e^{s_0 t}e^{(s-s_0)t}$, show that

$$\operatorname*{Res}_{s=s_0}[e^{st}F(s)] = e^{s_0 t}\left[b_1 + \frac{b_2}{1!}t + \cdots + \frac{b_{m-1}}{(m-2)!}t^{m-2} + \frac{b_m}{(m-1)!}t^{m-1}\right],$$

as stated at the beginning of Sec. 67.

13. Let the point $s_0 = \alpha + i\beta(\beta \neq 0)$ be a pole of order m of a function $F(s)$, which has a Laurent series representation

$$F(s) = \sum_{n=0}^{\infty} a_n(s - s_0)^n + \frac{b_1}{s - s_0} + \frac{b_2}{(s - s_0)^2} + \cdots + \frac{b_m}{(s - s_0)^m} \qquad (b_m \neq 0)$$

in the punctured disk $0 < |s - s_0| < R_2$. Also, assume that $\overline{F(s)} = F(\bar{s})$ at points s where $F(s)$ is analytic.

(*a*) With the aid of the result in Exercise 7, Sec. 43, point out how it follows that

$$F(\bar{s}) = \sum_{n=0}^{\infty} \bar{a}_n(\bar{s} - \bar{s}_0)^n + \frac{\bar{b}_1}{\bar{s} - \bar{s}_0} + \frac{\bar{b}_2}{(\bar{s} - \bar{s}_0)^2} + \cdots + \frac{\bar{b}_m}{(\bar{s} - \bar{s}_0)^m} \qquad (\bar{b}_m \neq 0)$$

when $0 < |\bar{s} - \bar{s}_0| < R_2$. Then replace $\bar{s}$ by s here to obtain a Laurent series representation for $F(s)$ in the punctured disk $0 < |s - \bar{s}_0| < R_2$, and conclude that $\bar{s}_0$ is a pole of order m of $F(s)$.

(b) Use results in Exercise 12 and part (a) above to show that

$$\operatorname*{Res}_{s=s_0}[e^{st}F(s)] + \operatorname*{Res}_{s=\bar{s}_0}[e^{st}F(s)] = 2e^{\alpha t} \operatorname{Re}\left\{ e^{i\beta t}\left[b_1 + \frac{b_2}{1!}t + \cdots + \frac{b_m}{(m-1)!}t^{m-1} \right] \right\}$$

when t is real, as stated at the beginning of Sec. 67.

14. Let $F(s)$ be the function in Exercise 13, and write the nonzero coefficient b_m there in exponential form as $b_m = r_m \exp(i\theta_m)$. Then use the main result in part (b) of Exercise 13 to show that when t is real, the sum of the residues of $e^{st}F(s)$ at $s_0 = \alpha + i\beta(\beta \neq 0)$ and $\bar{s}_0$ contains a term of the type

$$\frac{2r_m}{(m-1)!}t^{m-1}e^{\alpha t}\cos(\beta t + \theta_m).$$

Note that if $\alpha > 0$, the product $t^{m-1}e^{\alpha t}$ here tends to ∞ as t tends to ∞. When the inverse Laplace transform $f(t)$ is found by summing the residues of $e^{st}F(s)$, the term displayed above is, therefore, an *unstable* component of $f(t)$ if $\alpha > 0$; and it is said to be of *resonance* type. If $m \geq 2$ and $\alpha = 0$, the term is also of resonance type.

CHAPTER 8

MAPPING BY
ELEMENTARY
FUNCTIONS

The geometric interpretation of a function of a complex variable as a mapping, or transformation, was introduced in Sec. 10. We saw there how the nature of such a function can be displayed graphically, to some extent, by the manner in which it maps certain curves and regions.

In this chapter, we shall see further examples of how various curves and regions are mapped by elementary analytic functions. Applications of such results to physical problems are illustrated in Chaps. 10 and 11.

68. LINEAR TRANSFORMATIONS

To study the mapping

(1) $$w = Az,$$

where A is a nonzero complex constant and $z \neq 0$, we write A and z in exponential form:

$$A = ae^{i\alpha}, \qquad z = re^{i\theta}.$$

Then

(2) $$w = (ar)e^{i(\alpha+\theta)};$$

and we see from equation (2) that transformation (1) expands or contracts the radius vector representing z by the factor $a = |A|$ and rotates it through an angle $\alpha = \arg A$ about the origin. The image of a given region is, therefore, geometrically similar to that region.

The mapping

(3) $w = z + B,$

where B is any complex constant, is a translation by means of the vector representing B. That is, if

$$w = u + iv, \qquad z = x + iy, \qquad \text{and} \qquad B = b_1 + ib_2,$$

then the image of any point (x, y) in the z plane is the point

(4) $(u, v) = (x + b_1, y + b_2)$

in the w plane. Since each point in any given region of the z plane is mapped into the w plane in this manner, the image region is geometrically congruent to the original one.

The general (nonconstant) *linear transformation*

(5) $w = Az + B \qquad (A \neq 0),$

which is a composition of the transformations

$$Z = Az \quad (A \neq 0) \qquad \text{and} \qquad w = Z + B,$$

is evidently an expansion or contraction and a rotation, followed by a translation.

EXAMPLE.   The mapping

$$w = (1 + i)z + 2$$

transforms the rectangular region shown in the z plane of Fig. 75 into the rectangular region shown in the w plane there. This is seen by writing it as a composition of the transformations

$$Z = (1 + i)z \qquad \text{and} \qquad w = Z + 2.$$

Since $1 + i = \sqrt{2}\exp(i\pi/4)$, the first of these transformations is an expansion by the factor $\sqrt{2}$ and a rotation through the angle $\pi/4$. The second is a translation two units to the right.

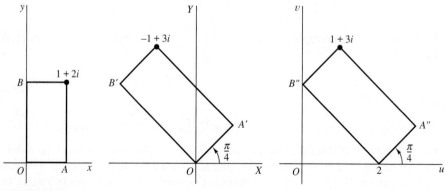

FIGURE 75
$w = (1 + i)z + 2.$

69.   THE TRANSFORMATION $w = 1/z$

The equation

(1)
$$w = \frac{1}{z}$$

establishes a one to one correspondence between the nonzero points of the z and the w planes. Since $z\bar{z} = |z|^2$, the mapping can be described by means of the successive transformations

(2)
$$Z = \frac{1}{|z|^2}z, \qquad w = \bar{Z}.$$

The first of these transformations is an inversion with respect to the unit circle $|z| = 1$. That is, the image of a nonzero point z is the point Z with the properties

$$|Z| = \frac{1}{|z|} \qquad \text{and} \qquad \arg Z = \arg z.$$

Thus the points exterior to the circle $|z| = 1$ are mapped onto the nonzero points interior to it (Fig. 76), and conversely. Any point on the circle is mapped onto itself. The second of transformations (2) is simply a reflection in the real axis.

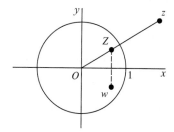

FIGURE 76

Because

$$\lim_{z \to 0} \frac{1}{z} = \infty \qquad \text{and} \qquad \lim_{z \to \infty} \frac{1}{z} = 0$$

(see Exercise 11, Sec. 14), it is natural to define a one to one transformation $w = T(z)$ from the extended z plane onto the extended w plane by writing

$$T(0) = \infty, \qquad T(\infty) = 0, \qquad \text{and} \qquad T(z) = \frac{1}{z}$$

for the remaining values of z. According to Exercise 12, Sec. 14, T is then continuous throughout the extended z plane. For this reason, when the point at infinity is involved in a discussion of the function $1/z$, it is tacitly assumed that $T(z)$ is intended.

When a point $w = u + iv$ is the image of a nonzero point $z = x + iy$ under the transformation $w = 1/z$, writing $w = \bar{z}/|z|^2$ reveals that

(3)
$$u = \frac{x}{x^2 + y^2}, \qquad v = \frac{-y}{x^2 + y^2}.$$

Also, since $z = 1/w = \overline{w}/|w|^2$,

(4) $$x = \frac{u}{u^2 + v^2}, \qquad y = \frac{-v}{u^2 + v^2}.$$

The following argument, based on these relations between coordinates, shows that *the mapping $w = 1/z$ transforms circles and lines into circles and lines.* When A, B, C, and D are all real numbers satisfying the condition $B^2 + C^2 > 4AD$, the equation

(5) $$A(x^2 + y^2) + Bx + Cy + D = 0$$

represents an arbitrary circle or line, where $A \neq 0$ for a circle and $A = 0$ for a line. The need for the condition $B^2 + C^2 > 4AD$ when $A \neq 0$ is evident if, by the method of completing the squares, we rewrite equation (5) as

$$\left(x + \frac{B}{2A}\right)^2 + \left(y + \frac{C}{2A}\right)^2 = \left(\frac{\sqrt{B^2 + C^2 - 4AD}}{2A}\right)^2.$$

When $A = 0$, the condition becomes $B^2 + C^2 > 0$, which means that B and C are not both zero. Returning to the verification of the statement in italics, we observe that if x and y satisfy equation (5), we can use relations (4) to substitute for those variables. After some simplifications, we find that u and v satisfy the equation (see also Exercise 20)

(6) $$D(u^2 + v^2) + Bu - Cv + A = 0,$$

which also represents a circle or line. Conversely, if u and v satisfy equation (6), it follows from relations (3) that x and y satisfy equation (5).

It is now clear from equations (5) and (6) that

(a) a circle ($A \neq 0$) not passing through the origin ($D \neq 0$) in the z plane is transformed into a circle not passing through the origin in the w plane;

(b) a circle ($A \neq 0$) through the origin ($D = 0$) in the z plane is transformed into a line that does not pass through the origin in the w plane;

(c) a line ($A = 0$) not passing through the origin ($D \neq 0$) in the z plane is transformed into a circle through the origin in the w plane;

(d) a line ($A = 0$) through the origin ($D = 0$) in the z plane is transformed into a line through the origin in the w plane.

EXAMPLE 1.   According to equations (5) and (6), a vertical line $x = c_1$ ($c_1 \neq 0$) is transformed by $w = 1/z$ into the circle $-c_1(u^2 + v^2) + u = 0$, or

(7) $$\left(u - \frac{1}{2c_1}\right)^2 + v^2 = \left(\frac{1}{2c_1}\right)^2,$$

which is centered on the u axis and tangent to the v axis. The image of a typical point (c_1, y) on the line is, by equations (3),

$$(u, v) = \left(\frac{c_1}{c_1^2 + y^2}, \frac{-y}{c_1^2 + y^2}\right).$$

If $c_1 > 0$, the circle (7) is evidently to the right of the v axis. As the point (c_1, y) moves up the entire line, its image traverses the circle once in the clockwise direction, the point at infinity in the extended z plane corresponding to the origin in the w plane. For if $y < 0$, then $v > 0$; and, as y increases through negative values to 0, u increases from 0 to $1/c_1$. Then, as y increases through positive values, v is negative and u decreases to 0.

If, on the other hand, $c_1 < 0$, the circle lies to the left of the v axis. As the point (c_1, y) moves upward, its image still makes one cycle, but in the counterclockwise direction. See Fig. 77, where the cases $c_1 = 1/3$ and $c_1 = -1/2$ are illustrated.

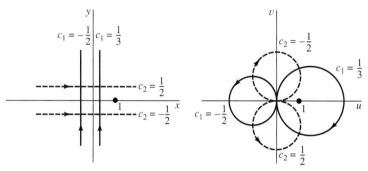

FIGURE 77
$w = 1/z$.

EXAMPLE 2. The transformation $w = 1/z$ maps a horizontal line $y = c_2$ ($c_2 \neq 0$) onto the circle

(8)
$$u^2 + \left(v + \frac{1}{2c_2}\right)^2 = \left(\frac{1}{2c_2}\right)^2,$$

which is centered on the v axis and tangent to the u axis. Two special cases are shown in Fig. 77, where the corresponding orientations of the lines and circles are also indicated.

EXAMPLE 3. When $w = 1/z$, the half plane $x \geq c_1$ ($c_1 > 0$) is mapped onto the disk

(9)
$$\left(u - \frac{1}{2c_1}\right)^2 + v^2 \leq \left(\frac{1}{2c_1}\right)^2.$$

For, according to Example 1, any line $x = c$ ($c \geq c_1$) is transformed into the circle

(10)
$$\left(u - \frac{1}{2c}\right)^2 + v^2 = \left(\frac{1}{2c}\right)^2.$$

Furthermore, as c increases through all values greater than c_1, the lines $x = c$ move to the right and the image circles (10) shrink in size. (See Fig. 78.) Since the lines $x = c$ pass through all points in the half plane $x \geq c_1$ and the circles (10) pass through all points in the disk (9), the mapping is established.

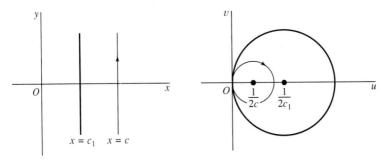

FIGURE 78
$w = 1/z$.

EXERCISES

*1. State why the transformation $w = iz$ is a rotation of the z plane through the angle $\pi/2$. Then find the image of the infinite strip $0 < x < 1$.

      Ans. $0 < v < 1$.

*2. Show that the transformation $w = iz + i$ maps the half plane $x > 0$ onto the half plane $v > 1$.

3. Find the region onto which the half plane $y > 0$ is mapped by the transformation $w = (1 + i)z$ by using (*a*) polar coordinates; (*b*) rectangular coordinates. Sketch this region.

      Ans. $v > u$.

4. Find the image of the half plane $y > 1$ under the transformation $w = (1 - i)z$.

*5. Find the image of the semi-infinite strip $x > 0, 0 < y < 2$ under the transformation $w = iz + 1$. Sketch the strip and its image.

      Ans. $-1 < u < 1, v < 0$.

6. Give a geometric description of the transformation $w = A(z + B)$, where A and B are complex constants and $A \neq 0$.

7. In Sec. 69, point out how it follows from the first of equations (4) that when $w = 1/z$, the inequality $x \geq c_1$ ($c_1 > 0$) is satisfied if and only if inequality (9) holds. Thus give an alternative verification of the mapping established in Example 3 in that section.

*8. Show that when $c_1 < 0$, the image of the half plane $x < c_1$ under the transformation $w = 1/z$ is the interior of a circle. What is the image when $c_1 = 0$?

*9. Show that the image of the half plane $y > c_2$ under the transformation $w = 1/z$ is the interior of a circle, provided $c_2 > 0$. Find the image when $c_2 < 0$; also find it when $c_2 = 0$.

10. Find the image of the infinite strip $0 < y < 1/(2c)$ under the transformation $w = 1/z$. Sketch the strip and its image.

      Ans. $u^2 + (v + c)^2 > c^2, v < 0$.

11. Find the image of the quadrant $x > 1, y > 0$ under the transformation $w = 1/z$.

      Ans. $(u - 1/2)^2 + v^2 < (1/2)^2, v < 0$.

12. Verify the mapping, where $w = 1/z$, of the regions and parts of the boundaries indicated in (*a*) Fig. 4, Appendix 2; (*b*) Fig. 5, Appendix 2.

*13. Describe geometrically the transformation $w = 1/(z - 1)$.

*14. Describe geometrically the transformation $w = i/z$. State why it transforms circles and lines into circles and lines.

15. Find the image of the semi-infinite strip $x > 0, 0 < y < 1$ under the transformation $w = i/z$. Sketch the strip and its image.

      Ans. $(u - 1/2)^2 + v^2 > (1/2)^2, u > 0, v > 0.$

16. By writing $w = \rho \exp(i\phi)$, show that the mapping $w = 1/z$ transforms the hyperbola $x^2 - y^2 = 1$ into the lemniscate $\rho^2 = \cos 2\phi$. (See Exercise 18, Sec. 4.)

17. Let the circle $|z| = 1$ have a positive, or counterclockwise, orientation. Determine the orientation of its image under the transformation $w = 1/z$.

18. Show that when a circle is transformed into a circle under the transformation $w = 1/z$, the center of the original circle is *never* mapped onto the center of the image circle.

*19. Using the exponential form $z = re^{i\theta}$ of z, show that the transformation

$$w = z + \frac{1}{z},$$

which is the sum of the identity transformation and the transformation discussed in Sec. 69, maps circles $r = r_0$ onto ellipses with parametric representations

$$u = \left(r_0 + \frac{1}{r_0}\right)\cos\theta, \qquad v = \left(r_0 - \frac{1}{r_0}\right)\sin\theta \qquad (0 \le \theta \le 2\pi)$$

and foci at the points $w = \pm 2$. Then show how it follows that this transformation maps the entire circle $|z| = 1$ onto the segment $-2 \le u \le 2$ of the u axis and the domain outside that circle onto the rest of the w plane.

20. (*a*) Write equation (5), Sec. 69, in the form

$$2Az\bar{z} + (B - Ci)z + (B + Ci)\bar{z} + 2D = 0,$$

    where $z = x + iy$.

    (*b*) Show that when $w = 1/z$, the result in part (*a*) becomes

$$2Dw\bar{w} + (B + Ci)w + (B - Ci)\bar{w} + 2A = 0.$$

Then show that if $w = u + iv$, this equation is the same as equation (6), Sec. 69.

70. LINEAR FRACTIONAL TRANSFORMATIONS

The transformation

$$(1) \qquad\qquad w = \frac{az + b}{cz + d} \qquad (ad - bc \ne 0),$$

where $a, b, c,$ and d are complex constants, is called a *linear fractional transformation*, or Möbius transformation. Observe that equation (1) can be written in the form

$$(2) \qquad\qquad Azw + Bz + Cw + D = 0 \qquad (AD - BC \ne 0);$$

and, conversely, any equation of type (2) can be put in the form (1). Since this alternative form is linear in z and linear in w, or bilinear in z and w, another name for a linear fractional transformation is *bilinear transformation*.

    When $c = 0$, the condition $ad - bc \ne 0$ with equation (1) becomes $ad \ne 0$; and we see that the transformation reduces to a nonconstant linear function. When

$c \neq 0$, equation (1) can be written

$$(3) \qquad w = \frac{a}{c} + \frac{bc - ad}{c} \cdot \frac{1}{cz + d} \qquad (ad - bc \neq 0).$$

So, once again, the condition $ad - bc \neq 0$ ensures that we do not have a constant function. The transformation $w = 1/z$ is, of course, a special case of transformation (1) when $c \neq 0$.

Equation (3) reveals that when $c \neq 0$, a linear fractional transformation is a composition of the mappings

$$Z = cz + d, \qquad W = \frac{1}{Z}, \qquad w = \frac{a}{c} + \frac{bc - ad}{c}W \qquad (ad - bc \neq 0).$$

It thus follows that *a linear fractional transformation always transforms circles and lines into circles and lines* because these special linear fractional transformations do this. (See Secs. 68 and 69.)

Solving equation (1) for z, we find that

$$(4) \qquad z = \frac{-dw + b}{cw - a} \qquad (ad - bc \neq 0).$$

Thus, when a given point w is the image of some point z under transformation (1), the point z is retrieved by means of equation (4). If $c = 0$, so that a and d are both nonzero, each point in the w plane is evidently the image of one and only one point in the z plane. The same is true if $c \neq 0$, except when $w = a/c$ since the denominator in equation (4) vanishes if w has that value. We can, however, enlarge the domain of definition of transformation (1) in order to define a linear fractional transformation T on the *extended* z plane such that the point $w = a/c$ is the image of $z = \infty$ when $c \neq 0$. We first write

$$(5) \qquad T(z) = \frac{az + b}{cz + d} \qquad (ad - bc \neq 0).$$

We then write

$$T(\infty) = \infty \qquad \text{if} \qquad c = 0$$

and

$$T(\infty) = \frac{a}{c} \quad \text{and} \quad T\left(-\frac{d}{c}\right) = \infty \qquad \text{if} \qquad c \neq 0.$$

In view of Exercise 10, Sec. 14, this makes T continuous on the extended z plane. It also agrees with the way in which we enlarged the domain of definition of the transformation $w = 1/z$ in Sec. 69.

When its domain of definition is enlarged in this way, the linear fractional transformation (5) is a *one to one* mapping of the extended z plane *onto* the extended w plane. That is, $T(z_1) \neq T(z_2)$ whenever $z_1 \neq z_2$; and, for each point w in the second plane, there is a point z in the first one such that $T(z) = w$. Hence, associated with the transformation T, there is an *inverse transformation* T^{-1}, which is defined on the extended w plane as follows:

$$T^{-1}(w) = z \qquad \text{if and only if} \qquad T(z) = w.$$

From equation (4), we see that

(6) $$T^{-1}(w) = \frac{-dw + b}{cw - a} \qquad (ad - bc \neq 0).$$

Evidently, T^{-1} is itself a linear fractional transformation, where

$$T^{-1}(\infty) = \infty \qquad \text{if} \qquad c = 0$$

and

$$T^{-1}\left(\frac{a}{c}\right) = \infty \quad \text{and} \quad T^{-1}(\infty) = -\frac{d}{c} \qquad \text{if} \qquad c \neq 0.$$

If T and S are two linear fractional transformations, then so is the composition $S[T(z)]$. This can be verified by combining expressions of the type (5). Note that, in particular, $T^{-1}[T(z)] = z$ for each point z in the extended plane.

There is always a linear fractional transformation that maps three given distinct points z_1, z_2, and z_3 onto three specified distinct points w_1, w_2, and w_3, respectively. Verification of this will appear in Sec. 71, where the image w of a point z under such a transformation is given implicitly in terms of z. We illustrate here a more direct approach to finding the desired transformation.

EXAMPLE 1. Let us find the special case of transformation (1) that maps the points

$$z_1 = -1, \qquad z_2 = 0, \qquad \text{and} \qquad z_3 = 1$$

onto the points

$$w_1 = -i, \qquad w_2 = 1, \qquad \text{and} \qquad w_3 = i.$$

Since 1 is the image of 0, expression (1) tells us that $1 = b/d$, or $d = b$. Thus

(7) $$w = \frac{az + b}{cz + b} \qquad [b(a - c) \neq 0].$$

Then, since -1 and 1 are transformed into $-i$ and i, respectively, it follows that

$$ic - ib = -a + b \qquad \text{and} \qquad ic + ib = a + b.$$

Adding corresponding sides of these equations, we find that $c = -ib$; and subtraction reveals that $a = ib$. Consequently,

$$w = \frac{ibz + b}{-ibz + b} = \frac{b(iz + 1)}{b(-iz + 1)}.$$

Since b is arbitrary and nonzero here, we may assign it the value unity (or cancel it out) and write

$$w = \frac{iz + 1}{-iz + 1} \cdot \frac{i}{i} = \frac{i - z}{i + z}.$$

EXAMPLE 2.   Suppose that the points

$$z_1 = 1, \qquad z_2 = 0, \qquad \text{and} \qquad z_3 = -1$$

are to be mapped onto

$$w_1 = i, \qquad w_2 = \infty, \qquad \text{and} \qquad w_3 = 1.$$

Since $w_2 = \infty$ corresponds to $z_2 = 0$, we require that $d = 0$ in expression (1); and so

$$(8) \qquad\qquad\qquad w = \frac{az + b}{cz} \qquad (bc \neq 0).$$

Because 1 is to be mapped onto i and -1 onto 1, we have the relations

$$ic = a + b \qquad \text{and} \qquad -c = -a + b;$$

and it follows that

$$b = \frac{i-1}{2}c, \qquad a = \frac{i+1}{2}c.$$

Finally, then, if we write $c = 2$, equation (8) becomes

$$w = \frac{(i+1)z + (i-1)}{2z}.$$

71.   AN IMPLICIT FORM

The equation

$$(1) \qquad\qquad \frac{(w - w_1)(w_2 - w_3)}{(w - w_3)(w_2 - w_1)} = \frac{(z - z_1)(z_2 - z_3)}{(z - z_3)(z_2 - z_1)}$$

defines (implicitly) a linear fractional transformation that maps distinct points z_1, z_2, and z_3 in the finite z plane onto distinct points w_1, w_2, and w_3, respectively, in the finite w plane.* To verify this, we write equation (1) as

$$(2) \quad (z - z_3)(w - w_1)(z_2 - z_1)(w_2 - w_3) = (z - z_1)(w - w_3)(z_2 - z_3)(w_2 - w_1).$$

If $z = z_1$, the right-hand side of equation (2) is zero; and it follows that $w = w_1$. Similarly, if $z = z_3$, the left-hand side is zero and, consequently, $w = w_3$. If $z = z_2$, we have the linear equation

$$(w - w_1)(w_2 - w_3) = (w - w_3)(w_2 - w_1),$$

---

*The two sides of equation (1) are *cross ratios*, which play an important role in more extensive developments of linear fractional transformations than in this book. See, for instance, R. P. Boas, "Invitation to Complex Analysis," pp. 192–196, 1993 or J. B. Conway, "Functions of One Complex Variable," 2d ed., pp. 48–55, 1988.

whose unique solution is $w = w_2$. One can see that the mapping defined by equation (1) is actually a linear fractional transformation by expanding the products in equation (2) and writing the result in the form (Sec. 70)

$$(3) \qquad Azw + Bz + Cw + D = 0.$$

The condition $AD - BC \neq 0$, which is needed with equation (3), is clearly satisfied since, as just demonstrated, equation (1) does not define a constant function. It is left to the reader (Exercise 10, Sec. 72) to show that equation (1) defines the *only* linear fractional transformation mapping the points z_1, z_2, and z_3 onto w_1, w_2, and w_3, respectively.

EXAMPLE 1. The transformation found in Example 1, Sec. 70, required that

$$z_1 = -1, \quad z_2 = 0, \quad z_3 = 1 \qquad \text{and} \qquad w_1 = -i, \quad w_2 = 1, \quad w_3 = i.$$

Using equation (1) to write

$$\frac{(w + i)(1 - i)}{(w - i)(1 + i)} = \frac{(z + 1)(0 - 1)}{(z - 1)(0 + 1)}$$

and then solving for w in terms of z, we arrive at the transformation

$$w = \frac{i - z}{i + z},$$

found earlier.

If equation (1) is modified properly, it can also be used when the point at infinity is one of the prescribed points in either the (extended) z or w plane. Suppose, for instance, that $z_1 = \infty$. Since any linear fractional transformation is continuous on the extended plane, we need only replace z_1 on the right-hand side of equation (1) by $1/z_1$, clear fractions, and let z_1 tend to zero:

$$\lim_{z_1 \to 0} \frac{(z - 1/z_1)(z_2 - z_3)}{(z - z_3)(z_2 - 1/z_1)} \cdot \frac{z_1}{z_1} = \lim_{z_1 \to 0} \frac{(z_1 z - 1)(z_2 - z_3)}{(z - z_3)(z_1 z_2 - 1)} = \frac{z_2 - z_3}{z - z_3}.$$

The desired modification of equation (1) is, then,

$$\frac{(w - w_1)(w_2 - w_3)}{(w - w_3)(w_2 - w_1)} = \frac{z_2 - z_3}{z - z_3}.$$

Note that this modification is obtained formally by simply deleting the factors involving z_1 in equation (1). It is easy to check that the same formal approach applies when any of the other prescribed points is ∞.

EXAMPLE 2. In Example 2, Sec. 70, the prescribed points were

$$z_1 = 1, \quad z_2 = 0, \quad z_3 = -1 \qquad \text{and} \qquad w_1 = i, \quad w_2 = \infty, \quad w_3 = 1.$$

In this case, we use the modification

$$\frac{w - w_1}{w - w_3} = \frac{(z - z_1)(z_2 - z_3)}{(z - z_3)(z_2 - z_1)}$$

of equation (1), which tells us that

$$\frac{w - i}{w - 1} = \frac{(z - 1)(0 + 1)}{(z + 1)(0 - 1)}.$$

Solving here for w, we arrive at the desired transformation:

$$w = \frac{(i + 1)z + (i - 1)}{2z}.$$

72. MAPPINGS OF THE UPPER HALF PLANE

Let us determine all linear fractional transformations that map the upper half plane $\operatorname{Im} z > 0$ onto the open disk $|w| < 1$ and the boundary $\operatorname{Im} z = 0$ onto the boundary $|w| = 1$ (Fig. 79).

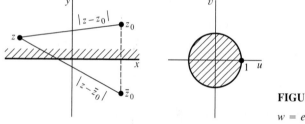

FIGURE 79

$w = e^{i\alpha} \dfrac{z - z_0}{z - \bar{z}_0}$ $(\operatorname{Im} z_0 > 0).$

Keeping in mind that points on the line $\operatorname{Im} z = 0$ are to be transformed into points on the circle $|w| = 1$, we start by selecting the points $z = 0, z = 1$, and $z = \infty$ on the line and determining conditions on a linear fractional transformation

$$(1) \qquad\qquad w = \frac{az + b}{cz + d} \qquad (ad - bc \neq 0)$$

which are necessary in order for the images of those points to have unit modulus.

We note from equation (1) that if $|w| = 1$ when $z = 0$, then $|b/d| = 1$; that is,

$$(2) \qquad\qquad |b| = |d| \neq 0.$$

Now, according to Sec. 70, the image w of the point $z = \infty$ is a finite number, namely $w = a/c$, only if $c \neq 0$. So the requirement that $|w| = 1$ when $z = \infty$ means that $|a/c| = 1$, or

$$(3) \qquad\qquad |a| = |c| \neq 0;$$

and the fact that a and c are nonzero enables us to rewrite equation (1) as

$$(4) \qquad\qquad w = \frac{a}{c} \cdot \frac{z + (b/a)}{z + (d/c)}.$$

Then, since $|a/c| = 1$ and

$$\left|\frac{b}{a}\right| = \left|\frac{d}{c}\right| \neq 0,$$

according to relations (2) and (3), equation (4) can be put in the form

$$(5) \qquad\qquad w = e^{i\alpha}\frac{z - z_0}{z - z_1} \qquad (|z_1| = |z_0| \neq 0),$$

where α is a real constant and z_0 and z_1 are (nonzero) complex constants.

Next, we impose on transformation (5) the condition that $|w| = 1$ when $z = 1$. This tells us that

$$|1 - z_1| = |1 - z_0|,$$

or

$$(1 - z_1)(1 - \bar{z}_1) = (1 - z_0)(1 - \bar{z}_0).$$

But $z_1\bar{z}_1 = z_0\bar{z}_0$ since $|z_1| = |z_0|$, and the above relation reduces to

$$z_1 + \bar{z}_1 = z_0 + \bar{z}_0;$$

that is, $\mathrm{Re}\, z_1 = \mathrm{Re}\, z_0$. It follows that either $z_1 = z_0$ or $z_1 = \bar{z}_0$, again since $|z_1| = |z_0|$. If $z_1 = z_0$, transformation (5) becomes the constant function $w = \exp(i\alpha)$; hence $z_1 = \bar{z}_0$.

Transformation (5), with $z_1 = \bar{z}_0$, maps the point z_0 onto the origin $w = 0$; and, since points interior to the circle $|w| = 1$ are to be the images of points *above* the real axis in the z plane, we may conclude that $\mathrm{Im}\, z_0 > 0$. Any linear fractional transformation having the mapping property stated in the first paragraph of this section must, therefore, be of the form

$$(6) \qquad\qquad w = e^{i\alpha}\frac{z - z_0}{z - \bar{z}_0} \qquad (\mathrm{Im}\, z_0 > 0),$$

where α is real.

It remains to show that, conversely, any linear fractional transformation of the form (6) has the desired mapping property. This is easily done by taking absolute values of each side of equation (6) and interpreting the resulting equation,

$$|w| = \frac{|z - z_0|}{|z - \bar{z}_0|},$$

geometrically. If a point z lies above the real axis, both it and the point z_0 lie on the same side of that axis, which is the perpendicular bisector of the line segment joining z_0 and $\bar{z}_0$. It follows that the distance $|z - z_0|$ is less than the distance $|z - \bar{z}_0|$ (Fig. 79); that is, $|w| < 1$. Likewise, if z lies below the real axis, the distance $|z - z_0|$ is greater than the distance $|z - \bar{z}_0|$; and so $|w| > 1$. Finally, if z is on the real axis, $|w| = 1$ because then $|z - z_0| = |z - \bar{z}_0|$. Since any linear fractional transformation is a one to one mapping of the extended z plane onto the extended w plane, this shows that *transformation* (6) *maps the half plane* Im $z > 0$ *onto the disk* $|w| < 1$ *and the boundary of the half plane onto the boundary of the disk.*

EXAMPLE 1.   The transformation $w = (i - z)/(i + z)$ in Examples 1 in Secs. 70 and 71 can be written

$$w = e^{i\pi} \frac{z - i}{z - \bar{i}}.$$

Hence it has the above mapping property.

Images of the upper half plane Im $z \geq 0$ under other types of linear fractional transformations are often fairly easy to determine by examining the particular transformation in question.

EXAMPLE 2.   By writing $z = x + iy$ and $w = u + iv$, we can readily show that the transformation

(7) $$w = \frac{z - 1}{z + 1}$$

maps the half plane $y > 0$ onto the half plane $v > 0$ and the x axis onto the u axis. We first note that when the number z is real, so is the number w. Consequently, since the image of the real axis $y = 0$ is either a circle or a line, it must be the real axis $v = 0$. Furthermore, for any point w in the finite w plane,

$$v = \operatorname{Im} w = \operatorname{Im} \frac{(z - 1)(\bar{z} + 1)}{(z + 1)(\bar{z} + 1)} = \frac{2y}{|z + 1|^2} \qquad (z \neq -1).$$

The numbers y and v thus have the same sign, and this means that points above the x axis correspond to points above the u axis and points below the x axis correspond to points below the u axis. Finally, since points on the x axis correspond to points on the u axis and since a linear fractional transformation is a one to one mapping of the extended plane onto the extended plane (Sec. 70), the stated mapping property of transformation (7) is established.

EXERCISES

1. Find the linear fractional transformation that maps the points $z_1 = 2, z_2 = i, z_3 = -2$ onto the points $w_1 = 1, w_2 = i, w_3 = -1$.
   Ans. $w = (3z + 2i)/(iz + 6)$.

2. Find the linear fractional transformation that maps the points $z_1 = -i, z_2 = 0, z_3 = i$ onto the points $w_1 = -1, w_2 = i, w_3 = 1$. Into what curve is the imaginary axis $x = 0$ transformed?

3. Find the bilinear transformation that maps the points $z_1 = \infty, z_2 = i, z_3 = 0$ onto the points $w_1 = 0, w_2 = i, w_3 = \infty$.
   Ans. $w = -1/z$.

4. Find the bilinear transformation that maps the distinct points z_1, z_2, z_3 onto the points $w_1 = 0, w_2 = 1, w_3 = \infty$.

$$Ans. \ w = \frac{(z - z_1)(z_2 - z_3)}{(z - z_3)(z_2 - z_1)}.$$

5. Show that a composition of two linear fractional transformations is again a linear fractional transformation, as stated in Sec. 70.

6. A *fixed point* of a transformation $w = f(z)$ is a point z_0 such that $f(z_0) = z_0$. Show that every linear fractional transformation, with the exception of the identity transformation $w = z$, has at most two fixed points in the extended plane.

7. Find the fixed points (Exercise 6) of the transformation
   (a) $w = (z - 1)/(z + 1)$;   (b) $w = (6z - 9)/z$.
       Ans. (a) $z = \pm i$;   (b) $z = 3$.

8. Modify equation (1), Sec. 71, for the case in which both z_2 and w_2 are the point at infinity. Then show that any linear fractional transformation must be of the form $w = az \ (a \neq 0)$ when its fixed points (Exercise 6) are 0 and ∞.

9. Prove that if the origin is a fixed point (Exercise 6) of a linear fractional transformation, then the transformation can be written in the form $w = z/(cz + d)$, where $d \neq 0$.

10. Show that there is only one linear fractional transformation that maps three given distinct points z_1, z_2, and z_3 in the extended z plane onto three specified distinct points w_1, w_2, and w_3 in the extended w plane.
       Suggestion: Let T and S be two such linear fractional transformations. Then, after pointing out why $S^{-1}[T(z_k)] = z_k \ (k = 1, 2, 3)$, use the results in Exercises 5 and 6 to show that $S^{-1}[T(z)] = z$ for all z. Thus show that $T(z) = S(z)$ for all z.

*11. Recall (Sec. 72) that the transformation $w = (i - z)/(i + z)$ maps the half plane $\mathrm{Im}\,z > 0$ onto the disk $|w| < 1$ and the boundary of the half plane onto the boundary of the disk. Show that a point $z = x$ is mapped onto the point

$$w = \frac{1 - x^2}{1 + x^2} + i\frac{2x}{1 + x^2},$$

and then complete the verification of the mapping illustrated in Fig. 13, Appendix 2, by showing that segments of the x axis are mapped as indicated there.

*12. Verify the mapping shown in Fig. 12, Appendix 2, where $w = (z - 1)/(z + 1)$.
       Suggestion: Write the given transformation as a composition of the mappings

$$Z = iz, \qquad W = \frac{i - Z}{i + Z}, \qquad w = -W.$$

   Then refer to the mapping verified in Exercise 11.

13. (a) By finding the inverse of the transformation $w = (i - z)/(i + z)$ and appealing to Fig. 13, Appendix 2, which was verified in Exercise 11, show that the transformation $w = i(1 - z)/(1 + z)$ maps the disk $|z| \leq 1$ onto the upper half plane $\mathrm{Im}\,w \geq 0$.
    (b) Show that the linear fractional transformation $w = (z - 2)/z$ can be written

$$Z = z - 1, \qquad W = i\frac{1 - Z}{1 + Z}, \qquad w = iW.$$

   Then, with the aid of the result in part (a), verify that the transformation $w = (z-2)/z$ maps the disk $|z - 1| \leq 1$ onto the left half plane $\mathrm{Re}\,w \leq 0$.

14. Transformation (6), Sec. 72, maps the point $z = \infty$ onto the point $w = \exp(i\alpha)$, which lies on the boundary of the disk $|w| \leq 1$. Show that if $0 < \alpha < 2\pi$ and the points $z = 0$

and $z = 1$ are to be mapped onto the points $w = 1$ and $w = \exp(i\alpha/2)$, respectively, then the transformation can be written

$$w = e^{i\alpha} \frac{z + \exp(-i\alpha/2)}{z + \exp(i\alpha/2)}.$$

15. Note that when $\alpha = \pi/2$, the transformation in Exercise 14 becomes

$$w = \frac{iz + \exp(i\pi/4)}{z + \exp(i\pi/4)}.$$

Verify that this special case maps points on the x axis in the manner indicated in Fig. 80.

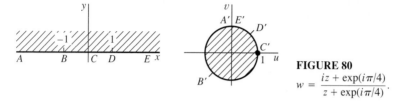

FIGURE 80
$$w = \frac{iz + \exp(i\pi/4)}{z + \exp(i\pi/4)}.$$

16. Show that when $\operatorname{Im} z_0 < 0$, transformation (6), Sec. 72, maps the lower half plane $\operatorname{Im} z \leq 0$ onto the unit disk $|w| \leq 1$.

17. With the aid of equation (1), Sec. 71, prove that if a linear fractional transformation maps each point of the x axis onto the u axis, then the coefficients in the transformation are all real, except possibly for a common complex factor. The converse statement is evident.

18. Let $T(z) = (az+b)/(cz+d)$, where $ad - bc \neq 0$, be any linear fractional transformation other than $T(z) = z$. Show that $T^{-1} = T$ if and only if $d = -a$.
    Suggestion: Write the equation $T^{-1}(z) = T(z)$ as

$$(a + d)[cz^2 + (d - a)z - b] = 0.$$

73. EXPONENTIAL AND LOGARITHMIC TRANSFORMATIONS

The transformation $w = e^z$ can be written $\rho e^{i\phi} = e^x e^{iy}$, where $z = x + iy$ and $w = \rho e^{i\phi}$. Thus (see Sec. 5) $\rho = e^x$ and $\phi = y + 2n\pi$, where n is any integer; and the above transformation from the z into the w plane can be expressed in the form

(1) $$\rho = e^x, \qquad \phi = y.$$

The image of a typical point $z = (c_1, y)$ on a vertical line $x = c_1$ has polar coordinates $\rho = \exp c_1$ and $\phi = y$ in the w plane. That image moves counterclockwise around the circle shown in Fig. 81 as z moves up the line. The image of the line is evidently the entire circle; and each point on the circle is the image of an infinite number of points, spaced 2π units apart, along the line.

A horizontal line $y = c_2$ is mapped in a one to one manner onto the ray $\phi = c_2$. As a point $z = (x, c_2)$ moves along that line from left to right, the coordinate $\rho = e^x$ of the image point on the ray increases through all positive values, as indicated in Fig. 81.

Vertical and horizontal line *segments* are mapped onto portions of circles and rays, respectively, and images of various regions are readily obtained from these observations.

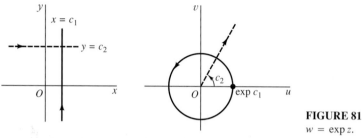

FIGURE 81
$w = \exp z$.

EXAMPLE 1. Let us show that the transformation $w = e^z$ maps the rectangular region $a \le x \le b, c \le y \le d$ onto the region $e^a \le \rho \le e^b, c \le \phi \le d$. The two regions and corresponding parts of their boundaries are indicated in Fig. 82. The vertical line segment AD is mapped onto the arc $\rho = e^a, c \le \phi \le d$, which is labeled $A'D'$. The images of vertical line segments to the right of AD and joining the horizontal parts of the boundary are larger arcs; eventually, the image of the line segment BC is the arc $\rho = e^b, c \le \phi \le d$, labeled $B'C'$. The mapping is one to one if $d - c < 2\pi$. In particular, if $c = 0$ and $d = \pi$, then $0 \le \phi \le \pi$; and the rectangular region is mapped onto half of a circular ring, as shown in Fig. 8, Appendix 2.

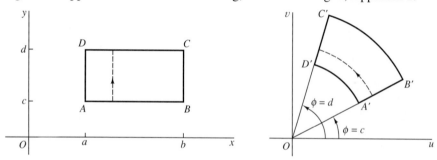

FIGURE 82
$w = \exp z$.

EXAMPLE 2. When $w = e^z$, the image of the infinite strip $a \le y \le b$ is the sector $a \le \phi \le b$ of the w plane (Fig. 83). This is seen by recalling from Fig. 81

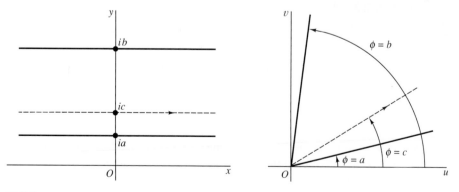

FIGURE 83
$w = \exp z$.

above how a horizontal line $y = c$ is transformed into a ray $\phi = c$ from the origin. The mapping of the strip here is evidently one to one if $b - a < 2\pi$; and the case in which $a = 0$ and $b = \pi$ is shown in Fig. 6 of Appendix 2, where corresponding points on the boundaries of the two regions are also indicated.

Any branch

$$(2) \qquad w = \log z = \ln r + i\theta \qquad (r > 0, \alpha < \theta < \alpha + 2\pi)$$

of the logarithmic function (see Sec. 26) maps its domain of definition in a one to one manner onto the strip $\alpha < v < \alpha + 2\pi$ in the w plane (Fig. 84). This is easy to see by letting a point $z = r \exp(i\theta_0)$, where $\alpha < \theta_0 < \alpha + 2\pi$, move outward from the origin along the ray $\theta = \theta_0$. Its image is evidently the point whose *rectangular* coordinates in the w plane are $(\ln r, \theta_0)$. Hence that image moves to the right along the entire length of the horizontal line $v = \theta_0$. Since these lines fill the strip $\alpha < \theta < \alpha + 2\pi$ as the choice of θ_0 varies between α and $\alpha + 2\pi$, the mapping of the strip is, indeed, one to one.

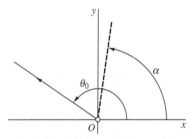 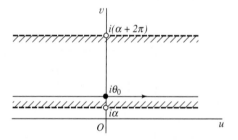

FIGURE 84
$w = \log z \quad (|z| > 0, \alpha < \arg z < \alpha + 2\pi)$.

Logarithmic mappings can be useful in applications, and we conclude this section with a mapping that we shall need later in Chap. 10 (Sec. 84).

EXAMPLE 3.   The transformation

$$(3) \qquad w = \operatorname{Log} \frac{z - 1}{z + 1},$$

where the principal branch of the logarithmic function is used, is a composition of the functions

$$(4) \qquad Z = \frac{z - 1}{z + 1} \qquad \text{and} \qquad w = \operatorname{Log} Z.$$

We know from Example 2 in Sec. 72 that the first of transformations (4) maps the upper half plane $y > 0$ onto the upper half plane $Y > 0$, where $z = x + iy$ and $Z = X + iY$. Furthermore, from Fig. 84 in this section, it is easy to see that the second of transformations (4) maps the half plane $Y > 0$ onto the strip $0 < v < \pi$, where

$w = u + iv$. Consequently, transformation (3) maps the half plane $y > 0$ onto the strip $0 < v < \pi$. Corresponding boundary points are shown in Fig. 19, Appendix 2.

74.   THE TRANSFORMATION $w = \sin z$

Since (Sec. 24)

$$\sin z = \sin x \cosh y + i \cos x \sinh y,$$

the transformation $w = \sin z$ can be written

(1) $u = \sin x \cosh y, \qquad v = \cos x \sinh y.$

The following examples illustrate various mapping properties of this transformation.

EXAMPLE 1.   The transformation $w = \sin z$ is a one to one mapping of the semi-infinite strip $-\pi/2 \le x \le \pi/2, y \ge 0$ in the z plane onto the upper half $v \ge 0$ of the w plane.

To verify this, we first show that the boundary of the strip is mapped in a one to one manner onto the real axis in the w plane, as indicated in Fig. 85. The image of the line segment BA is found by writing $x = \pi/2$ in equations (1) and restricting y to be nonnegative. Since $u = \cosh y$ and $v = 0$ when $x = \pi/2$, a typical point $(\pi/2, y)$ on BA is mapped onto the point $(\cosh y, 0)$ in the w plane; and that image must move to the right from B' along the u axis as $(\pi/2, y)$ moves upward from B. A point $(x, 0)$ on the horizonal segment DB has image $(\sin x, 0)$, which moves to the right from D' to B' as x increases from $x = -\pi/2$ to $x = \pi/2$, or as $(x, 0)$ goes from D to B. Finally, as a point $(-\pi/2, y)$ on the line segment DE moves upward from D, its image $(-\cosh y, 0)$ moves to the left from D'.

One way to see how the interior of the strip is mapped onto the upper half $v > 0$ of the w plane is to examine the images of certain vertical half lines. If $0 < c_1 < \pi/2$, points on the line $x = c_1$ are transformed into points on the curve

(2) $u = \sin c_1 \cosh y, \qquad v = \cos c_1 \sinh y \qquad (-\infty < y < \infty),$

which is the right-hand branch of the hyperbola

(3) $$\frac{u^2}{\sin^2 c_1} - \frac{v^2}{\cos^2 c_1} = 1$$

with foci at the points

$$w = \pm \sqrt{\sin^2 c_1 + \cos^2 c_1} = \pm 1.$$

The second of equations (2) shows that as a point on the line moves upward, its image on the hyperbola also moves upward. In particular, there is a one to one mapping of the top half ($y > 0$) of the line onto the top half ($v > 0$) of the hyperbola's branch. Such a half line L and its image L' are shown in Fig. 85. If $-\pi/2 < c_1 < 0$, the line $x = c_1$ is mapped onto the left-hand branch of the same hyperbola; and, as before, there is a one to one correspondence between points on the top half of the line and

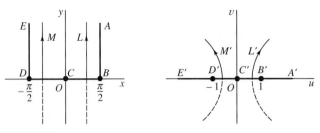

FIGURE 85
$w = \sin z$.

those on the top half of the hyperbola's branch. That half line and its image are labeled M and M' in Fig. 85.

The line $x = 0$, or the y axis, needs to be considered separately. According to equations (1), the image of each point $(0, y)$ is $(0, \sinh y)$. Hence the y axis is mapped onto the v axis in a one to one manner, the positive y axis corresponding to the positive v axis.

Now each point in the interior $-\pi/2 < x < \pi/2, y > 0$ of the strip lies on one of the above-mentioned half lines, and it is important to notice that the images of those half lines are distinct and constitute the entire half plane $v > 0$. More precisely, if the upper half L of a line $x = c_1$ $(0 < c_1 < \pi/2)$ is thought of as moving to the left toward the positive y axis, the right-hand branch of the hyperbola containing L' is opening up wider and its vertex $(\sin c_1, 0)$ is tending toward the origin $w = 0$. Hence L' tends to become the positive v axis, which we found in the preceding paragraph to be the image of the positive y axis. On the other hand, as L approaches the segment BA of the boundary of the strip, the branch of the hyperbola closes down around the segment $B'A'$ of the u axis and its vertex $(\sin c_1, 0)$ tends toward the point $w = 1$. Similar statements can be made regarding the half line M and its image M' in Fig. 85. We may conclude that the image of each point in the interior of the strip lies in the upper half plane $v > 0$ and, furthermore, that each point in the half plane is the image of exactly one point in the interior of the strip.

This completes our demonstration that the transformation $w = \sin z$ is a one to one mapping of the strip $-\pi/2 \le x \le \pi/2, y \ge 0$ onto the half plane $v \ge 0$. The final result is shown in Fig. 9, Appendix 2. The right-hand half of the strip is evidently mapped onto the first quadrant of the w plane, as shown in Fig. 10, Appendix 2.

Another convenient way to find the images of certain regions when $w = \sin z$ is to consider the images of *horizontal* line segments $y = c_2$ $(-\pi \le x \le \pi)$, where $c_2 > 0$. According to equations (1), the image of such a line segment is the curve with parametric representation

(4) $u = \sin x \cosh c_2,$ $v = \cos x \sinh c_2$ $(-\pi \le x \le \pi)$.

That curve is readily seen to be the ellipse

(5) $$\frac{u^2}{\cosh^2 c_2} + \frac{v^2}{\sinh^2 c_2} = 1,$$

whose foci lie at the points

$$w = \pm\sqrt{\cosh^2 c_2 - \sinh^2 c_2} = \pm 1.$$

The image of a point (x, c_2) moving to the right from point A to point E in Fig. 86 makes one circuit around the ellipse in the clockwise direction. Note that when smaller values of the positive number c_2 are taken, the ellipse becomes smaller but retains the same foci $(\pm 1, 0)$. In the limiting case $c_2 = 0$, equations (4) become

$$u = \sin x, \qquad v = 0 \qquad (-\pi \leq x \leq \pi);$$

and we find that the interval $-\pi \leq x \leq \pi$ of the x axis is mapped onto the interval $-1 \leq u \leq 1$ of the u axis. The mapping is not, however, one to one, as it is when $c_2 > 0$.

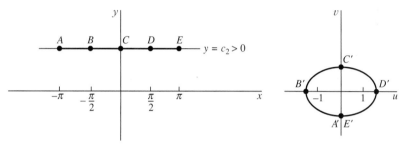

FIGURE 86
$w = \sin z.$

We now show how these observations can be used.

EXAMPLE 2. The rectangular region $-\pi/2 \leq x \leq \pi/2, 0 \leq y \leq b$ is mapped by $w = \sin z$ in a one to one manner onto the semi-elliptical region shown in Fig. 87, where corresponding boundary points are also indicated. For if L is a line segment $y = c_2 (-\pi/2 \leq x \leq \pi/2)$, where $0 < c_2 \leq b$, its image L' is the top half of the ellipse (5). As c_2 decreases, L moves downward toward the x axis and the semi-ellipse L' also moves downward and tends to become the line segment $E'F'A'$ from $w = -1$ to $w = 1$. In fact, when $c_2 = 0$, equations (4) become

$$u = \sin x, \qquad v = 0 \qquad \left(-\frac{\pi}{2} \leq x \leq \frac{\pi}{2}\right);$$

FIGURE 87
$w = \sin z.$

and this is clearly a one to one mapping of the segment EFA onto $E'F'A'$. Inasmuch as any point in the semi-elliptical region in the w plane lies on one and only one of the semi-ellipses, or on the limiting case $E'F'A'$, that point is the image of exactly one point in the rectangular region in the z plane. The desired mapping, which is also shown in Fig. 11 of Appendix 2, is now established.

Mappings by various other functions closely related to the sine function are easily obtained once mappings by the sine function are known.

EXAMPLE 3.   We need only recall the identity (Sec. 24)

$$\cos z = \sin\left(z + \frac{\pi}{2}\right)$$

to see that the transformation

$$w = \cos z$$

can be written successively as

$$Z = z + \frac{\pi}{2}, \qquad w = \sin Z.$$

Hence the cosine transformation is the same as the sine transformation preceded by a translation to the right through $\pi/2$ units.

EXAMPLE 4.   According to Sec. 25, the transformation

$$w = \sinh z$$

can be written $w = -i\sin(iz)$, or

$$Z = iz, \qquad W = \sin Z, \qquad w = -iW.$$

It is, therefore, a combination of the sine transformation and rotations through right angles. The transformation

$$w = \cosh z$$

is, likewise, essentially a cosine transformation since $\cosh z = \cos(iz)$.

EXERCISES

*1. Show that the lines $ay = x$ $(a \neq 0)$ are mapped onto the spirals $\rho = \exp(a\phi)$ under the transformation $w = \exp z$, where $w = \rho\exp(i\phi)$.

*2. By considering the images of horizontal line segments, verify that the image of the rectangular region $a \leq x \leq b, c \leq y \leq d$ under the transformation $w = \exp z$ is the region $e^a \leq \rho \leq e^b, c \leq \phi \leq d$, as shown in Fig. 82 (Sec. 73).

3. Verify the mapping of the region and boundary shown in Fig. 7 of Appendix 2, where the transformation is $w = \exp z$.

*4. Find the image of the semi-infinite strip $x \geq 0, 0 \leq y \leq \pi$ under the transformation $w = \exp z$, and label corresponding portions of the boundaries.

5. The equation $w = \log(z - 1)$ can be written

$$Z = z - 1, \qquad w = \log Z.$$

Find a branch of $\log Z$ such that the cut z plane consisting of all points except those on the segment $x \geq 1$ of the real axis is mapped by $w = \log(z - 1)$ onto the strip $0 < v < 2\pi$ in the w plane.

*6. Show that the transformation $w = \sin z$ maps the top half $(y > 0)$ of the line $x = c_1 \, (-\pi/2 < c_1 < 0)$ in a one to one manner onto the top half $(v > 0)$ of the left-hand branch of hyperbola (3), Sec. 74, as indicated in Fig. 85 of that section.

*7. Show that under the transformation $w = \sin z$, a line $x = c_1 \, (\pi/2 < c_1 < \pi)$ is mapped onto the right-hand branch of hyperbola (3), Sec. 74. Note that the mapping is one to one and that the upper and lower halves of the line are mapped onto the *lower* and *upper* halves, respectively, of the branch.

8. Vertical half lines were used in Example 1, Sec. 74, to show that the transformation $w = \sin z$ is a one to one mapping of the open region $-\pi/2 < x < \pi/2, y > 0$ onto the half plane $v > 0$. Verify that result by using, instead, horizontal line segments $y = c_2 \, (-\pi/2 < x < \pi/2)$, where $c_2 > 0$.

9. (*a*) Show that under the transformation $w = \sin z$, the images of the line segments forming the boundary of the rectangular region $0 \leq x \leq \pi/2, 0 \leq y \leq 1$ are the line segments and the arc $D'E'$ indicated in Fig. 88. The arc $D'E'$ is a quarter of the ellipse

$$\frac{u^2}{\cosh^2 1} + \frac{v^2}{\sinh^2 1} = 1.$$

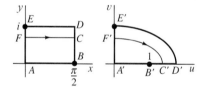

FIGURE 88
$w = \sin z$.

(*b*) Complete the mapping indicated in Fig. 88 by using images of horizontal line segments to prove that the transformation $w = \sin z$ establishes a one to one correspondence between the interior points of the regions $ABDE$ and $A'B'D'E'$.

*10. Verify that the interior of a rectangular region $-\pi \leq x \leq \pi, a \leq y \leq b$ lying above the x axis is mapped by $w = \sin z$ onto the interior of an elliptical ring which has a cut along the segment $-\sinh b \leq v \leq -\sinh a$ of the negative real axis, as indicated in Fig. 89. Note that, while the mapping of the interior of the rectangular region is one to one, the mapping of its boundary is *not*.

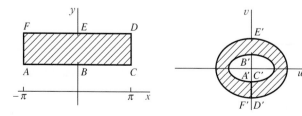

FIGURE 89
$w = \sin z$.

*11.(a) Show that the equation $w = \cosh z$ can be written

$$Z = iz + \frac{\pi}{2}, \qquad w = \sin Z.$$

(b) Use the result in part (a), together with the mapping by $\sin z$ shown in Fig. 10, Appendix 2, to verify that the transformation $w = \cosh z$ maps the semi-infinite strip $x \geq 0, 0 \leq y \leq \pi/2$ in the z plane onto the first quadrant $u \geq 0, v \geq 0$ of the w plane. Indicate corresponding parts of the boundaries of the two regions.

*12. Observe that the transformation $w = \cosh z$ can be expressed as a composition of the mappings

$$Z = e^z, \qquad W = Z + \frac{1}{Z}, \qquad w = \frac{1}{2}W.$$

Then, by referring to Figs. 7 and 16 in Appendix 2, show that when $w = \cosh z$, the semi-infinite strip $x \leq 0, 0 \leq y \leq \pi$ in the z plane is mapped onto the lower half $v \leq 0$ of the w plane. Indicate corresponding parts of the boundaries.

*13.(a) Verify that the equation $w = \sin z$ can be written

$$Z = i\left(z + \frac{\pi}{2}\right), \qquad W = \cosh Z, \qquad w = -W.$$

(b) Use the result in part (a) here and the one in Exercise 12 to show that the transformation $w = \sin z$ maps the semi-infinite strip $-\pi/2 \leq x \leq \pi/2, y \geq 0$ onto the half plane $v \geq 0$, as shown in Fig. 9, Appendix 2. (This mapping was verified in a different way in Example 1, Sec. 74.)

75. MAPPINGS BY BRANCHES OF $z^{1/2}$

As in Sec. 7, the values of $z^{1/2}$ are the two square roots of z when $z \neq 0$. According to that section, if polar coordinates are used and $z = r \exp(i\Theta) \, (r > 0, -\pi < \Theta \leq \pi)$, then

$$(1) \qquad z^{1/2} = \sqrt{r} \exp \frac{i(\Theta + 2k\pi)}{2} \qquad (k = 0, 1),$$

the principal root occurring when $k = 0$. In Sec. 27, we saw that $z^{1/2}$ can also be written

$$(2) \qquad z^{1/2} = \exp\left(\frac{1}{2} \log z\right) \qquad (z \neq 0).$$

The *principal branch* $F_0(z)$ of the double-valued function $z^{1/2}$ is then obtained by taking the principal branch of $\log z$ and writing (see Sec. 28)

$$F_0(z) = \exp\left(\frac{1}{2} \operatorname{Log} z\right) \qquad (|z| > 0, -\pi < \operatorname{Arg} z < \pi),$$

or

$$(3) \qquad F_0(z) = \sqrt{r} \exp \frac{i\Theta}{2} \qquad (r > 0, -\pi < \Theta < \pi).$$

The right-hand side of this equation is, of course, the same as the right-hand side of equation (1) when $k = 0$ and $-\pi < \Theta < \pi$ there. The origin and the ray $\Theta = \pi$ form the branch cut for F_0, and the origin is the branch point.

Images of curves and regions under the transformation $w = F_0(z)$ may be obtained by writing $w = \rho \exp(i\phi)$, where $\rho = \sqrt{r}$ and $\phi = \Theta/2$. Arguments are evidently halved by this transformation, and it is understood that $w = 0$ when $z = 0$.

EXAMPLE 1. It is easy to verify that $w = F_0(z)$ is a one to one mapping of the quarter disk $0 \le r \le 2, 0 \le \theta \le \pi/2$ in the z plane onto the sector $0 \le \rho \le \sqrt{2}, 0 \le \phi \le \pi/4$ in the w plane (Fig. 90). To do this, we observe that as a point $z = r \exp(i\theta_1)$ $(0 \le \theta_1 \le \pi/2)$ moves outward from the origin along a radius R_1 of length 2 and with angle of inclination θ_1, its image $w = \sqrt{r} \exp(i\theta_1/2)$ moves outward from the origin in the w plane along a radius R'_1 whose length is $\sqrt{2}$ and angle of inclination is $\theta_1/2$. See Fig. 90, where another radius R_2 and its image R'_2 are also shown. It is now clear from the figure that if the region in the z plane is thought of as being swept out by a radius, starting with DA and ending with DC, then the region in the w plane is swept out by the corresponding radius, starting with $D'A'$ and ending with $D'C'$. This establishes a one to one correspondence between points in the two regions.

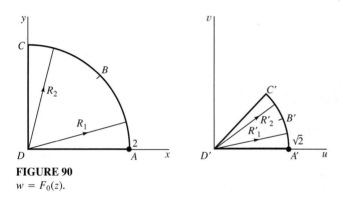

FIGURE 90
$w = F_0(z)$.

EXAMPLE 2. The transformation $w = F_0(\sin z)$ can be written

$$Z = \sin z, \quad w = F_0(Z) \quad (|Z| > 0, -\pi < \operatorname{Arg} Z < \pi).$$

As noted at the end of Example 1 in Sec. 74, the first transformation maps the semi-infinite strip $0 \le x \le \pi/2, y \ge 0$ onto the first quadrant $X \ge 0, Y \ge 0$ in the Z plane. The second transformation, with the understanding that $F_0(0) = 0$, maps that quadrant onto an octant in the w plane. These successive transformations are illustrated in Fig. 91, where corresponding boundary points are shown.

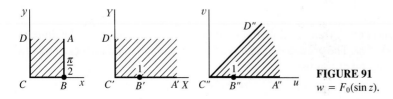

FIGURE 91
$w = F_0(\sin z)$.

When $-\pi < \Theta < \pi$ and the branch

$$\log z = \ln r + i(\Theta + 2\pi)$$

of the logarithmic function is used, equation (2) yields the branch

(4) $F_1(z) = \sqrt{r}\exp\dfrac{i(\Theta + 2\pi)}{2}$ $(r > 0, -\pi < \Theta < \pi)$

of $z^{1/2}$, which corresponds to the choice $k = 1$ in equation (1). Since $\exp(i\pi) = -1$, it follows that $F_1(z) = -F_0(z)$. The values $\pm F_0(z)$ thus represent the totality of values of $z^{1/2}$ at all points in the domain $r > 0, -\pi < \Theta < \pi$. If, by means of expression (3), we extend the domain of definition of F_0 to include the ray $\Theta = \pi$ and if we write $F_0(0) = 0$, then the values $\pm F_0(z)$ represent the totality of values of $z^{1/2}$ in the entire z plane.

Other branches of $z^{1/2}$ are obtained by using other branches of $\log z$ in expression (2). A branch where the ray $\theta = \alpha$ is used to form the branch cut is given by the equation

(5) $f_\alpha(z) = \sqrt{r}\exp\dfrac{i\theta}{2}$ $(r > 0, \alpha < \theta < \alpha + 2\pi)$.

Observe that when $\alpha = -\pi$, we have the branch $F_0(z)$ and that when $\alpha = \pi$, we have the branch $F_1(z)$. Just as in the case of F_0, the domain of definition of f_α can be extended to the entire complex plane by using expression (5) to define f_α at the nonzero points on the branch cut and by writing $f_\alpha(0) = 0$. Such extensions are, however, never continuous in the entire complex plane.

Finally, suppose that n is any positive integer, where $n \geq 2$. The values of $z^{1/n}$ are the nth roots of z when $z \neq 0$; and, according to Secs. 7 and 27, the multiple-valued function $z^{1/n}$ can be written

(6) $z^{1/n} = \exp\left(\dfrac{1}{n}\log z\right) = \sqrt[n]{r}\exp\dfrac{i(\Theta + 2k\pi)}{n}$ $(k = 0, 1, 2, \ldots, n - 1)$,

where $r = |z|$ and $\Theta = \operatorname{Arg} z$. The case $n = 2$ has just been considered. In the general case, each of the n functions

(7) $F_k(z) = \sqrt[n]{r}\exp\dfrac{i(\Theta + 2k\pi)}{n}$ $(k = 0, 1, 2, \ldots, n - 1)$

is a branch of $z^{1/n}$, defined on the domain $r > 0, -\pi < \Theta < \pi$. When $w = \rho e^{i\phi}$, the transformation $w = F_k(z)$ is a one to one mapping of that domain onto the domain

$$\rho > 0, \qquad \dfrac{(2k - 1)\pi}{n} < \phi < \dfrac{(2k + 1)\pi}{n}.$$

These n branches of $z^{1/n}$ yield the n distinct nth roots of z at any point z in the domain $r > 0, -\pi < \Theta < \pi$. The principal branch occurs when $k = 0$, and further branches of the type (5) are readily constructed.

76.   SQUARE ROOTS OF POLYNOMIALS

We now consider some mappings that are compositions of polynomials and square roots of z.

EXAMPLE 1.   Branches of the double-valued function $(z - z_0)^{1/2}$ can be obtained by noting that it is a composition of the translation $Z = z - z_0$ with the double-valued function $Z^{1/2}$. Each branch of $Z^{1/2}$ yields a branch of $(z - z_0)^{1/2}$. When $Z = Re^{i\theta}$, branches of $Z^{1/2}$ are

$$Z^{1/2} = \sqrt{R}\exp\frac{i\theta}{2} \qquad (R > 0, \alpha < \theta < \alpha + 2\pi).$$

Hence if we write

$$R = |z - z_0|, \qquad \Theta = \operatorname{Arg}(z - z_0), \qquad \text{and} \qquad \theta = \arg(z - z_0),$$

two branches of $(z - z_0)^{1/2}$ are

(1) $$G_0(z) = \sqrt{R}\exp\frac{i\Theta}{2} \qquad (R > 0, -\pi < \Theta < \pi)$$

and

(2) $$g_0(z) = \sqrt{R}\exp\frac{i\theta}{2} \qquad (R > 0, 0 < \theta < 2\pi).$$

The branch of $Z^{1/2}$ used in writing $G_0(z)$ is defined at all points in the Z plane except for the origin and points on the ray $\operatorname{Arg} Z = \pi$. The transformation $w = G_0(z)$ is, therefore, a one to one mapping of the domain

$$|z - z_0| > 0, \quad -\pi < \operatorname{Arg}(z - z_0) < \pi$$

onto the right half $\operatorname{Re} w > 0$ of the w plane (Fig. 92). The transformation $w = g_0(z)$ maps the domain

$$|z - z_0| > 0, \quad 0 < \arg(z - z_0) < 2\pi$$

in a one to one manner onto the upper half plane $\operatorname{Im} w > 0$.

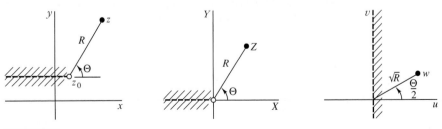

FIGURE 92
$w = G_0(z)$.

EXAMPLE 2.   For an instructive but less elementary example, we now consider the double-valued function $(z^2 - 1)^{1/2}$. Using established properties of logarithms, we can write

$$(z^2 - 1)^{1/2} = \exp\left[\frac{1}{2}\log(z^2 - 1)\right] = \exp\left[\frac{1}{2}\log(z - 1) + \frac{1}{2}\log(z + 1)\right],$$

or

(3) $$(z^2 - 1)^{1/2} = (z - 1)^{1/2}(z + 1)^{1/2} \qquad (z \neq \pm 1).$$

Thus, if $f_1(z)$ is a branch of $(z - 1)^{1/2}$ defined on a domain D_1 and $f_2(z)$ is a branch of $(z + 1)^{1/2}$ defined on a domain D_2, the product $f(z) = f_1(z)f_2(z)$ is a branch of $(z^2 - 1)^{1/2}$ defined at all points lying in both D_1 and D_2.

In order to obtain a specific branch of $(z^2 - 1)^{1/2}$, we use the branch of $(z - 1)^{1/2}$ and the branch of $(z + 1)^{1/2}$ given by equation (2). If we write

$$r_1 = |z - 1| \qquad \text{and} \qquad \theta_1 = \arg(z - 1),$$

that branch of $(z - 1)^{1/2}$ is

$$f_1(z) = \sqrt{r_1}\exp\frac{i\theta_1}{2} \qquad (r_1 > 0, 0 < \theta_1 < 2\pi).$$

The branch of $(z + 1)^{1/2}$ given by equation (2) is

$$f_2(z) = \sqrt{r_2}\exp\frac{i\theta_2}{2} \qquad (r_2 > 0, 0 < \theta_2 < 2\pi),$$

where

$$r_2 = |z + 1| \qquad \text{and} \qquad \theta_2 = \arg(z + 1).$$

The product of these two branches is, therefore, the branch f of $(z^2 - 1)^{1/2}$ defined by the equation

(4) $$f(z) = \sqrt{r_1 r_2}\exp\frac{i(\theta_1 + \theta_2)}{2},$$

where

$$r_k > 0, \qquad 0 < \theta_k < 2\pi \qquad (k = 1, 2).$$

As illustrated in Fig. 93, the branch f is defined everywhere in the z plane except on the ray $r_2 \geq 0, \theta_2 = 0$, which is the portion $x \geq -1$ of the x axis.

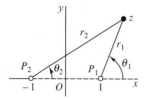

FIGURE 93

The branch f of $(z^2 - 1)^{1/2}$ given in equation (4) can be extended to a function

(5) $$F(z) = \sqrt{r_1 r_2}\, \exp \frac{i(\theta_1 + \theta_2)}{2},$$

where

$$r_k > 0, \quad 0 \le \theta_k < 2\pi \quad (k = 1, 2) \qquad \text{and} \qquad r_1 + r_2 > 2.$$

As we shall now see, this function is analytic everywhere in its domain of definition, which is the entire z plane except for the segment $-1 \le x \le 1$ of the x axis.

Since $F(z) = f(z)$ for all z in the domain of definition of F except on the ray $r_1 > 0, \theta_1 = 0$, we need only show that F is analytic on that ray. To do this, we form the product of the branches of $(z - 1)^{1/2}$ and $(z + 1)^{1/2}$ which are given by equation (1). That is, we consider the function

$$G(z) = \sqrt{r_1 r_2}\, \exp \frac{i(\Theta_1 + \Theta_2)}{2},$$

where

$$r_1 = |z - 1|, \quad r_2 = |z + 1|, \quad \Theta_1 = \operatorname{Arg}(z - 1), \quad \Theta_2 = \operatorname{Arg}(z + 1)$$

and where

$$r_k > 0, \qquad -\pi < \Theta_k < \pi \qquad (k = 1, 2).$$

Observe that G is analytic in the entire z plane except for the ray $r_1 \ge 0, \Theta_1 = \pi$. Now $F(z) = G(z)$ when the point z lies above or on the ray $r_1 > 0, \Theta_1 = 0$; for then $\theta_k = \Theta_k$ $(k = 1, 2)$. When z lies below that ray, $\theta_k = \Theta_k + 2\pi$ $(k = 1, 2)$. Consequently, $\exp(i\theta_k/2) = -\exp(i\Theta_k/2)$; and this means that

$$\exp \frac{i(\theta_1 + \theta_2)}{2} = \left(\exp \frac{i\theta_1}{2}\right)\left(\exp \frac{i\theta_2}{2}\right) = \exp \frac{i(\Theta_1 + \Theta_2)}{2}.$$

So, again, $F(z) = G(z)$. Since $F(z) = G(z)$ in a domain containing the ray $r_1 > 0, \Theta_1 = 0$ and since $G(z)$ is analytic in that domain, $F(z)$ is analytic there. Hence *$F(z)$ is analytic everywhere except on the line segment $P_2 P_1$ in Fig. 93.*

The function F defined by equation (5) cannot itself be extended to a function which is analytic at points on the line segment $P_2 P_1$; for the value on the right in equation (5) jumps from $i\sqrt{r_1 r_2}$ to numbers near $-i\sqrt{r_1 r_2}$ as the point z moves downward across that line segment. Hence the extension would not even be continuous there.

The transformation $w = F(z)$ is, as we shall see, a one to one mapping of the domain D_z consisting of all points in the z plane except those on the line segment $P_2 P_1$ onto the domain D_w consisting of the entire w plane with the exception of the segment $-1 \le v \le 1$ of the v axis (Fig. 94).

Before verifying this, we note that if $z = iy$ $(y > 0)$, then

$$r_1 = r_2 > 1 \qquad \text{and} \qquad \theta_1 + \theta_2 = \pi;$$

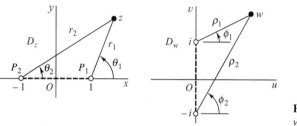

FIGURE 94
$w = F(z)$.

hence the positive y axis is mapped by $w = F(z)$ onto that part of the v axis for which $v > 1$. The negative y axis is, moreover, mapped onto that part of the v axis for which $v < -1$. Each point in the upper half $y > 0$ of the domain D_z is mapped into the upper half $v > 0$ of the w plane, and each point in the lower half $y < 0$ of the domain D_z is mapped into the lower half $v < 0$ of the w plane. The ray $r_1 > 0, \theta_1 = 0$ is mapped onto the positive real axis in the w plane, and the ray $r_2 > 0, \theta_2 = \pi$ is mapped onto the negative real axis there.

To show that the transformation $w = F(z)$ is *one to one*, we observe that if $F(z_1) = F(z_2)$, then $z_1^2 - 1 = z_2^2 - 1$. From this, it follows that $z_1 = z_2$ or $z_1 = -z_2$. However, because of the manner in which F maps the upper and lower halves of the domain D_z, as well as the portions of the real axis lying in D_z, the case $z_1 = -z_2$ is impossible. Thus, if $F(z_1) = F(z_2)$, then $z_1 = z_2$; and F is one to one.

We can show that F maps the domain D_z *onto* the domain D_w by finding a function H mapping D_w into D_z with the property that if $z = H(z)$, then $w = F(z)$. This will show that, for any point w in D_w, there exists a point z in D_z such that $F(z) = w$; that is, the mapping F is onto. The mapping H will be the inverse of F.

To find H, we first note that if w is a value of $(z^2 - 1)^{1/2}$ for a specific z, then $w^2 = z^2 - 1$; and z is, therefore, a value of $(w^2 + 1)^{1/2}$ for that w. The function H will be a branch of the double-valued function

$$(w^2 + 1)^{1/2} = (w - i)^{1/2}(w + i)^{1/2} \qquad (w \neq \pm i).$$

Following our procedure for obtaining the function $F(z)$, we write $w - i = \rho_1 \exp(i\phi_1)$ and $w + i = \rho_2 \exp(i\phi_2)$. (See Fig. 94.) With the restrictions

$$\rho_k > 0, \qquad \frac{-\pi}{2} \le \phi_k < \frac{3\pi}{2} \quad (k = 1, 2) \qquad \text{and} \qquad \rho_1 + \rho_2 > 2,$$

we then write

(6)
$$H(w) = \sqrt{\rho_1 \rho_2} \exp \frac{i(\phi_1 + \phi_2)}{2},$$

the domain of definition being D_w. The transformation $z = H(w)$ maps points of D_w lying above or below the u axis onto points above or below the x axis, respectively. It maps the positive u axis into that part of the positive x axis where $x > 1$ and the negative u axis into that part of the negative x axis where $x < -1$. If $z = H(w)$, then $z^2 = w^2 + 1$; and so $w^2 = z^2 - 1$. Since z is in D_z and since $F(z)$ and $-F(z)$ are the two values of $(z^2 - 1)^{1/2}$ for a point in D_z, we see that $w = F(z)$ or $w = -F(z)$.

But it is evident from the manner in which F and H map the upper and lower halves of their domains of definition, including the portions of the real axes lying in those domains, that $w = F(z)$.

   Mappings by branches of double-valued functions

(7) $w = (z^2 + Az + B)^{1/2} = [(z - z_0)^2 - z_1^2]^{1/2}$ $(z_1 \neq 0)$,

where $A = -2z_0$ and $B = z_0^2 - z_1^2$, can be treated with the aid of the results found for the function F in Example 2 and the successive transformations

(8) $Z = \dfrac{z - z_0}{z_1},$ $W = (Z^2 - 1)^{1/2},$ $w = z_1 W.$

EXERCISES

*1. By referring to Fig. 10, Appendix 2, show that the transformation $w = \sin^2 z$ maps the strip $0 \leq x \leq \pi/2, y \geq 0$ onto the half plane $v \geq 0$. Indicate corresponding parts of the boundaries.

2. Use Fig. 9, Appendix 2, to show that under the transformation $w = (\sin z)^{1/4}$, where the principal branch of the fractional power is taken, the semi-infinite strip $-\pi/2 < x < \pi/2, y > 0$ is mapped onto the part of the first quadrant lying between the line $v = u$ and the u axis. Label corresponding parts of the boundaries.

3. According to Example 2, Sec. 72, the linear fractional transformation $Z = (z-1)/(z+1)$ maps the x axis onto the X axis and the half planes $y > 0$ and $y < 0$ onto the half planes $Y > 0$ and $Y < 0$, respectively. Show that, in particular, it maps the segment $-1 \leq x \leq 1$ of the x axis onto the segment $X \leq 0$ of the X axis. Then show that when the principal branch of the square root is used, the composite function

$$w = Z^{1/2} = \left(\frac{z - 1}{z + 1}\right)^{1/2}$$

   maps the z plane, except for the segment $-1 \leq x \leq 1$ of the x axis, onto the half plane $u > 0$.

4. Determine the image of the domain $r > 0, -\pi < \Theta < \pi$ in the z plane under each of the transformations $w = F_k(z)$ $(k = 0, 1, 2, 3)$, where $F_k(z)$ are the four branches of $z^{1/4}$ given by equation (7), Sec. 75, when $n = 4$. Use these branches to determine the fourth roots of i.

5. The branch F of $(z^2 - 1)^{1/2}$ in Example 2, Sec. 76, was defined in terms of the coordinates $r_1, r_2, \theta_1, \theta_2$. Explain geometrically why the conditions $r_1 > 0, 0 < \theta_1 + \theta_2 < \pi$ describe the quadrant $x > 0, y > 0$ of the z plane. Then show that the transformation $w = F(z)$ maps that quadrant onto the quadrant $u > 0, v > 0$ of the w plane.
   Suggestion: To show that the quadrant $x > 0, y > 0$ in the z plane is described, note that $\theta_1 + \theta_2 = \pi$ at each point on the positive y axis and that $\theta_1 + \theta_2$ decreases as a point z moves to the right along a ray $\theta_2 = c$ $(0 < c < \pi/2)$.

6. For the transformation $w = F(z)$ of the first quadrant of the z plane onto the first quadrant of the w plane (Exercise 5), show that

$$u = \frac{1}{\sqrt{2}} \sqrt{r_1 r_2 + x^2 - y^2 - 1} \quad \text{and} \quad v = \frac{1}{\sqrt{2}} \sqrt{r_1 r_2 - x^2 + y^2 + 1},$$

where $r_1^2 r_2^2 = (x^2 + y^2 + 1)^2 - 4x^2$, and that the image of the portion of the hyperbola $x^2 - y^2 = 1$ in the first quadrant is the ray $v = u \ (u > 0)$.

7. Show that in Exercise 6 the domain D that lies under the hyperbola and in the first quadrant of the z plane is described by the conditions $r_1 > 0, 0 < \theta_1 + \theta_2 < \pi/2$. Then show that the image of D is the octant $0 < v < u$. Sketch the domain D and its image.

8. Let F be the branch of $(z^2 - 1)^{1/2}$ defined in Example 2, Sec. 76, and let $z_0 = r_0 \exp(i\theta_0)$ be a fixed point, where $r_0 > 0$ and $0 \le \theta_0 < 2\pi$. Show that a branch F_0 of $(z^2 - z_0^2)^{1/2}$, whose branch cut is the line segment between the points z_0 and $-z_0$, can be written $F_0(z) = z_0 F(Z)$, where $Z = z/z_0$.

9. Write $z - 1 = r_1 \exp(i\theta_1)$ and $z + 1 = r_2 \exp(i\Theta_2)$, where

$$0 < \theta_1 < 2\pi \qquad \text{and} \qquad -\pi < \Theta_2 < \pi,$$

to define a branch of the function

(a) $(z^2 - 1)^{1/2}$;   (b) $\left(\dfrac{z-1}{z+1}\right)^{1/2}$.

In each case, the branch cut should consist of the two rays $\theta_1 = 0$ and $\Theta_2 = \pi$.

10. Using the notation in Sec. 76, show that the function

$$w = \left(\frac{z-1}{z+1}\right)^{1/2} = \sqrt{\frac{r_1}{r_2}} \exp \frac{i(\theta_1 - \theta_2)}{2}$$

is a branch with the same domain of definition D_z and the same branch cut as the function $w = F(z)$ in that section. Show that this transformation maps D_z onto the right half plane $\rho > 0, -\pi/2 < \phi < \pi/2$, where the point $w = 1$ is the image of the point $z = \infty$. Also, show that the inverse transformation is

$$z = \frac{1 + w^2}{1 - w^2} \qquad (\operatorname{Re} w > 0).$$

(Compare Exercise 3.)

11. Show that the transformation in Exercise 10 maps the region outside the unit circle $|z| = 1$ in the upper half of the z plane onto the region in the first quadrant of the w plane between the line $v = u$ and the u axis. Sketch the two regions.

12. Write $z = r \exp(i\Theta), z - 1 = r_1 \exp(i\Theta_1)$, and $z + 1 = r_2 \exp(i\Theta_2)$, where the values of all three arguments lie between $-\pi$ and π. Then define a branch of the function $[z(z^2 - 1)]^{1/2}$ whose branch cut consists of the two segments $x \le -1$ and $0 \le x \le 1$ of the x axis.

77. RIEMANN SURFACES

The remaining two sections of this chapter constitute a brief introduction to the concept of a mapping defined on a *Riemann surface*, which is a generalization of the complex plane consisting of more than one sheet. The theory rests on the fact that at each point on such a surface only one value of a given multiple-valued function is assigned. The material in these two sections will not be used in the chapters to follow, and the reader may skip to Chap. 9 without disruption.

Once a Riemann surface is devised for a given function, the function is single-valued on the surface and the theory of single-valued functions applies there. Complexities arising because the function is multiple-valued are thus relieved by a geometric device. However, the description of those surfaces and the arrangement of proper connections between the sheets can become quite involved. We limit our attention to fairly simple examples and begin with a surface for log z.

EXAMPLE 1. Corresponding to each nonzero number z, the multiple-valued function

$$(1) \qquad\qquad \log z = \ln r + i\theta$$

has infinitely many values. To describe log z as a single-valued function, we replace the z plane, with the origin deleted, by a surface on which a new point is located whenever the argument of the number z is increased or decreased by 2π, or an integral multiple of 2π.

We treat the z plane, with the origin deleted, as a thin sheet R_0 which is cut along the positive half of the real axis. On that sheet, let θ range from 0 to 2π. Let a second sheet R_1 be cut in the same way and placed in front of the sheet R_0. The lower edge of the slit in R_0 is then joined to the upper edge of the slit in R_1. On R_1, the angle θ ranges from 2π to 4π; so, when z is represented by a point on R_1, the imaginary component of log z ranges from 2π to 4π.

A sheet R_2 is then cut in the same way and placed in front of R_1. The lower edge of the slit in R_1 is joined to the upper edge of the slit in this new sheet, and similarly for sheets $R_3, R_4, \ldots$. A sheet R_{-1} on which θ varies from 0 to -2π is cut and placed behind R_0, with the lower edge of its slit connected to the upper edge of the slit in R_0; the sheets $R_{-2}, R_{-3}, \ldots$ are constructed in like manner. The coordinates r and θ of a point on any sheet can be considered as polar coordinates of the projection of the point onto the original z plane, the angular coordinate θ being restricted to a definite range of 2π radians on each sheet.

Consider any continuous curve on this connected surface of infinitely many sheets. As a point z describes that curve, the values of log z vary continuously since θ, in addition to r, varies continuously; and log z now assumes just one value corresponding to each point on the curve. For example, as the point makes a complete cycle around the origin on the sheet R_0 over the path indicated in Fig. 95, the angle changes from 0 to 2π. As it moves across the ray $\theta = 2\pi$, the point passes to the sheet R_1 of the surface. As the point completes a cycle in R_1, the angle θ varies from 2π to 4π; and, as it crosses the ray $\theta = 4\pi$, the point passes to the sheet R_2.

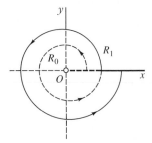

FIGURE 95

The surface described here is a Riemann surface for $\log z$. It is a connected surface of infinitely many sheets, arranged so that $\log z$ is a single-valued function of points on it.

The transformation $w = \log z$ maps the whole Riemann surface in a one to one manner onto the entire w plane. The image of the sheet R_0 is the strip $0 \leq v \leq 2\pi$. As a point z moves onto the sheet R_1 over the arc shown in Fig. 96, its image w moves upward across the line $v = 2\pi$, as indicated in that figure.

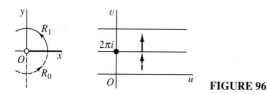

FIGURE 96

Note that $\log z$, defined on the sheet R_1, represents the analytic continuation (Sec. 52) of the single-valued analytic function

$$f(z) = \ln r + i\theta \qquad (0 < \theta < 2\pi)$$

upward across the positive real axis. In this sense, $\log z$ is not only a single-valued function of all points z on the Riemann surface but also an *analytic* function at all points there.

The sheets could, of course, be cut along the negative real axis, or along any other ray from the origin, and properly joined along the slits to form other Riemann surfaces for $\log z$.

EXAMPLE 2.   Corresponding to each point in the z plane other than the origin, the square root function

(2) $$z^{1/2} = \sqrt{r}e^{i\theta/2}$$

has two values. A Riemann surface for $z^{1/2}$ is obtained by replacing the z plane with a surface made up of two sheets R_0 and R_1, each cut along the positive real axis and with R_1 placed in front of R_0. The lower edge of the slit in R_0 is joined to the upper edge of the slit in R_1, and the lower edge of the slit in R_1 is joined to the upper edge of the slit in R_0.

As a point z starts from the upper edge of the slit in R_0 and describes a continuous circuit around the origin in the counterclockwise direction (Fig. 97), the angle θ increases from 0 to 2π. The point then passes from the sheet R_0 to the sheet R_1, where θ increases from 2π to 4π. As the point moves still further, it passes back to the sheet R_0, where the values of θ can vary from 4π to 6π or from 0 to 2π, a choice that does not affect the value of $z^{1/2}$, etc. Note that the value of $z^{1/2}$ at a point where the circuit passes from the sheet R_0 to the sheet R_1 is different from the value of $z^{1/2}$ at a point where the circuit passes from the sheet R_1 to the sheet R_0.

We have thus constructed a Riemann surface on which $z^{1/2}$ is single-valued for each nonzero z. In that construction, the edges of the sheets R_0 and R_1 are joined

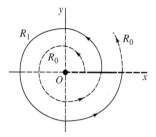

FIGURE 97

in pairs in such a way that the resulting surface is closed and connected. The points where two of the edges are joined are distinct from the points where the other two edges are joined. Thus it is physically impossible to build a model of that Riemann surface. In visualizing a Riemann surface, it is important to understand how we are to proceed when we arrive at an edge of a slit.

The origin is a special point on this Riemann surface. It is common to both sheets, and a curve around the origin on the surface must wind around it twice in order to be a closed curve. A point of this kind on a Riemann surface is called a *branch point*.

The image of the sheet R_0 under the transformation $w = z^{1/2}$ is the upper half of the w plane since the argument of w is $\theta/2$ on R_0, where $0 \le \theta/2 \le \pi$. Likewise, the image of the sheet R_1 is the lower half of the w plane. As defined on either sheet, the function is the analytic continuation, across the cut, of the function defined on the other sheet. In this respect, the single-valued function $z^{1/2}$ of points on the Riemann surface is analytic at all points except the origin.

78. SURFACES FOR RELATED FUNCTIONS

EXAMPLE 1. Let us describe a Riemann surface for the double-valued function

$$(1) \qquad f(z) = (z^2 - 1)^{1/2} = \sqrt{r_1 r_2} \exp \frac{i(\theta_1 + \theta_2)}{2},$$

where $z - 1 = r_1 \exp(i\theta_1)$ and $z + 1 = r_2 \exp(i\theta_2)$. A branch of this function, with the line segment $P_1 P_2$ between the branch points $z = \pm 1$ as a branch cut (Fig. 98), was described in Example 2, Sec. 76. That branch is as written above, with the restrictions $r_k > 0, 0 \le \theta_k < 2\pi \ (k = 1, 2)$ and $r_1 + r_2 > 2$. The branch is not defined on the segment $P_1 P_2$.

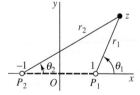

FIGURE 98

A Riemann surface for the double-valued function (1) must consist of two sheets of R_0 and R_1. Let both sheets be cut along the segment P_1P_2. The lower edge of the slit in R_0 is then joined to the upper edge of the slit in R_1, and the lower edge in R_1 is joined to the upper edge in R_0.

On the sheet R_0, let the angles θ_1 and θ_2 range from 0 to 2π. If a point on the sheet R_0 describes a simple closed curve that encloses the segment P_1P_2 once in the counterclockwise direction, then both θ_1 and θ_2 change by the amount 2π upon the return of the point to its original position. The change in $(\theta_1 + \theta_2)/2$ is also 2π, and the value of f is unchanged. If a point starting on the sheet R_0 describes a path that passes twice around just the branch point $z = 1$, it crosses from the sheet R_0 onto the sheet R_1 and then back onto the sheet R_0 before it returns to its original position. In this case, the value of θ_1 changes by the amount 4π, while the value of θ_2 does not change at all. Similarly, for a circuit twice around the point $z = -1$, the value of θ_2 changes by 4π, while the value of θ_1 remains unchanged. Again, the change in $(\theta_1 + \theta_2)/2$ is 2π; and the value of f is unchanged. Thus, on the sheet R_0, the range of the angles θ_1 and θ_2 may be extended by changing both θ_1 and θ_2 by the same integral multiple of 2π or by changing just one of the angles by a multiple of 4π. In either case, the total change in both angles is an even integral multiple of 2π.

To obtain the range of values for θ_1 and θ_2 on the sheet R_1, we note that if a point starts on the sheet R_0 and describes a path around just one of the branch points once, it crosses onto the sheet R_1 and does not return to the sheet R_0. In this case, the value of one of the angles is changed by 2π, while the value of the other remains unchanged. Hence on the sheet R_1 one angle can range from 2π to 4π, while the other ranges from 0 to 2π. Their sum then ranges from 2π to 4π, and the value of $(\theta_1 + \theta_2)/2$, which is the argument of $f(z)$, ranges from π to 2π. Again, the range of the angles is extended by changing the value of just one of the angles by an integral multiple of 4π or by changing the value of both angles by the same integral multiple of 2π.

The double-valued function (1) may now be considered as a single-valued function of the points on the Riemann surface just constructed. The transformation $w = f(z)$ maps each of the sheets used in the construction of that surface onto the entire w plane.

EXAMPLE 2. Consider the double-valued function

$$(2) \qquad f(z) = [z(z^2 - 1)]^{1/2} = \sqrt{rr_1r_2}\exp\frac{i(\theta + \theta_1 + \theta_2)}{2}$$

(Fig. 99). The points $z = 0, \pm1$ are branch points of this function. We note that if the point z describes a circuit that includes all three of those points, the argument of $f(z)$ changes by the angle 3π and the value of the function thus changes. Consequently, a branch cut must run from one of those branch points to the point at infinity in order to describe a single-valued branch of f. Hence the point at infinity is also a branch point, as one can show by noting that the function $f(1/z)$ has a branch point at $z = 0$.

Let two sheets be cut along the line segment L_2 from $z = -1$ to $z = 0$ and along the part L_1 of the real axis to the right of the point $z = 1$. We specify that each

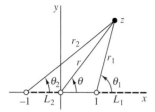

FIGURE 99

of the three angles θ, θ_1, and θ_2 may range from 0 to 2π on the sheet R_0 and from 2π to 4π on the sheet R_1. We also specify that the angles corresponding to a point on either sheet may be changed by integral multiples of 2π in such a way that the sum of the three angles changes by an integral multiple of 4π. The value of the function f is, therefore, unaltered.

A Riemann surface for the double-valued function (2) is obtained by joining the lower edges in R_0 of the slits along L_1 and L_2 to the upper edges in R_1 of the slits along L_1 and L_2, respectively. The lower edges in R_1 of the slits along L_1 and L_2 are then joined to the upper edges in R_0 of the slits along L_1 and L_2, respectively. It is readily verified with the aid of Fig. 99 that one branch of the function is represented by its values at points on R_0 and the other branch at points on R_1.

EXERCISES

1. Describe a Riemann surface for the triple-valued function $w = (z - 1)^{1/3}$, and point out which third of the w plane represents the image of each sheet of that surface.

2. Describe the Riemann surface for $\log z$ obtained by cutting the z plane along the negative real axis. Compare this Riemann surface with the one obtained in Example 1, Sec. 77.

3. Determine the image under the transformation $w = \log z$ of the sheet R_n, where n is an arbitrary integer, of the Riemann surface for $\log z$ given in Example 1, Sec. 77.

4. Verify that, under the transformation $w = z^{1/2}$, the sheet R_1 of the Riemann surface for $z^{1/2}$ given in Example 2, Sec. 77, is mapped onto the lower half of the w plane.

5. Describe the curve, on a Riemann surface for $z^{1/2}$, whose image is the entire circle $|w| = 1$ under the transformation $w = z^{1/2}$.

6. Corresponding to each point on the Riemann surface described in Example 2, Sec. 78, for the function $w = f(z)$ in that example, there is just one value of w. Show that, corresponding to each value of w, there are, in general, three points on the surface.

7. Describe a Riemann surface for the multiple-valued function

$$f(z) = \left(\frac{z - 1}{z}\right)^{1/2}.$$

8. Let C denote the positively oriented circle $|z - 2| = 1$ on the Riemann surface described in Example 2, Sec. 77, for $z^{1/2}$, where the upper half of that circle lies on the sheet R_0 and the lower half on R_1. Note that, for each point z on C, one can write

$$z^{1/2} = \sqrt{r}e^{i\theta/2} \qquad \text{where} \qquad 4\pi - \frac{\pi}{2} < \theta < 4\pi + \frac{\pi}{2}.$$

State why it follows that

$$\int_C z^{1/2}\, dz = 0.$$

Generalize this result to fit the case of the other simple closed curves that cross from one sheet to another without enclosing the branch points. Generalize to other functions, thus extending the Cauchy-Goursat theorem to integrals of multiple-valued functions.

9. Note that the Riemann surface described in Example 1, Sec. 78, for $(z^2 - 1)^{1/2}$ is also a Riemann surface for the function

$$g(z) = z + (z^2 - 1)^{1/2}.$$

Let f_0 denote the branch of $(z^2 - 1)^{1/2}$ defined on the sheet R_0, and show that the branches g_0 and g_1 of g on the two sheets are given by the equations

$$g_0(z) = \frac{1}{g_1(z)} = z + f_0(z).$$

10. In Exercise 9, the branch f_0 of $(z^2 - 1)^{1/2}$ can be described by the equation

$$f_0(z) = \sqrt{r_1 r_2}\left(\exp \frac{i\theta_1}{2}\right)\left(\exp \frac{i\theta_2}{2}\right),$$

where θ_1 and θ_2 range from 0 to 2π and

$$z - 1 = r_1 \exp(i\theta_1), \qquad z + 1 = r_2 \exp(i\theta_2).$$

Note that $2z = r_1 \exp(i\theta_1) + r_2 \exp(i\theta_2)$, and show that the branch g_0 of the function $g(z) = z + (z^2 - 1)^{1/2}$ can be written in the form

$$g_0(z) = \frac{1}{2}\left(\sqrt{r_1}\exp \frac{i\theta_1}{2} + \sqrt{r_2}\exp \frac{i\theta_2}{2}\right)^2.$$

Find $g_0(z)\overline{g_0(z)}$, and note that $r_1 + r_2 \geq 2$ and $\cos[(\theta_1 - \theta_2)/2] \geq 0$ for all z, to prove that $|g_0(z)| \geq 1$. Then show that the transformation $w = z + (z^2 - 1)^{1/2}$ maps the sheet R_0 of the Riemann surface onto the region $|w| \geq 1$, the sheet R_1 onto the region $|w| \leq 1$, and the branch cut between the points $z = \pm 1$ onto the circle $|w| = 1$. Note that the transformation used here is an inverse of the transformation

$$z = \frac{1}{2}\left(w + \frac{1}{w}\right).$$

CHAPTER
9

CONFORMAL MAPPING

In this chapter, we introduce and develop the concept of a conformal mapping, with emphasis on connections between such mappings and harmonic functions. Applications to physical problems will follow in the next chapter.

79. PRESERVATION OF ANGLES

Let C be a smooth arc (Sec. 31), represented by the equation $z = z(t)$ ($a \leq t \leq b$), and let $f(z)$ be a function defined at all points on C. The equation $w = f[z(t)]$ ($a \leq t \leq b$) is a parametric representation of the image Γ of C under the transformation $w = f(z)$.

Suppose that C passes through a point $z_0 = z(t_0)$ ($a < t_0 < b$) at which f is analytic and that $f'(z_0) \neq 0$. According to the chain rule given in Exercise 11, Sec. 31, if $w(t) = f[z(t)]$, then

$$(1) \qquad w'(t_0) = f'[z(t_0)]z'(t_0);$$

and this means that (see Sec. 6)

$$(2) \qquad \arg w'(t_0) = \arg f'[z(t_0)] + \arg z'(t_0).$$

Statement (2) is useful in relating the directions of C and Γ at the points z_0 and $w_0 = f(z_0)$, respectively.

To be specific, let ψ_0 denote a value of $\arg f'(z_0)$, and let θ_0 be the angle of inclination of a directed line tangent to C at z_0 (Fig. 100). According to Sec. 31, θ_0 is a value of $\arg z'(t_0)$; and it follows from statement (2) that the quantity

$$\phi_0 = \psi_0 + \theta_0$$

is a value of $\arg w'(t_0)$ and is, therefore, the angle of inclination of a directed line tangent to Γ at the point $w_0 = f(z_0)$. Hence the angle of inclination of the directed line at w_0 differs from the angle of inclination of the directed line at z_0 by the *angle of rotation*

(3) $$\psi_0 = \arg f'(z_0).$$

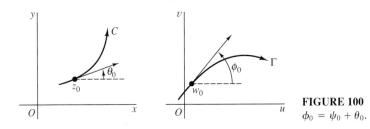

FIGURE 100
$\phi_0 = \psi_0 + \theta_0.$

Now let C_1 and C_2 be two smooth arcs passing through z_0, and let θ_1 and θ_2 be angles of inclination of directed lines tangent to C_1 and C_2, respectively, at z_0. We know from the preceding paragraph that the quantities

$$\phi_1 = \psi_0 + \theta_1 \qquad \text{and} \qquad \phi_2 = \psi_0 + \theta_2$$

are angles of inclination of directed lines tangent to the image curves Γ_1 and Γ_2, respectively, at the point $w_0 = f(z_0)$. Thus $\phi_2 - \phi_1 = \theta_2 - \theta_1$; that is, the angle $\phi_2 - \phi_1$ from Γ_1 to Γ_2 is the same in *magnitude* and *sense* as the angle $\theta_2 - \theta_1$ from C_1 to C_2. Those angles are denoted by α in Fig. 101.

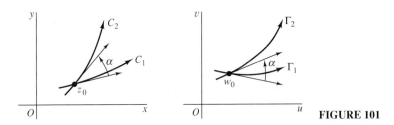

FIGURE 101

Because of this angle-preserving property, a transformation $w = f(z)$ is said to be *conformal* at a point z_0 if f is analytic there and $f'(z_0) \neq 0$. Such a transformation is actually conformal at each point in a neighborhood of z_0. For f must be analytic in a neighborhood of z_0 (Sec. 20); and, since f' is continuous at z_0 (Sec. 40), it follows from Sec. 14 that there is also a neighborhood of that point throughout which $f'(z) \neq 0$.

A transformation $w = f(z)$, defined on a domain D, is referred to as a *conformal transformation*, or *conformal mapping*, when it is conformal at each point in D. That is, the mapping is conformal in D if f is analytic in D and its derivative f' has no zeros there. Each of the elementary functions studied in Chap. 3 can be used to define a transformation that is conformal in some domain.

EXAMPLE 1.   The mapping $w = e^z$ is conformal throughout the entire z plane since $(e^z)' = e^z \neq 0$ for each z. Consider any two lines $x = c_1$ and $y = c_2$ in the z plane, the first directed upward and the second directed to the right. According to Sec. 73, their images under the mapping $w = e^z$ are a positively oriented circle centered at the origin and a ray from the origin, respectively. As illustrated in Fig. 81 (Sec. 73), the angle between the lines at their point of intersection is a right angle in the negative direction, and the same is true of the angle between the circle and the ray at the corresponding point in the w plane. The conformality of the mapping $w = e^z$ is also illustrated in Figs. 7 and 8 of Appendix 2.

EXAMPLE 2.   Consider two smooth arcs which are level curves $u(x, y) = c_1$ and $v(x, y) = c_2$ of the real and imaginary components, respectively, of a function

$$f(z) = u(x, y) + iv(x, y),$$

and suppose that they intersect at a point z_0 where f is analytic and $f'(z_0) \neq 0$. The transformation $w = f(z)$ is conformal at z_0 and maps these arcs into the lines $u = c_1$ and $v = c_2$, which are orthogonal at the point $w_0 = f(z_0)$. According to our theory, then, the arcs must be orthogonal at z_0. This has already been verified and illustrated in Exercises 14 through 18 of Sec. 22.

A mapping that preserves the magnitude of the angle between two smooth arcs but not necessarily the sense is called an *isogonal mapping*.

EXAMPLE 3.   The transformation $w = \bar{z}$, which is a reflection in the real axis, is isogonal but not conformal. If it is followed by conformal transformation, the resulting transformation $w = f(\bar{z})$ is also isogonal but not conformal.

Suppose that f is not a constant function and is analytic at a point z_0. If $f'(z_0) = 0$, then z_0 is called a *critical point* of the transformation $w = f(z)$.

EXAMPLE 4.   The point $z = 0$ is a critical point of the transformation $w = 1 + z^2$, which is a composition of the mappings $Z = z^2$ and $w = 1 + Z$. A ray $\theta = \alpha$ from the point $z = 0$ is evidently mapped onto the ray from the point $w = 1$ whose angle of inclination is 2α. Moreover, the angle between any two rays drawn from the critical point $z = 0$ is doubled by the transformation.

More generally, it can be shown that if z_0 is a critical point of a transformation $w = f(z)$, there is an integer m ($m \geq 2$) such that the angle between any two smooth arcs passing through z_0 is multiplied by m under that transformation. The integer m is the smallest positive integer such that $f^{(m)}(z_0) \neq 0$. Verification of these facts is left to the exercises.

80.   FURTHER PROPERTIES

Another property of a transformation $w = f(z)$ that is conformal at a point z_0 is obtained by considering the modulus of $f'(z_0)$. From the definition of derivative and property (12), Sec. 12, of limits, we know that

$$(1) \qquad |f'(z_0)| = \left| \lim_{z \to z_0} \frac{f(z) - f(z_0)}{z - z_0} \right| = \lim_{z \to z_0} \frac{|f(z) - f(z_0)|}{|z - z_0|}.$$

Now $|z - z_0|$ is the length of a line segment joining z_0 and z, and $|f(z) - f(z_0)|$ is the length of the line segment joining the points $f(z_0)$ and $f(z)$ in the w plane. Evidently, then, if z is near the point z_0, the ratio

$$\frac{|f(z) - f(z_0)|}{|z - z_0|}$$

of the two lengths is approximately the number $|f'(z_0)|$. Note that $|f'(z_0)|$ represents an expansion if it is greater than unity and a contraction if it is less than unity.

Although the angle of rotation $\arg f'(z)$ (Sec. 79) and the *scale factor* $|f'(z)|$ vary, in general, from point to point, it follows from the continuity of f' that their values are approximately $\arg f'(z_0)$ and $|f'(z_0)|$ at points z near z_0. Hence the image of a small region in a neighborhood of z_0 *conforms* to the original region in the sense that it has approximately the same shape. A large region may, however, be transformed into a region that bears no resemblance to the original one.

EXAMPLE 1. When $f(z) = z^2$, the transformation

$$w = f(z) = x^2 - y^2 + i2xy$$

is conformal at the point $z = 1 + i$, where the half lines $y = x$ $(x \geq 0)$ and $x = 1$ $(y \geq 0)$ intersect. We denote those half lines by C_1 and C_2, with positive sense upward, and observe that the angle from C_1 to C_2 is $\pi/4$ at their point of intersection (Fig. 102). Since the image of a point (x, y) in the z plane is a point in the w plane whose rectangular coordinates are

$$u = x^2 - y^2 \qquad \text{and} \qquad v = 2xy,$$

the half line C_1 is transformed into the curve Γ_1 with parametric representation

$$(2) \qquad u = 0, \qquad v = 2x^2 \qquad (0 \leq x < \infty).$$

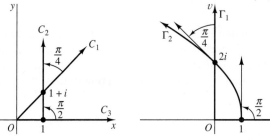

FIGURE 102
$w = z^2$.

Thus Γ_1 is the upper half $v \geq 0$ of the v axis. The half line C_2 is transformed into the curve Γ_2 represented by the equations

$$(3) \qquad\qquad u = 1 - y^2, \qquad v = 2y \qquad (0 \leq y < \infty).$$

Hence Γ_2 is the upper half of the parabola $v^2 = -4(u - 1)$. Note that, in each case, the positive sense of the image curve is upward.

If u and v are the variables in representation (3) for the image curve Γ_2, then

$$\frac{dv}{du} = \frac{dv/dy}{du/dy} = \frac{2}{-2y} = -\frac{2}{v}.$$

In particular, $dv/du = -1$ when $v = 2$. Consequently, the angle from the image curve Γ_1 to the image curve Γ_2 at the point $w = f(1 + i) = 2i$ is $\pi/4$, as required by the conformality of the mapping there. As anticipated, the angle of rotation $\pi/4$ at the point $z = 1 + i$ is a value of

$$\arg[f'(1 + i)] = \arg[2(1 + i)] = \frac{\pi}{4} + 2n\pi \qquad (n = 0, \pm 1, \pm 2, \ldots).$$

The scale factor at that point is the number

$$|f'(1 + i)| = |2(1 + i)| = 2\sqrt{2}.$$

To illustrate how the angle of rotation and the scale factor can change from point to point, we note that they are 0 and 2, respectively, at the point $z = 1$ since $f'(1) = 2$. See Fig. 102, where the curves C_2 and Γ_2 are the same as above and where the nonnegative x axis C_3 is transformed into the nonnegative u axis Γ_3.

A transformation $w = f(z)$ that is conformal at a point z_0 has a *local inverse* there. That is, if $w_0 = f(z_0)$, then there exists a unique transformation $z = g(w)$, defined and analytic in a neighborhood N of w_0, such that $g(w_0) = z_0$ and $f[g(w)] = w$ for all points w in N. The derivative of g is, in fact,

$$(4) \qquad\qquad g'(w) = \frac{1}{f'(z)}.$$

We note from expression (4) that the transformation $z = g(w)$ is itself conformal at w_0.

Let us verify the existence of such an inverse, which is a direct consequence of results in advanced calculus.* As noted in Sec. 79, the conformality of the transformation $w = f(z)$ at z_0 implies that there is some neighborhood of z_0 where the transformation is conformal and, consequently, where f is analytic. Hence, if we write $z = x + iy$, $z_0 = x_0 + iy_0$, and

$$f(z) = u(x, y) + iv(x, y),$$

---

*The results from advanced calculus to be used here appear in, for instance, A. E. Taylor and W. R. Mann, "Advanced Calculus," 3d ed., pp. 241–247, 1983.

we know that there is a neighborhood of the point (x_0, y_0) throughout which the functions $u(x, y)$ and $v(x, y)$, along with their partial derivatives of all orders, are continuous (see Sec. 40).

Now the pair of equations

$$(5) \qquad u = u(x, y), \qquad v = v(x, y)$$

represents a transformation from the neighborhood just mentioned into the uv plane. Moreover, the determinant

$$J = \begin{vmatrix} u_x & u_y \\ v_x & v_y \end{vmatrix} = u_x v_y - v_x u_y,$$

which is known as the *Jacobian* of the transformation, is nonzero at the point (x_0, y_0). For, in view of the Cauchy-Riemann equations $u_x = v_y$ and $u_y = -v_x$, one can write J as

$$J = (u_x)^2 + (v_x)^2 = |f'(z)|^2;$$

and $f'(z_0) \neq 0$ since the transformation $w = f(z)$ is conformal at z_0. The above continuity conditions on the functions $u(x, y)$ and $v(x, y)$ and their derivatives, together with this condition on the Jacobian, are sufficient to ensure the existence of a local inverse of transformation (5) at (x_0, y_0). That is, if

$$(6) \qquad u_0 = u(x_0, y_0) \qquad \text{and} \qquad v_0 = v(x_0, y_0),$$

then there is a unique continuous transformation

$$(7) \qquad x = x(u, v), \qquad y = y(u, v),$$

defined on a neighborhood N of the point (u_0, v_0) and mapping that point onto (x_0, y_0), such that equations (5) hold when equations (7) hold. Also, in addition to being continuous, the functions (7) have continuous first-order partial derivatives satisfying the equations

$$(8) \qquad x_u = \frac{1}{J} v_y, \qquad x_v = -\frac{1}{J} u_y, \qquad y_u = -\frac{1}{J} v_x, \qquad y_v = \frac{1}{J} u_x$$

throughout N.

If we write $w = u + iv$, $w_0 = u_0 + iv_0$, and

$$(9) \qquad g(w) = x(u, v) + i y(u, v),$$

the transformation $z = g(w)$ is evidently the local inverse of the transformation $w = f(z)$ at z_0. Transformations (5) and (7) can be written

$$u + iv = u(x, y) + iv(x, y) \qquad \text{and} \qquad x + iy = x(u, v) + iy(u, v);$$

and these last two equations are the same as

$$w = f(z) \qquad \text{and} \qquad z = g(w),$$

where g has the desired properties. Equations (8) can be used to show that g is analytic in N. Details are left to the exercises, where expression (4) for $g'(w)$ is also derived.

EXAMPLE 2. If $f(z) = e^z$, the transformation $w = f(z)$ is conformal every-where in the z plane and, in particular, at the point $z_0 = 2\pi i$. The image of this choice of z_0 is the point $w_0 = 1$. When points in the w plane are expressed in the form $w = \rho e^{i\phi}$, the local inverse at z_0 can be obtained by writing $g(w) = \log w$, where $\log w$ denotes the branch

$$\log w = \ln \rho + i\phi \qquad (\rho > 0, \pi < \phi < 3\pi)$$

of the logarithmic function, restricted to any neighborhood of w_0 that does not contain the origin. Observe that

$$g(1) = \ln 1 + i2\pi = 2\pi i$$

and that, when w is in that neighborhood,

$$f[g(w)] = \exp(\log w) = w.$$

Also,

$$g'(w) = \frac{d}{dw} \log w = \frac{1}{w} = \frac{1}{\exp z},$$

in accordance with equation (4).

Note that, if the point $z_0 = 0$ is chosen, one can use the principal branch

$$\text{Log } w = \ln \rho + i\phi \qquad (\rho > 0, -\pi < \phi < \pi)$$

of the logarithmic function to define g. In this case, $g(1) = 0$.

EXERCISES

1. Determine the angle of rotation at the point $z = 2 + i$ when the transformation is $w = z^2$, and illustrate it for some particular curve. Show that the scale factor of the transformation at that point is $2\sqrt{5}$.

2. What angle of rotation is produced by the transformation $w = 1/z$ at the point
   (a) $z = 1$; (b) $z = i$?
   Ans. (a) π; (b) 0.

*3. Show that under the transformation $w = 1/z$, the images of the lines $y = x - 1$ and $y = 0$ are the circle $u^2 + v^2 - u - v = 0$ and the line $v = 0$, respectively. Sketch all four curves, determine corresponding directions along them, and verify the conformality of the mapping at the point $z = 1$.

4. Show that the angle of rotation at a nonzero point $z_0 = r_0 \exp(i\theta_0)$ under the transformation $w = z^n$ $(n = 1, 2, \ldots)$ is $(n-1)\theta_0$. Determine the scale factor of the transformation at that point.
   Ans. nr_0^{n-1}.

5. Show that the transformation $w = \sin z$ is conformal at all points except

$$z = \left(\frac{\pi}{2} + n\pi\right) \qquad (n = 0, \pm1, \pm2, \ldots).$$

Note that this is in agreement with the mapping of directed line segments shown in Figs. 9, 10, and 11 of Appendix 2.

6. Find the local inverse of the transformation $w = z^2$ at the point
   (a) $z_0 = 2$;   (b) $z_0 = -2$;   (c) $z_0 = -i$.
        Ans. (a) $w^{1/2} = \sqrt{\rho}\,e^{i\phi/2}$ $(\rho > 0, -\pi < \phi < \pi)$;
             (c) $w^{1/2} = \sqrt{\rho}\,e^{i\phi/2}$ $(\rho > 0, 2\pi < \phi < 4\pi)$.

7. In Sec. 80, it was pointed out that the components $x(u, v)$ and $y(u, v)$ of the inverse function $g(w)$ defined by equation (9) are continuous and have continuous first-order partial derivatives in the neighborhood N. Use equations (8), Sec. 80, to show that the Cauchy-Riemann equations $x_u = y_v$, $x_v = -y_u$ hold in N. Then conclude that $g(w)$ is analytic in that neighborhood.

8. Show that if $z = g(w)$ is the local inverse of a conformal transformation $w = f(z)$ at a point z_0, then

$$g'(w) = \frac{1}{f'(z)}$$

   at points w in the neighborhood N where g is analytic (Exercise 7).
        Suggestion: Start with the fact that $f[g(w)] = w$, and apply the chain rule for differentiating composite functions.

9. Let C be a smooth arc lying in a domain D throughout which a transformation $w = f(z)$ is conformal, and let Γ denote the image of C under that transformation. Show that Γ is also a smooth arc.

10. Suppose that a function f is analytic at z_0 and that

$$f'(z_0) = f''(z_0) = \cdots = f^{(m-1)}(z_0) = 0, \qquad f^{(m)}(z_0) \neq 0$$

   for some positive integer m $(m \geq 1)$. Also, write $w_0 = f(z_0)$.

   (a) Use the Taylor series for f about the point z_0 to show that there is a neighborhood of z_0 in which the difference $f(z) - w_0$ can be written

$$f(z) - w_0 = (z - z_0)^m \frac{f^{(m)}(z_0)}{m!}[1 + g(z)],$$

   where $g(z)$ is continuous at z_0 and $g(z_0) = 0$.

   (b) Let Γ be the image of a smooth arc C under the transformation $w = f(z)$, as shown in Fig. 100 (Sec. 79), and note that the angles of inclination θ_0 and ϕ_0 in that figure are limits of $\arg(z - z_0)$ and $\arg[f(z) - w_0]$, respectively, as z approaches z_0 along the arc C. Then use the result in part (a) to show that θ_0 and ϕ_0 are related by the equation

$$\phi_0 = m\theta_0 + \arg f^{(m)}(z_0).$$

   (c) Let α denote the angle between two smooth arcs C_1 and C_2 passing through z_0, as shown on the left in Fig. 101 (Sec. 79). Show how it follows from the relation obtained in part (b) that the corresponding angle between the image curves Γ_1 and Γ_2 at the point $w_0 = f(z_0)$ is $m\alpha$. (Note that the transformation is conformal at z_0 when $m = 1$ and that z_0 is a critical point when $m \geq 2$.)

81. HARMONIC CONJUGATES

We saw in Sec. 22 that if a function

$$f(z) = u(x, y) + iv(x, y)$$

is analytic in a domain D, then the real-valued functions u and v are harmonic in that domain. That is, they have continuous partial derivatives of the first and second order in D and satisfy Laplace's equation there:

$$(1) \qquad\qquad u_{xx} + u_{yy} = 0, \qquad v_{xx} + v_{yy} = 0.$$

We had seen earlier that the first-order partial derivatives of u and v satisfy the Cauchy-Riemann equations

$$(2) \qquad\qquad u_x = v_y, \qquad u_y = -v_x;$$

and, as pointed out in Sec. 22, v is called a harmonic conjugate of u.

Suppose now that $u(x, y)$ is any given harmonic function defined on a *simply connected* (Sec. 38) domain D. In this section, we show that $u(x, y)$ always has a harmonic conjugate $v(x, y)$ in D by deriving an expression for $v(x, y)$.

To accomplish this, we first recall some important facts about line integrals in advanced calculus.* Suppose that $P(x, y)$ and $Q(x, y)$ have continuous first-order partial derivatives in a simply connected domain D of the xy plane, and let (x_0, y_0) and (x, y) be any two points in D. If $P_y = Q_x$ everywhere in D, then the line integral

$$\int_C P(s, t)\, ds + Q(s, t)\, dt$$

from (x_0, y_0) to (x, y) is independent of the contour C that is taken as long as the contour lies entirely in D. Furthermore, when the point (x_0, y_0) is kept fixed and (x, y) is allowed to vary throughout D, the integral represents a single-valued function

$$(3) \qquad\qquad F(x, y) = \int_{(x_0, y_0)}^{(x, y)} P(s, t)\, ds + Q(s, t)\, dt$$

of x and y whose first-order partial derivatives are given by the equations

$$(4) \qquad\qquad F_x(x, y) = P(x, y), \qquad F_y(x, y) = Q(x, y).$$

Note that the value of F is changed by an additive constant when a different point (x_0, y_0) is taken.

Returning to the given harmonic function $u(x, y)$, observe how it follows from Laplace's equation $u_{xx} + u_{yy} = 0$ that

$$(-u_y)_y = (u_x)_x$$

everywhere in D. Also, the second-order partial derivatives of u are continuous in D; and this means that the first-order partial derivatives of $-u_y$ and u_x are continuous there. Thus, if (x_0, y_0) is a fixed point in D, the function

$$(5) \qquad\qquad v(x, y) = \int_{(x_0, y_0)}^{(x, y)} -u_t(s, t)\, ds + u_s(s, t)\, dt$$

---

*See, for example, W. Kaplan, "Advanced Mathematics for Engineers," pp. 546–550, 1981.

is well defined for all (x, y) in D; and, according to equations (4),

(6) $$v_x(x, y) = -u_y(x, y), \qquad v_y(x, y) = u_x(x, y).$$

These are the Cauchy-Riemann equations. Since the first-order partial derivatives of u are continuous, it is evident from equations (6) that those derivatives of v are also continuous. Hence (Sec. 18) $u(x, y) + iv(x, y)$ is an analytic function in D; and v is, therefore, a harmonic conjugate of u.

The function v defined by equation (5) is, of course, not the only harmonic conjugate of u. The function $v(x, y) + c$, where c is any real constant, is also a harmonic conjugate of u. [Recall Exercise 11(a), Sec. 22.]

   EXAMPLE.   Consider the function $u(x, y) = xy$, which is harmonic throughout the entire xy plane. According to equation (5), the function

$$v(x, y) = \int_{(0,0)}^{(x,y)} -s \, ds + t \, dt$$

is a harmonic conjugate of $u(x, y)$. The integral here is readily evaluated by inspection. It can also be evaluated by integrating first along the horizontal path from the point $(0, 0)$ to the point $(x, 0)$ and then along the vertical path from $(x, 0)$ to the point (x, y). The result is

$$v(x, y) = -\frac{1}{2}x^2 + \frac{1}{2}y^2,$$

and the corresponding analytic function is

$$f(z) = xy - \frac{i}{2}(x^2 - y^2) = -\frac{i}{2}z^2.$$

82.   TRANSFORMATIONS OF HARMONIC FUNCTIONS

The problem of finding a function that is harmonic in a specified domain and satisfies prescribed conditions on the boundary of the domain is prominent in applied mathematics. If the values of the function are prescribed along the boundary, the problem is known as a boundary value problem of the first kind, or a *Dirichlet problem*. If the values of the normal derivative of the function are prescribed on the boundary, the boundary value problem is one of the second kind, or a *Neumann problem*. Modifications and combinations of those types of boundary conditions also arise.

   The domains most frequently encountered in the applications are simply connected; and, since a function that is harmonic in a simply connected domain always has a harmonic conjugate (Sec. 81), solutions of boundary value problems for such domains are the real or imaginary parts of analytic functions.

   EXAMPLE 1.   In Example 1, Sec. 22, we saw that the function $T(x, y) = e^{-y} \sin x$ satisfies a certain Dirichlet problem for the strip $0 < x < \pi$, $y > 0$ and noted that it represents a solution of a temperature problem. The function $T(x, y)$, which is actually harmonic throughout the xy plane, is evidently the real part of the entire function

$$-ie^{iz} = e^{-y} \sin x - ie^{-y} \cos x.$$

It is also the imaginary part of the entire function e^{iz}.

Sometimes a solution of a given boundary value problem can be *discovered* by identifying it as the real or imaginary part of an analytic function. But the success of that procedure depends on the simplicity of the problem and on one's familiarity with the real and imaginary parts of a variety of analytic functions. The following theorem is an important aid.

Theorem. *Suppose that an analytic function*

(1) $$w = f(z) = u(x, y) + iv(x, y)$$

maps a domain D_z in the z plane onto a domain D_w in the w plane. If $h(u, v)$ is a harmonic function defined on D_w, then the function

(2) $$H(x, y) = h[u(x, y), v(x, y)]$$

is harmonic in D_z.

We first prove the theorem for the case in which the domain D_w is simply connected. According to Sec. 81, that property of D_w ensures that the given harmonic function $h(u, v)$ has a harmonic conjugate $g(u, v)$. Hence the function

(3) $$\Phi(w) = h(u, v) + ig(u, v)$$

is analytic in D_w. Since the function $f(z)$ is analytic in D_z, the composite function $\Phi[f(z)]$ is also analytic in D_z. Consequently, the real part $h[u(x, y), v(x, y)]$ of this composition is harmonic in D_z.

If D_w is *not* simply connected, we observe that each point w_0 in D_w has a neighborhood $|w - w_0| < \varepsilon$ lying entirely in D_w. Since that neighborhood is simply connected, a function of the type (3) is analytic in it. Furthermore, since f is continuous at a point z_0 in D_z whose image is w_0, there is a neighborhood $|z - z_0| < \delta$ whose image is contained in the neighborhood $|w - w_0| < \varepsilon$. Hence it follows that the composition $\Phi[f(z)]$ is analytic in the neighborhood $|z - z_0| < \delta$, and we may conclude that $h[u(x, y), v(x, y)]$ is harmonic there. Finally, since w_0 was arbitrarily chosen in D_w and since each point in D_z is mapped onto such a point under the transformation $w = f(z)$, the function $h[u(x, y), v(x, y)]$ must be harmonic throughout D_z.

The proof of the theorem for the general case in which D_w is not necessarily simply connected can also be accomplished directly by means of the chain rule for partial derivatives. The computations are, however, somewhat involved (see Exercise 8, Sec. 83).

EXAMPLE 2. The function $h(u, v) = e^{-v} \sin u$ is harmonic in the domain D_w consisting of all points in the upper half plane $v > 0$ (see Example 1). If the transformation is $w = z^2$, then $u(x, y) = x^2 - y^2$ and $v(x, y) = 2xy$; moreover, the domain D_z in the z plane consisting of the points in the first quadrant $x > 0, y > 0$ is mapped onto the domain D_w, as shown in Example 3, Sec. 10. Hence the function

$$H(x, y) = e^{-2xy} \sin(x^2 - y^2)$$

is harmonic in D_z.

EXAMPLE 3. Consider the function $h(u, v) = \operatorname{Im} w = v$, which is harmonic in the horizontal strip $-\pi/2 < v < \pi/2$. It follows readily from the discussion following Example 2 in Sec. 73 that the transformation $w = \operatorname{Log} z$ maps the right half plane $x > 0$ onto that strip. By writing

$$\operatorname{Log} z = \ln \sqrt{x^2 + y^2} + i \arctan \frac{y}{x},$$

where $-\pi/2 < \arctan t < \pi/2$, we find that the function

$$H(x, y) = \arctan \frac{y}{x}$$

is harmonic in the half plane $x > 0$.

83. TRANSFORMATIONS OF BOUNDARY CONDITIONS

The conditions that a function or its normal derivative have prescribed values along the boundary of a domain in which it is harmonic are the most common, although not the only, important types of boundary conditions. In this section, we show that certain of these conditions remain unaltered under the change of variables associated with a conformal transformation. These results will be used in Chap. 10 to solve boundary value problems. The basic technique there is to transform a given boundary value problem in the xy plane into a simpler one in the uv plane and then to use the theorems of this and the preceding section to write the solution of the original problem in terms of the solution obtained for the simpler one.

Theorem. Suppose that a transformation

(1) $$w = f(z) = u(x, y) + iv(x, y)$$

is conformal on a smooth arc C, and let Γ be the image of C under that transformation. If, along Γ, a function $h(u, v)$ satisfies either of the conditions

(2) $$h = h_0 \qquad \text{or} \qquad \frac{dh}{dn} = 0,$$

where h_0 is a real constant and dh/dn denotes derivatives normal to Γ, then, along C, the function

(3) $$H(x, y) = h[u(x, y), v(x, y)]$$

satisfies the corresponding condition

$$(4) \qquad\qquad H = h_0 \qquad or \qquad \frac{dH}{dN} = 0,$$

where dh/dN denotes derivatives normal to C.

To show that the condition $h = h_0$ on Γ implies that $H = h_0$ on C, we note from equation (3) that the value of H at any point (x, y) on C is the same as the value of h at the image (u, v) of (x, y) under transformation (1). Since the image point (u, v) lies on Γ and since $h = h_0$ along that curve, it follows that $H = h_0$ along C.

Suppose, on the other hand, that $dh/dn = 0$ on Γ. From calculus, we know that

$$(5) \qquad\qquad \frac{dh}{dn} = (\text{grad } h) \cdot \mathbf{n},$$

where grad h denotes the gradient of h at a point (u, v) on Γ and $\mathbf{n}$ is a unit vector normal to Γ at (u, v). Since $dh/dn = 0$ at (u, v), equation (5) tells us that grad h is orthogonal to $\mathbf{n}$ at (u, v). That is, grad h is tangent to Γ there (Fig. 103). But gradients are orthogonal to level curves; and, because grad h is tangent to Γ, we see that Γ is orthogonal to a level curve $h(u, v) = c$ passing through (u, v).

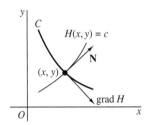 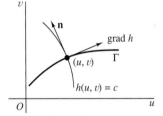

FIGURE 103

Now, according to equation (3), the level curve $H(x, y) = c$ in the z plane can be written

$$h[u(x, y), v(x, y)] = c;$$

and so it is evidently transformed into the level curve $h(u, v) = c$ under transformation (1). Furthermore, since C is transformed into Γ and Γ is orthogonal to the level curve $h(u, v) = c$, as demonstrated in the preceding paragraph, it follows from the conformality of transformation (1) on C that C is orthogonal to the level curve $H(x, y) = c$ at the point (x, y) corresponding to (u, v). Because gradients are orthogonal to level curves, this means that grad H is tangent to C at (x, y) (see Fig. 103). Consequently, if $\mathbf{N}$ denotes a unit vector normal to C at (x, y), grad H is orthogonal to $\mathbf{N}$. That is,

$$(6) \qquad\qquad (\text{grad } H) \cdot \mathbf{N} = 0.$$

Finally, since

$$\frac{dH}{dN} = (\text{grad}\,H) \cdot \mathbf{N},$$

we may conclude from equation (6) that $dH/dN = 0$ at points on C.

In this discussion, we have tacitly assumed that $\text{grad}\,h \neq \mathbf{0}$. If $\text{grad}\,h = \mathbf{0}$, it follows from the identity

$$|\,\text{grad}\,H(x, y)| = |\,\text{grad}\,h(u, v)|\,|f'(z)|,$$

derived in Exercise 10(a) below, that $\text{grad}\,H = \mathbf{0}$; hence dh/dn and the corresponding normal derivative dH/dN are both zero. We also assumed that

(a) $\text{grad}\,h$ and $\text{grad}\,H$ always exist;

(b) the level curve $H(x, y) = c$ is a smooth arc at the point (x, y) when $\text{grad}\,h \neq \mathbf{0}$ at (u, v).

Condition (b) ensures that angles between arcs are preserved by transformation (1) when it is conformal. In all of our applications, both conditions (a) and (b) will be satisfied.

EXAMPLE.   Consider, for instance, the function $h(u, v) = v + 2$. The transformation

$$w = iz^2 = -2xy + i(x^2 - y^2)$$

is conformal when $z \neq 0$. It maps the half line $y = x$ $(x > 0)$ onto the negative u axis, where $h = 2$, and the positive x axis onto the positive v axis, where the normal derivative h_u is 0 (Fig. 104). According to the above theorem, the function

$$H(x, y) = x^2 - y^2 + 2$$

must satisfy the condition $H = 2$ along the half line $y = x$ $(x > 0)$ and $H_y = 0$ along the positive x axis, as one can verify directly.

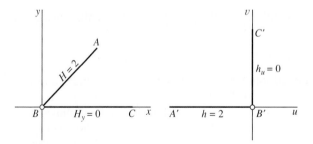

FIGURE 104

A boundary condition that is not of one of the two types mentioned in the theorem may be transformed into a condition that is substantially different from the original one (see Exercise 6). New boundary conditions for the transformed problem can be obtained for a particular transformation in any case. It is interesting to note that, under a conformal transformation, the ratio of a directional derivative of H along a smooth arc C in the z plane to the directional derivative of h along the image curve Γ at the corresponding point in the w plane is $|f'(z)|$; usually, this ratio is not constant along a given arc. (See Exercise 10.)

EXERCISES

1. Use expression (5), Sec. 81, to find a harmonic conjugate of the harmonic function $u(x, y) = x^3 - 3xy^2$. Write the resulting analytic function in terms of the complex variable z.

2. Let $u(x, y)$ be harmonic in a simply connected domain D. By appealing to results in Secs. 81 and 40, show that its partial derivatives of all orders are continuous throughout D.

3. The transformation $w = \exp z$ maps the horizontal strip $0 < y < \pi$ onto the upper half plane $v > 0$, as shown in Fig. 6 of Appendix 2; and the function

$$h(u, v) = \operatorname{Re}(w^2) = u^2 - v^2$$

is harmonic in that half plane. With the aid of the theorem in Sec. 82, show that the function $H(x, y) = e^{2x} \cos 2y$ is harmonic in the strip. Verify this result directly.

4. Under the transformation $w = \exp z$, the image of the segment $0 \le y \le \pi$ of the y axis is the semicircle $u^2 + v^2 = 1, v \ge 0$. Also, the function

$$h(u, v) = \operatorname{Re}\left(2 - w + \frac{1}{w}\right) = 2 - u + \frac{u}{u^2 + v^2}$$

is harmonic everywhere in the w plane except for the origin; and it assumes the value $h = 2$ on the semicircle. Write an explicit expression for the function $H(x, y)$ defined in the theorem of Sec. 83. Then illustrate the theorem by showing directly that $H = 2$ along the segment $0 \le y \le \pi$ of the y axis.

5. The transformation $w = z^2$ maps the positive x and y axes and the origin in the z plane onto the u axis in the w plane. Consider the harmonic function

$$h(u, v) = \operatorname{Re}(e^{-w}) = e^{-u} \cos v,$$

and observe that its normal derivative h_v along the u axis is zero. Then illustrate the theorem in Sec. 83 when $f(z) = z^2$ by showing directly that the normal derivative of the function $H(x, y)$ defined in that theorem is zero along both positive axes in the z plane. (Note that the transformation $w = z^2$ is not conformal at the origin.)

6. Replace the function $h(u, v)$ in Exercise 5 by the harmonic function

$$h(u, v) = \operatorname{Re}(-2iw + e^{-w}) = 2v + e^{-u} \cos v.$$

Then show that $h_v = 2$ along the u axis but that $H_y = 4x$ along the positive x axis and $H_x = 4y$ along the positive y axis. This illustrates how a condition of the type

$$\frac{dh}{dn} = h_0 \ne 0$$

is *not* necessarily transformed into a condition of the type $dH/dN = h_0$.

7. Show that if a function $H(x, y)$ is a solution of a Neumann problem (Sec. 82), then $H(x, y) + A$, where A is any real constant, is also a solution of that problem.

8. Suppose that an analytic function $w = f(z) = u(x, y) + iv(x, y)$ maps a domain D_z in the z plane onto a domain D_w in the w plane; and let a function $h(u, v)$, with continuous partial derivatives of the first and second order, be defined on D_w. Use the chain rule for partial derivatives to show that if $H(x, y) = h[u(x, y), v(x, y)]$, then

$$H_{xx}(x, y) + H_{yy}(x, y) = [h_{uu}(u, v) + h_{vv}(u, v)] |f'(z)|^2.$$

Conclude that the function $H(x, y)$ is harmonic in D_z when $h(u, v)$ is harmonic in D_w. This is an alternative proof of the theorem in Sec. 82, even when the domain D_w is multiply connected.

     Suggestion: In the simplifications, it is important to note that since f is analytic, the Cauchy-Riemann equations $u_x = v_y$, $u_y = -v_x$ hold and that the functions u and v both satisfy Laplace's equation. Also, the continuity conditions on the derivatives of h ensure that $h_{vu} = h_{uv}$.

9. Let $p(u, v)$ be a function that has continuous partial derivatives of the first and second order and satisfies *Poisson's equation*

$$p_{uu}(u, v) + p_{vv}(u, v) = \Phi(u, v)$$

in a domain D_w of the w plane, where Φ is a prescribed function. Show how it follows from the identity obtained in Exercise 8 that if an analytic function

$$w = f(z) = u(x, y) + iv(x, y)$$

maps a domain D_z onto the domain D_w, then the function

$$P(x, y) = p[u(x, y), v(x, y)]$$

satisfies the Poisson equation

$$P_{xx}(x, y) + P_{yy}(x, y) = \Phi[u(x, y), v(x, y)] |f'(z)|^2$$

in D_z.

10. Suppose that $w = f(z) = u(x, y) + iv(x, y)$ is a conformal mapping of a smooth arc C onto a smooth arc Γ in the w plane. Let the function $h(u, v)$ be defined on Γ, and write

$$H(x, y) = h[u(x, y), v(x, y)].$$

(a) From calculus, we know that the x and y components of grad H are the partial derivatives H_x and H_y, respectively; likewise, grad h has components h_u and h_v. By applying the chain rule for partial derivatives and using the Cauchy-Riemann equations, show that if (x, y) is a point on C and (u, v) is its image on Γ, then

$$|\text{grad}\, H(x, y)| = |\text{grad}\, h(u, v)| |f'(z)|.$$

(b) Show that the angle from the arc C to grad H at point (x, y) on C is equal to the angle from Γ to grad h at the image (u, v) of the point (x, y).

(c) Let s and σ denote distance along the arcs C and Γ, respectively; and let $\mathbf{t}$ and $\boldsymbol{\tau}$ denote unit tangent vectors at a point (x, y) on C and its image (u, v), in the direction of increasing distance. With the aid of the results in parts (a) and (b) and using the fact that

$$\frac{dH}{ds} = (\text{grad}\, H) \cdot \mathbf{t} \qquad \text{and} \qquad \frac{dh}{d\sigma} = (\text{grad}\, h) \cdot \boldsymbol{\tau},$$

show that the directional derivative along the arc Γ is transformed as follows:

$$\frac{dH}{ds} = \frac{dh}{d\sigma} |f'(z)|.$$

CHAPTER
10

APPLICATIONS
OF CONFORMAL
MAPPING

We now use conformal mapping to solve a number of physical problems involving Laplace's equation in two independent variables. Problems in heat conduction, electrostatic potential, and fluid flow will be treated. Since these problems are intended to illustrate methods, they will be kept on a fairly elementary level.

84. STEADY TEMPERATURES

In the theory of heat conduction, the *flux* across a surface within a solid body at a point on that surface is the quantity of heat flowing in a specified direction normal to the surface per unit time per unit area at the point. Flux is, therefore, measured in such units as calories per second per square centimeter. It is denoted here by Φ, and it varies with the normal derivative of the temperature T at the point on the surface:

$$(1) \qquad \qquad \Phi = -K\frac{dT}{dN} \qquad (K > 0).$$

The constant K is known as the *thermal conductivity* of the material of the solid, which is assumed to be homogeneous.

The points in the solid are assigned rectangular coordinates in three-dimensional space, and we restrict our attention to those cases in which the temperature T varies with only the x and y coordinates. Since T does not vary with the coordinate along the axis perpendicular to the xy plane, the flow of heat is, then, two-dimensional and parallel to that plane. We assume, moreover, that the flow is in a steady state; that is, T does not vary with time.

It is assumed that no thermal energy is created or destroyed within the solid. That is, no heat sources or sinks are present there. Also, the temperature function $T(x, y)$ and its partial derivatives of the first and second order are continuous at each point interior to the solid. This statement and expression (1) for the flux of heat are postulates for the mathematical theory of heat conduction, postulates that also apply at points within a solid containing a continuous distribution of sources or sinks.

Consider now an element of volume that is interior to the solid and that has the shape of a rectangular prism of unit height perpendicular to the xy plane, with base Δx by Δy in that plane (Fig. 105). The time rate of flow of heat toward the right across the left-hand face is $-KT_x(x, y)\Delta y$; and, toward the right across the right-hand face, it is $-KT_x(x + \Delta x, y)\Delta y$. Subtracting the first rate from the second, we obtain the net rate of heat loss from the element through those two faces. This resultant rate can be written

$$-K\left[\frac{T_x(x + \Delta x, y) - T_x(x, y)}{\Delta x}\right]\Delta x\Delta y,$$

or

(2) $$-KT_{xx}(x, y)\Delta x\Delta y$$

if Δx is very small. Expression (2) is, of course, an approximation whose accuracy increases as Δx and Δy are made smaller.

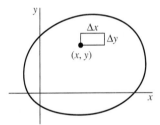

FIGURE 105

In like manner, the resultant rate of heat loss through the other faces perpendicular to the xy plane is found to be

(3) $$-KT_{yy}(x, y)\Delta x\Delta y.$$

Heat enters or leaves the element only through these four faces, and the temperatures within the element are steady. Hence the sum of expressions (2) and (3) is zero; that is,

(4) $$T_{xx}(x, y) + T_{yy}(x, y) = 0.$$

The temperature function thus satisfies Laplace's equation at each interior point of the solid.

In view of equation (4) and the continuity of the temperature function and its partial derivatives, T is a *harmonic function of x and y* in the domain representing the interior of the solid body.

The surfaces $T(x, y) = c_1$, where c_1 is any real constant, are the *isotherms* within the solid. They can also be considered as curves in the xy plane; then $T(x, y)$ can be interpreted as the temperature at a point (x, y) in a thin sheet of material in that plane, with the faces of the sheet thermally insulated. The isotherms are the level curves of the function T.

The gradient of T is perpendicular to the isotherm at each point, and the maximum flux at a point is in the direction of the gradient there. If $T(x, y)$ denotes temperatures in a thin sheet and if S is a harmonic conjugate of the function T, then a curve $S(x, y) = c_2$ has the gradient of T as a tangent vector at each point where the analytic function $T(x, y) + iS(x, y)$ is conformal. The curves $S(x, y) = c_2$ are called *lines of flow*.

If the normal derivative dT/dN is zero along any part of the boundary of the sheet, then the flux of heat across that part is zero. That is, the part is thermally insulated and is, therefore, a line of flow.

The function T may also denote the concentration of a substance that is diffusing through a solid. In that case, K is the diffusion constant. The above discussion and the derivation of equation (4) apply as well to steady-state diffusion.

85. STEADY TEMPERATURES IN A HALF PLANE

Let us find an expression for the steady temperatures $T(x, y)$ in a thin semi-infinite plate $y \geq 0$ whose faces are insulated and whose edge $y = 0$ is kept at temperature zero except for the segment $-1 < x < 1$, where it is kept at temperature unity (Fig. 106). The function $T(x, y)$ is to be bounded; this condition is natural if we consider the given plate as the limiting case of the plate $0 \leq y \leq y_0$ whose upper edge is kept at a fixed temperature as y_0 is increased. In fact, it would be physically reasonable to stipulate that $T(x, y)$ approach zero as y tends to infinity.

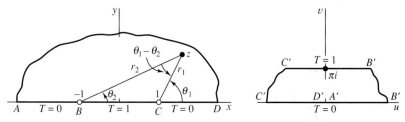

FIGURE 106

$$w = \log \frac{z - 1}{z + 1} \quad \left(\frac{r_1}{r_2} > 0, -\frac{\pi}{2} < \theta_1 - \theta_2 < \frac{3\pi}{2} \right).$$

The boundary value problem to be solved can be written

$$(1) \qquad T_{xx}(x, y) + T_{yy}(x, y) = 0 \qquad (-\infty < x < \infty, y > 0),$$

$$(2) \qquad T(x, 0) = \begin{cases} 1 & \text{when } |x| < 1, \\ 0 & \text{when } |x| > 1; \end{cases}$$

also, $|T(x, y)| < M$ where M is some positive constant. This is a Dirichlet problem for the upper half of the xy plane. Our method of solution will be to obtain a new Dirichlet problem for a region in the uv plane. That region will be the image of the half plane under a transformation $w = f(z)$ that is analytic in the domain $y > 0$ and that is conformal along the boundary $y = 0$ except at the points $(\pm 1, 0)$, where it is undefined. It will be a simple matter to discover a bounded harmonic function satisfying the new problem. The two theorems in Chap. 9 will then be applied to transform the solution of the problem in the uv plane into a solution of the original problem in the xy plane. Specifically, a harmonic function of u and v will be transformed into a harmonic function of x and y, and the boundary conditions in the uv plane will be preserved on corresponding portions of the boundary in the xy plane. There should be no confusion if we use the same symbol T to denote the different temperature functions in the two planes.

Let us write

$$z - 1 = r_1 \exp(i\theta_1) \qquad \text{and} \qquad z + 1 = r_2 \exp(i\theta_2),$$

where $0 \le \theta_k \le \pi$ ($k = 1, 2$). The transformation

$$(3) \quad w = \log \frac{z - 1}{z + 1} = \ln \frac{r_1}{r_2} + i(\theta_1 - \theta_2) \qquad \left(\frac{r_1}{r_2} > 0, -\frac{\pi}{2} < \theta_1 - \theta_2 < \frac{3\pi}{2} \right)$$

is defined on the upper half plane $y \ge 0$, except for the points $z = \pm 1$, since $0 \le \theta_1 - \theta_2 \le \pi$ in that region. (See Fig. 106.) Now the value of the logarithm is the principal value when $0 \le \theta_1 - \theta_2 \le \pi$, and we note from Fig. 19 of Appendix 2 that the upper half plane $y > 0$ is mapped onto the strip $0 < v < \pi$ in the w plane. Indeed, it was that figure which suggested transformation (3) here. The segment of the x axis between $z = -1$ and $z = 1$, where $\theta_1 - \theta_2 = \pi$, is mapped onto the upper edge of the strip; and the rest of the x axis, where $\theta_1 - \theta_2 = 0$, is mapped onto the lower edge. The required analyticity and conformality conditions are evidently satisfied by transformation (3).

A bounded harmonic function of u and v that is zero on the edge $v = 0$ of the strip and unity on the edge $v = \pi$ is clearly

$$(4) \qquad\qquad\qquad T = \frac{1}{\pi} v;$$

it is harmonic since it is the imaginary part of the entire function $(1/\pi)w$. Changing to x and y coordinates by means of the equation

$$(5) \qquad\qquad\qquad w = \ln \left| \frac{z - 1}{z + 1} \right| + i \arg \left(\frac{z - 1}{z + 1} \right),$$

we find that

$$v = \arg \left[\frac{(z - 1)(\bar{z} + 1)}{(z + 1)(\bar{z} + 1)} \right] = \arg \left[\frac{x^2 + y^2 - 1 + i2y}{(x + 1)^2 + y^2} \right],$$

or

$$v = \arctan \left(\frac{2y}{x^2 + y^2 - 1} \right).$$

The range of the arctangent function here is from 0 to π since

$$\arg\left(\frac{z-1}{z+1}\right) = \theta_1 - \theta_2$$

and $0 \leq \theta_1 - \theta_2 \leq \pi$. Expression (4) now takes the form

(6) $$T = \frac{1}{\pi} \arctan\left(\frac{2y}{x^2+y^2-1}\right) \qquad (0 \leq \arctan t \leq \pi).$$

Since the function (4) is harmonic in the strip $0 < v < \pi$ and since transformation (3) is analytic in the half plane $y > 0$, we may apply the theorem in Sec. 82 to conclude that the function (6) is harmonic in that half plane. The boundary conditions for the two harmonic functions are the same on corresponding parts of the boundaries because they are of the type $h = h_0$, treated in the theorem of Sec. 83. The bounded function (6) is, therefore, the desired solution of the original problem. One can, of course, verify directly that the function (6) satisfies Laplace's equation and has the values tending to those indicated on the left in Fig. 106 as the point (x, y) approaches the x axis from above.

The isotherms $T(x, y) = c_1$ $(0 < c_1 < 1)$ are arcs of the circles

$$x^2 + (y - \cot \pi c_1)^2 = \csc^2 \pi c_1,$$

passing through the points $(\pm 1, 0)$ and with centers on the y axis.

Finally, we note that since the product of a harmonic function by a constant is also harmonic, the function

$$T = \frac{T_0}{\pi} \arctan\left(\frac{2y}{x^2+y^2-1}\right) \qquad (0 \leq \arctan t \leq \pi)$$

represents the steady temperatures in the given half plane when the temperature $T = 1$ along the segment $-1 < x < 1$ of the x axis is replaced by any fixed temperature $T = T_0$.

86. A RELATED PROBLEM

Consider a semi-infinite slab in three-dimensional space, bounded by the planes $x = \pm \pi/2$ and $y = 0$, when the first two surfaces are kept at temperature zero and the third at temperature unity. We wish to find an expression for the temperature $T(x, y)$ at any interior point of the slab. The problem is also that of finding temperatures in a thin plate having the form of a semi-infinite strip $-\pi/2 \leq x \leq \pi/2, y \geq 0$ when the faces of the plate are perfectly insulated (Fig. 107).

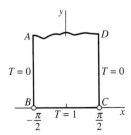

FIGURE 107

The boundary value problem here is

(1) $$T_{xx}(x, y) + T_{yy}(x, y) = 0 \qquad \left(-\frac{\pi}{2} < x < \frac{\pi}{2}, y > 0\right),$$

(2) $$T\left(-\frac{\pi}{2}, y\right) = T\left(\frac{\pi}{2}, y\right) = 0 \qquad (y > 0),$$

(3) $$T(x, 0) = 1 \qquad \left(-\frac{\pi}{2} < x < \frac{\pi}{2}\right),$$

where $T(x, y)$ is bounded.

In view of Example 1 in Sec. 74, as well as Fig. 9 of Appendix 2, the mapping

(4) $$w = \sin z$$

transforms this boundary value problem into the one posed in Sec. 85 (Fig. 106). Hence, according to solution (6) in that section,

(5) $$T = \frac{1}{\pi} \arctan\left(\frac{2v}{u^2 + v^2 - 1}\right) \qquad (0 \le \arctan t \le \pi).$$

The change of variables indicated in equation (4) can be written

$$u = \sin x \cosh y, \qquad v = \cos x \sinh y;$$

and the harmonic function (5) becomes

$$T = \frac{1}{\pi} \arctan\left(\frac{2 \cos x \sinh y}{\sin^2 x \cosh^2 y + \cos^2 x \sinh^2 y - 1}\right).$$

Since the denominator here reduces to $\sinh^2 y - \cos^2 x$, the quotient can be put in the form

$$\frac{2 \cos x \sinh y}{\sinh^2 y - \cos^2 x} = \frac{2(\cos x / \sinh y)}{1 - (\cos x / \sinh y)^2} = \tan 2\alpha,$$

where $\tan \alpha = \cos x / \sinh y$. Hence $T = (2/\pi)\alpha$; that is,

(6) $$T = \frac{2}{\pi} \arctan\left(\frac{\cos x}{\sinh y}\right) \qquad \left(0 \le \arctan t \le \frac{\pi}{2}\right).$$

This arctangent function has the range 0 to $\pi/2$ since its argument is nonnegative.

Since $\sin z$ is entire and the function (5) is harmonic in the half plane $v > 0$, the function (6) is harmonic in the strip $-\pi/2 < x < \pi/2$, $y > 0$. Also, the function (5) satisfies the boundary condition $T = 1$ when $|u| < 1$ and $v = 0$, as well as the condition $T = 0$ when $|u| > 1$ and $v = 0$. The function (6) thus satisfies boundary conditions (2) and (3). Moreover, $|T(x, y)| \le 1$ throughout the strip. Expression (6) is, therefore, the temperature formula that is sought.

The isotherms $T(x, y) = c_1$ $(0 < c_1 < 1)$ are the portions of the surfaces

$$\cos x = \tan\left(\frac{\pi c_1}{2}\right) \sinh y$$

within the slab, each surface passing through the points $(\pm\pi/2, 0)$ in the xy plane. If K is the thermal conductivity, the flux of heat into the slab through the surface lying in the plane $y = 0$ is

$$-KT_y(x,0) = \frac{2K}{\pi \cos x} \qquad \left(-\frac{\pi}{2} < x < \frac{\pi}{2}\right).$$

The flux outward through the surface lying in the plane $x = \pi/2$ is

$$-KT_x\left(\frac{\pi}{2}, y\right) = \frac{2K}{\pi \sinh y} \qquad (y > 0).$$

The boundary value problem posed in this section can also be solved by the *method of separation of variables*. That method is more direct, but it gives the solution in the form of an infinite series.*

87. TEMPERATURES IN A QUADRANT

Let us find the steady temperatures in a thin plate having the form of a quadrant if a segment at the end of one edge is insulated, if the rest of that edge is kept at a fixed temperature, and if the second edge is kept at another fixed temperature. The surfaces are insulated, and so the problem is two-dimensional.

The temperature scale and the unit of length can be chosen so that the boundary value problem for the temperature function T becomes

(1) $$T_{xx}(x, y) + T_{yy}(x, y) = 0 \qquad (x > 0, y > 0),$$

(2) $$\begin{cases} T_y(x,0) = 0 & \text{when } 0 < x < 1, \\ T(x,0) = 1 & \text{when } x > 1, \end{cases}$$

(3) $$T(0, y) = 0 \qquad (y > 0),$$

where $T(x, y)$ is bounded in the quadrant. The plate and its boundary conditions are shown on the left in Fig. 108. Conditions (2) prescribe the values of the normal derivative of the function T over a part of a boundary line and the values of the function itself over the rest of that line. The separation-of-variables method mentioned at the end of Sec. 86 is not adapted to such problems with different types of conditions along the same boundary line.

As indicated in Fig. 10 of Appendix 2, the transformation

(4) $$z = \sin w$$

is a one to one mapping of the strip $0 \le u \le \pi/2, v \ge 0$ onto the quadrant $x \ge 0, y \ge 0$. Observe now that the existence of an inverse is ensured by the fact that the

---

*A similar problem is treated in the authors' "Fourier Series and Boundary Value Problems," 5th ed., Problem 7, p. 150, 1993. Also, a short discussion of the uniqueness of solutions to boundary value problems can be found in Chap. 9 of that book.

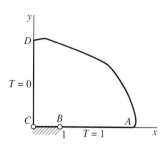
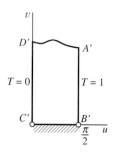

FIGURE 108

given transformation is both one to one and onto. Since transformation (4) is conformal throughout the strip except at the point $w = \pi/2$, the inverse transformation must be conformal throughout the quadrant except at the point $z = 1$. That inverse transformation maps the segment $0 < x < 1$ of the x axis onto the base of the strip and the rest of the boundary onto the sides of the strip as shown in Fig. 108.

Since the inverse of transformation (4) is conformal in the quadrant, except when $z = 1$, the solution to the given problem can be obtained by finding a function that is harmonic in the strip and satisfies the boundary conditions shown on the right in Fig. 108. Observe that these boundary conditions are of the types $h = h_0$ and $dh/dn = 0$ in the theorem of Sec. 83.

The required temperature function T for the new boundary value problem is clearly

$$(5) \qquad\qquad T = \frac{2}{\pi}u,$$

the function $(2/\pi)u$ being the real part of the entire function $(2/\pi)w$. We must now express T in terms of x and y.

To obtain u in terms of x and y, we first note that, according to equation (4),

$$(6) \qquad\qquad x = \sin u \cosh v, \qquad y = \cos u \sinh v.$$

When $0 < u < \pi/2$, both $\sin u$ and $\cos u$ are nonzero; and, consequently,

$$(7) \qquad\qquad \frac{x^2}{\sin^2 u} - \frac{y^2}{\cos^2 u} = 1.$$

Now it is convenient to observe that, for each fixed u, hyperbola (7) has foci at the points

$$z = \pm\sqrt{\sin^2 u + \cos^2 u} = \pm 1$$

and that the length of the transverse axis, which is the line segment joining the two vertices, is $2 \sin u$. Thus the absolute value of the difference of the distances between the foci and a point (x, y) lying on the part of the hyperbola in the first quadrant is

$$\sqrt{(x+1)^2 + y^2} - \sqrt{(x-1)^2 + y^2} = 2 \sin u.$$

It follows directly from equations (6) that this relation also holds when $u = 0$ or $u = \pi/2$. In view of equation (5), then, the required temperature function is

$$
(8) \qquad T = \frac{2}{\pi} \arcsin \frac{1}{2} \left[\sqrt{(x+1)^2 + y^2} - \sqrt{(x-1)^2 + y^2} \right],
$$

where, since $0 \le u \le \pi/2$, the arcsine function has the range 0 to $\pi/2$.

If we wish to verify that this function satisfies boundary conditions (2), we must remember that $\sqrt{(x-1)^2}$ denotes $x - 1$ when $x > 1$ and $1 - x$ when $0 < x < 1$, the square roots being positive. Note, too, that the temperature at any point along the insulated part of the lower edge of the plate is

$$
T(x,0) = \frac{2}{\pi} \arcsin x \qquad (0 < x < 1).
$$

It can be seen from equation (5) that the isotherms $T(x, y) = c_1$ $(0 < c_1 < 1)$ are the parts of the confocal hyperbolas (7), where $u = \pi c_1 / 2$, which lie in the first quadrant. Since the function $(2/\pi)v$ is a harmonic conjugate of the function (5), the lines of flow are quarters of the confocal ellipses obtained by holding v constant in equations (6).

EXERCISES

1. In the problem of the semi-infinite plate shown on the left in Fig. 106 (Sec. 85), obtain a harmonic conjugate of the temperature function $T(x, y)$ from equation (5), Sec. 85, and find the lines of flow of heat. Show that those lines of flow consist of the upper half of the y axis and the upper halves of certain circles on either side of that axis, the centers of the circles lying on the segment AB or CD of the x axis.

2. Show that if the function T in Sec. 85 is not required to be bounded, then the harmonic function (4) in that section can be replaced by the harmonic function

$$
T = \mathrm{Im}\left(\frac{1}{\pi} w + A \cosh w \right) = \frac{1}{\pi} v + A \sinh u \sin v,
$$

where A is an arbitrary real constant. Conclude that the solution of the Dirichlet problem for the strip in the uv plane (Fig. 106) would not, then, be unique.

3. Suppose that the condition that T be bounded is omitted from the problem for temperatures in the semi-infinite slab of Sec. 86 (Fig. 107). Show that an infinite number of solutions are then possible by noting the effect of adding to the solution found there the imaginary part of the function $A \sin z$, where A is an arbitrary real constant.

4. Use the function Log z to find an expression for the bounded steady temperatures in a plate having the form of a quadrant $x \ge 0$, $y \ge 0$ (Fig. 109) if its faces are perfectly insulated

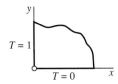

$T = 1$

$T = 0$ x **FIGURE 109**

and its edges have temperatures $T(x,0) = 0$ and $T(0, y) = 1$. Find the isotherms and lines of flow, and draw some of them.

Ans. $T = (2/\pi)\arctan(y/x)$.

5. Find the steady temperatures in a solid whose shape is that of a long cylindrical wedge if its boundary planes $\theta = 0$ and $\theta = \theta_0$ $(0 < r < r_0)$ are kept at constant temperatures zero and T_0, respectively, and if its surface $r = r_0$ $(0 < \theta < \theta_0)$ is perfectly insulated (Fig. 110).

Ans. $T = (T_0/\theta_0)\arctan(y/x)$.

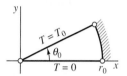

FIGURE 110

6. Find the bounded steady temperatures $T(x, y)$ in the semi-infinite solid $y \geq 0$ if $T = 0$ on the part $x < -1$ $(y = 0)$ of the boundary, if $T = 1$ on the part $x > 1$ $(y = 0)$, and if the strip $-1 < x < 1$ $(y = 0)$ of the boundary is insulated (Fig. 111).

$$Ans. \ T = \frac{1}{2} + \frac{1}{\pi}\arcsin\frac{1}{2}\left[\sqrt{(x+1)^2 + y^2} - \sqrt{(x-1)^2 + y^2}\right]$$

$(-\pi/2 \leq \arcsin t \leq \pi/2)$.

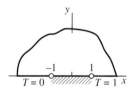

FIGURE 111

7. Find the bounded steady temperatures in the solid $x \geq 0$, $y \geq 0$ when the boundary surfaces are kept at fixed temperatures except for insulated strips of equal width at the corner, as shown in Fig. 112.

Suggestion: This problem can be transformed into the one of Exercise 6.

$$Ans. \ T = \frac{1}{2} + \frac{1}{\pi}\arcsin\frac{1}{2}\left[\sqrt{(x^2 - y^2 + 1)^2 + 4x^2y^2} - \sqrt{(x^2 - y^2 - 1)^2 + 4x^2y^2}\right]$$

$(-\pi/2 \leq \arctan t \leq \pi/2)$.

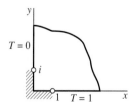

FIGURE 112

8. Solve the following Dirichlet problem for a semi-infinite strip (Fig.113):

$$H_{xx}(x, y) + H_{yy}(x, y) = 0 \qquad (0 < x < \pi/2, y > 0),$$

$$H(x, 0) = 0 \qquad (0 < x < \pi/2),$$

$$H(0, y) = 1, \quad H(\pi/2, y) = 0 \qquad (y > 0),$$

where $0 \le H(x, y) \le 1$.

Suggestion: This problem can be transformed into the one of Exercise 4.

$$Ans. \ H = \frac{2}{\pi} \arctan \left(\frac{\tanh y}{\tan x} \right).$$

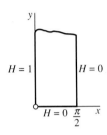

$H = 1$ $H = 0$

$H = 0$ $\frac{\pi}{2}$ x

FIGURE 113

9. Derive an expression for the temperatures $T(r, \theta)$ in a semicircular plate $r \le 1, 0 \le \theta \le \pi$ with insulated faces if $T = 1$ along the radial edge $\theta = 0$ $(0 < r < 1)$ and $T = 0$ on the rest of the boundary.

Suggestion: This problem can be transformed into the one of Exercise 8.

$$Ans. \ T = \frac{2}{\pi} \arctan \left(\frac{1 - r}{1 + r} \cot \frac{\theta}{2} \right).$$

10. Solve the boundary value problem for the plate $x \ge 0, y \ge 0$ in the z plane when the faces are insulated and the boundary conditions are those indicated in Fig. 114.

Suggestion: Use the mapping $w = i/z = i\bar{z}/|z|^2$ to transform this problem into the one posed in Sec. 87 (Fig. 108).

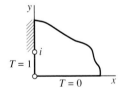

$T = 1$

$T = 0$ x

FIGURE 114

πi $T = 0$

$T = 1$ x

FIGURE 115

11. The portions $x < 0$ $(y = 0)$ and $x < 0$ $(y = \pi)$ of the edges of an infinite plate $0 \le y \le \pi$ are thermally insulated, as are the faces of the plate. The conditions $T(x, 0) = 1$ and $T(x, \pi) = 0$ are maintained when $x > 0$ (Fig. 115). Find the steady temperatures in the plate.

Suggestion: This problem can be transformed into the one of Exercise 6.

12. Consider a thin plate, with insulated faces, whose shape is the upper half of the region enclosed by an ellipse with foci $(\pm1,0)$. The temperature on the elliptical part of its boundary is $T = 1$. The temperature along the segment $-1 < x < 1$ of the x axis is $T = 0$, and the rest of the boundary along the x axis is insulated. With the aid of Fig. 11 in Appendix 2, find the lines of flow of heat.

13. According to Sec. 42 and Exercise 7 of that section, if a function $f(z) = u(x, y) + iv(x, y)$ is continuous in a closed bounded region R and analytic and not constant in the interior of R, then the function $u(x, y)$ reaches its maximum and minimum values on the boundary of R, and never in the interior. By interpreting $u(x, y)$ as a steady temperature, state a physical reason why that property of maximum and minimum values should hold true.

88. ELECTROSTATIC POTENTIAL

In an electrostatic force field, the *field intensity* at a point is a vector representing the force exerted on a unit positive charge placed at that point. The electrostatic *potential* is a scalar function of the space coordinates such that, at each point, its directional derivative in any direction is the negative of the component of the field intensity in that direction.

For two stationary charged particles, the magnitude of the force of attraction or repulsion exerted by one particle on the other is directly proportional to the product of the charges and inversely proportional to the square of the distance between those particles. From this inverse-square law, it can be shown that the potential at a point due to a single particle in space is inversely proportional to the distance between the point and the particle. In any region free of charges, the potential due to a distribution of charges outside that region can be shown to satisfy Laplace's equation for three-dimensional space.

If conditions are such that the potential V is the same in all planes parallel to the xy plane, then in regions free of charges V is a harmonic function of just the two variables x and y:

$$V_{xx}(x, y) + V_{yy}(x, y) = 0.$$

The field intensity vector at each point is parallel to the xy plane, with x and y components $-V_x(x, y)$ and $-V_y(x, y)$, respectively. That vector is, therefore, the negative of the gradient of $V(x, y)$.

A surface along which $V(x, y)$ is constant is an equipotential surface. The tangential component of the field intensity vector at a point on a conducting surface is zero in the static case since charges are free to move on such a surface. Hence $V(x, y)$ is constant along the surface of a conductor, and that surface is an *equipotential*.

If U is a harmonic conjugate of V, the curves $U(x, y) = c_2$ in the xy plane are called *flux lines*. When such a curve intersects an equipotential curve $V(x, y) = c_1$ at a point where the derivative of the analytic function $V(x, y) + iU(x, y)$ is not zero, the two curves are orthogonal at that point and the field intensity is tangent to the flux line there.

Boundary value problems for the potential V are the same mathematical problems as those for steady temperatures T; and, as in the case of steady temperatures, the methods of complex variables are limited to two-dimensional problems.

The problem posed in Sec. 86 (Fig. 107), for instance, can be interpreted as that of finding the two-dimensional electrostatic potential in the empty space $-\pi/2 < x < \pi/2$, $y > 0$ bounded by the conducting planes $x = \pm\pi/2$ and $y = 0$, insulated at their intersections, when the first two surfaces are kept at potential zero and the third at potential unity.

The potential in the steady flow of electricity in a plane conducting sheet is also a harmonic function at points free from sources and sinks. Gravitational potential is a further example of a harmonic function in physics.

89. POTENTIAL IN A CYLINDRICAL SPACE

A long hollow circular cylinder is made out of a thin sheet of conducting material, and the cylinder is split lengthwise to form two equal parts. Those parts are separated by slender strips of insulating material and are used as electrodes, one of which is grounded at potential zero and the other kept at a different fixed potential. We take the coordinate axes and units of length and potential difference as indicated on the left in Fig. 116. We then interpret the electrostatic potential $V(x, y)$ over any cross section of the enclosed space that is distant from the ends of the cylinder as a harmonic function inside the circle $x^2 + y^2 = 1$ in the xy plane. Note that $V = 0$ on the upper half of the circle and that $V = 1$ on the lower half.

A linear fractional transformation that maps the upper half plane onto the interior of the unit circle centered at the origin, the positive real axis onto the upper half of the circle, and the negative real axis onto the lower half of the circle is verified in Exercise 11, Sec. 72. The result is given in Fig. 13 of Appendix 2; interchanging z and w there, we find that the inverse of the transformation

$$(1) \qquad\qquad z = \frac{i - w}{i + w}$$

gives us a new problem for V in a half plane, indicated on the right in Fig. 116.

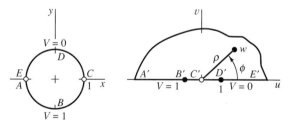

FIGURE 116

Now the imaginary part of the function

$$(2) \qquad\qquad \frac{1}{\pi}\mathrm{Log}\, w = \frac{1}{\pi}\ln\rho + \frac{i}{\pi}\phi \qquad (\rho > 0, 0 \le \phi \le \pi)$$

is a bounded function of u and v that assumes the required constant values on the two parts $\phi = 0$ and $\phi = \pi$ of the u axis. Hence the desired harmonic function for the half plane is

$$(3) \qquad V = \frac{1}{\pi} \arctan\left(\frac{v}{u}\right),$$

where the values of the arctangent function range from 0 to π.

The inverse of transformation (1) is

$$(4) \qquad w = i\frac{1-z}{1+z},$$

from which u and v can be expressed in terms of x and y. Equation (3) then becomes

$$(5) \qquad V = \frac{1}{\pi} \arctan\left(\frac{1-x^2-y^2}{2y}\right) \qquad (0 \le \arctan t \le \pi).$$

The function (5) is the potential function for the space enclosed by the cylindrical electrodes since it is harmonic inside the circle and assumes the required values on the semicircles. If we wish to verify this solution, we must note that

$$\lim_{\substack{t\to 0 \\ t>0}} \arctan t = 0 \qquad \text{and} \qquad \lim_{\substack{t\to 0 \\ t<0}} \arctan t = \pi.$$

The equipotential curves $V(x,y) = c_1$ $(0 < c_1 < 1)$ in the circular region are arcs of the circles

$$x^2 + (y + \tan \pi c_1)^2 = \sec^2 \pi c_1,$$

with each circle passing through the points $(\pm 1, 0)$. Also, the segment of the x axis between those points is the equipotential $V(x,y) = 1/2$. A harmonic conjugate U of V is $-(1/\pi)\ln \rho$, or the imaginary part of the function $-(i/\pi)\operatorname{Log} w$. In view of equation (4), U may be written

$$U = -\frac{1}{\pi} \ln\left|\frac{1-z}{1+z}\right|.$$

From this equation, it can be seen that the flux lines $U(x,y) = c_2$ are arcs of circles with centers on the x axis. The segment of the y axis between the electrodes is also a flux line.

EXERCISES

1. The harmonic function (3) of Sec. 89 is bounded in the half plane $v \ge 0$ and satisfies the boundary conditions indicated on the right in Fig. 116. Show that if the imaginary part of Ae^w, where A is any real constant, is added to that function, then the resulting function satisfies all of the requirements except for the boundedness condition.

2. Show that transformation (4) of Sec. 89 maps the upper half of the circular region shown on the left in Fig. 116 onto the first quadrant of the w plane and the diameter CE onto the positive v axis. Then find the electrostatic potential V in the space enclosed by the half cylinder $x^2 + y^2 = 1, y \ge 0$ and the plane $y = 0$ when $V = 0$ on the cylindrical surface

and $V = 1$ on the planar surface (Fig. 117).

$$Ans.\ V = \frac{2}{\pi} \arctan\left(\frac{1 - x^2 - y^2}{2y}\right).$$

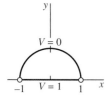

FIGURE 117

3. Find the electrostatic potential $V(r, \theta)$ in the space $0 < r < 1, 0 < \theta < \pi/4$, bounded by the half planes $\theta = 0$ and $\theta = \pi/4$ and the portion $0 \leq \theta \leq \pi/4$ of the cylindrical surface $r = 1$, when $V = 1$ on the planar surfaces and $V = 0$ on the cylindrical one. (See Exercise 2.) Verify that the function obtained satisfies the boundary conditions.

4. Note that all branches of $\log z$ have the same real component, which is harmonic everywhere except at the origin. Then write an expression for the electrostatic potential $V(x, y)$ in the space between two coaxial conducting cylindrical surfaces $x^2 + y^2 = 1$ and $x^2 + y^2 = r_0^2$ $(r_0 \neq 1)$ when $V = 0$ on the first surface and $V = 1$ on the second.

$$Ans.\ V = \frac{\ln(x^2 + y^2)}{2 \ln r_0}.$$

5. Find the bounded electrostatic potential $V(x, y)$ in the space $y > 0$ bounded by an infinite conducting plane $y = 0$ one strip $(-a < x < a, y = 0)$ of which is insulated from the rest of the plane and kept at potential $V = 1$, while $V = 0$ on the rest (Fig. 118). Verify that the function obtained satisfies the boundary conditions.

$$Ans.\ V = \frac{1}{\pi} \arctan\left(\frac{2ay}{x^2 + y^2 - a^2}\right) \qquad (0 \leq \arctan t \leq \pi).$$

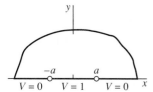

FIGURE 118

6. Derive an expression for the electrostatic potential in the semi-infinite space indicated in Fig. 119, bounded by two half planes and a half cylinder, when $V = 1$ on the cylindrical

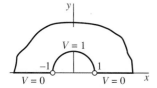

FIGURE 119

surface and $V = 0$ on the planar surfaces. Draw some of the equipotential curves in the xy plane.

$$\textit{Ans. } V = \frac{2}{\pi} \arctan\left(\frac{2y}{x^2 + y^2 - 1}\right).$$

7. Find the potential V in the space between the planes $y = 0$ and $y = \pi$ when $V = 0$ on the part of each of those planes where $x > 0$ and $V = 1$ on the parts where $x < 0$ (Fig. 120).
Check the result with the boundary conditions.

$$\textit{Ans. } V = \frac{1}{\pi} \arctan\left(\frac{\sin y}{\sinh x}\right) \qquad (0 \le \arctan t \le \pi).$$

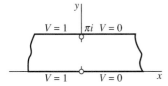

$$\textbf{FIGURE 120}$$

8. Derive an expression for the electrostatic potential V in the space interior to a long cylinder $r = 1$ when $V = 0$ on the first quadrant ($r = 1, 0 < \theta < \pi/2$) of the cylindrical surface and $V = 1$ on the rest ($r = 1, \pi/2 < \theta < 2\pi$) of that surface. (See Exercise 15, Sec. 72, and Fig. 80 there.) Show that $V = 3/4$ on the axis of the cylinder. Check the result with the boundary conditions.

9. Using Fig. 20 of Appendix 2, find a temperature function $T(x, y)$ that is harmonic in the shaded domain of the xy plane shown there and assumes the values $T = 0$ along the arc ABC and $T = 1$ along the line segment DEF. Verify that the function obtained satisfies the required boundary conditions. (See Exercise 2.)

10. The Dirichlet problem

$$V_{xx}(x, y) + V_{yy}(x, y) = 0 \qquad (0 < x < a, 0 < y < b),$$
$$V(x, 0) = 0, \qquad V(x, b) = 1 \qquad (0 < x < a),$$
$$V(0, y) = V(a, y) = 0 \qquad (0 < y < b)$$

for $V(x, y)$ in a rectangle can be solved by the method of separation of variables.* The solution is

$$V = \frac{4}{\pi} \sum_{n=1}^{\infty} \frac{\sinh(m\pi y/a)}{m \sinh(m\pi b/a)} \sin \frac{m\pi x}{a} \qquad (m = 2n - 1).$$

Accepting this result and adapting it to a problem in the uv plane, find the potential $V(r, \theta)$ in the space $1 < r < r_0, 0 < \theta < \pi$ when $V = 1$ on the part of the boundary where $\theta = \pi$ and $V = 0$ on the rest of the boundary. (See Fig. 121.)

$$\textit{Ans. } V = \frac{4}{\pi} \sum_{n=1}^{\infty} \frac{\sinh(\alpha_n \theta)}{\sinh(\alpha_n \pi)} \cdot \frac{\sin(\alpha_n \ln r)}{2n - 1} \qquad \left[\alpha_n = \frac{(2n - 1)\pi}{\ln r_0} \right].$$

---

*See the authors' "Fourier Series and Boundary Value Problems," 5th ed., pp. 143–145 and 198–199, 1993.

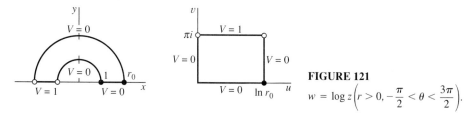

FIGURE 121

$$w = \log z \left(r > 0, -\frac{\pi}{2} < \theta < \frac{3\pi}{2} \right).$$

11. With the aid of the solution of the Dirichlet problem for the rectangle $0 \le x \le a, 0 \le y \le b$ that was used in Exercise 10, find the potential function $V(r, \theta)$ for the space $1 < r < r_0, 0 < \theta < \pi$ when $V = 1$ on the part of the boundary $r = r_0, 0 < \theta < \pi$ and $V = 0$ on the rest of the boundary (Fig. 122).

$$Ans. \ V = \frac{4}{\pi} \sum_{n=1}^{\infty} \left(\frac{r^m - r^{-m}}{r_0^m - r_0^{-m}} \right) \frac{\sin m\theta}{m} \qquad (m = 2n - 1).$$

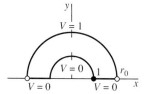

FIGURE 122

90. TWO-DIMENSIONAL FLUID FLOW

Harmonic functions play an important role in hydrodynamics and aerodynamics. Again, we consider only the two-dimensional steady-state type of problem. That is, the motion of the fluid is assumed to be the same in all planes parallel to the xy plane, the velocity being parallel to that plane and independent of time. It is, then, sufficient to consider the motion of a sheet of fluid in the xy plane.

We let the vector representing the complex number

$$V = p + iq$$

denote the velocity of a particle of the fluid at any point (x, y); hence the x and y components of the velocity vector are $p(x, y)$ and $q(x, y)$, respectively. At points interior to a region of flow in which no sources or sinks of the fluid occur, the real-valued functions $p(x, y)$ and $q(x, y)$ and their first-order partial derivatives are assumed to be continuous.

The *circulation* of the fluid along any contour C is defined as the line integral with respect to arc length σ of the tangential component $V_T(x, y)$ of the velocity vector along C:

(1)
$$\int_C V_T(x, y) d\sigma.$$

The ratio of the circulation along C to the length of C is, therefore, a mean speed of the fluid along that contour. It is shown in advanced calculus that such an integral can be written*

(2) $$\int_C V_T(x, y)d\sigma = \int_C p(x, y)dx + q(x, y)dy.$$

When C is a positively oriented simple closed contour lying in a simply connected domain of flow containing no sources or sinks, Green's theorem (see Sec. 36) enables us to write

$$\int_C p(x, y)\, dx + q(x, y)\, dy = \int\int_R [q_x(x, y) - p_y(x, y)]\, dA,$$

where R is the closed region consisting of points interior to and on C. Thus

(3) $$\int_C V_T(x, y)d\sigma = \int\int_R [q_x(x, y) - p_y(x, y)]\, dA$$

for such a contour.

A physical interpretation of the integrand on the right in expression (3) for the circulation along the simple closed contour C is readily given. We let C denote a circle of radius r which is centered at a point (x_0, y_0) and taken counterclockwise. The mean speed along C is then found by dividing the circulation by the circumference $2\pi r$, and the corresponding mean angular speed of the fluid about the center of the circle is obtained by dividing that mean speed by r:

$$\frac{l}{\pi r^2} \int\int_R \frac{1}{2}[q_x(x, y) - p_y(x, y)]\, dA.$$

Now this is also an expression for the mean value of the function

(4) $$\omega(x, y) = \frac{1}{2}[q_x(x, y) - p_y(x, y)]$$

over the circular region R bounded by C. Its limit as r tends to zero is the value of ω at the point (x_0, y_0). Hence the function $\omega(x, y)$, called the *rotation* of the fluid, represents the limiting angular speed of a circular element of the fluid as the circle shrinks to its center (x, y), the point at which ω is evaluated.

If $\omega(x, y) = 0$ at each point in some simply connected domain, the flow is *irrotational* in that domain. We consider only irrotational flows here, and we also assume that the fluid is *incompressible* and *free from viscosity*. Under our assumption of steady irrotational flow of fluids with uniform density ρ, it can be shown that the fluid pressure $P(x, y)$ satisfies the following special case of *Bernoulli's equation*:

$$\frac{P}{\rho} + \frac{1}{2}|V|^2 = \text{constant}.$$

Note that the pressure is greatest where the speed $|V|$ is least.

---

*Properties of line integrals in advanced calculus that are used in this and the following section are to be found in, for instance, W. Kaplan, "Advanced Mathematics for Engineers," Chap. 10, 1981.

Let D be a simply connected domain in which the flow is irrotational. According to equation (4), $p_y = q_x$ throughout D. This relation between partial derivatives implies that the line integral

$$\int_C p(s,t)\,ds + q(s,t)\,dt$$

along a contour C lying entirely in D and joining any two points (x_0, y_0) and (x, y) in D is actually independent of path. Thus, if (x_0, y_0) is fixed, the function

$$(5) \qquad \phi(x, y) = \int_{(x_0,y_0)}^{(x,y)} p(s,t)\,ds + q(s,t)\,dt$$

is well defined on D; and, by taking partial derivatives on each side of this equation, we find that

$$(6) \qquad \phi_x(x, y) = p(x, y), \qquad \phi_y(x, y) = q(x, y).$$

From equations (6), we see that the velocity vector $V = p + iq$ is the gradient of ϕ; and the directional derivative of ϕ in any direction represents the component of the velocity of flow in that direction.

The function $\phi(x, y)$ is called the *velocity potential*. From equation (5), it is evident that $\phi(x, y)$ changes by an additive constant when the reference point (x_0, y_0) is changed. The level curves $\phi(x, y) = c_1$ are called *equipotentials*. Because it is the gradient of $\phi(x, y)$, the velocity vector V is normal to an equipotential at any point where V is not the zero vector.

Just as in the case of the flow of heat, the condition that the incompressible fluid enter or leave an element of volume only by flowing through the boundary of that element requires that $\phi(x, y)$ must satisfy Laplace's equation

$$\phi_{xx}(x, y) + \phi_{yy}(x, y) = 0$$

in a domain where the fluid is free from sources or sinks. In view of equations (6) and the continuity of the functions p and q and their first-order partial derivatives, it follows that the partial derivatives of the first and second order of ϕ are continuous in such a domain. Hence the velocity potential ϕ is a *harmonic* function in that domain.

91. THE STREAM FUNCTION

According to Sec. 90, the velocity vector

$$(1) \qquad V = p(x, y) + iq(x, y)$$

for a simply connected domain in which the flow is irrotational can be written

$$(2) \qquad V = \phi_x(x, y) + i\phi_y(x, y) = \operatorname{grad} \phi(x, y),$$

where ϕ is the velocity potential. When the velocity vector is not the zero vector, it is normal to an equipotential passing through the point (x, y). If, moreover, $\psi(x, y)$ denotes a harmonic conjugate of $\phi(x, y)$ (see Sec. 81), the velocity vector is tangent to a curve $\psi(x, y) = c_2$. The curves $\psi(x, y) = c_2$ are called the *streamlines* of the flow, and the function ψ is the *stream function*. In particular, a boundary across which fluid cannot flow is a streamline.

The analytic function

$$F(z) = \phi(x, y) + i\psi(x, y)$$

is called the *complex potential* of the flow. Note that

$$F'(z) = \phi_x(x, y) + i\psi_x(x, y),$$

or, in view of the Cauchy-Riemann equations,

$$F'(z) = \phi_x(x, y) - i\phi_y(x, y).$$

Expression (2) for the velocity thus becomes

(3) $$V = \overline{F'(z)}.$$

The speed, or magnitude of the velocity, is obtained by writing

$$|V| = |F'(z)|.$$

According to equation (5), Sec. 81, if ϕ is harmonic in a simply connected domain D, a harmonic conjugate of ϕ there can be written

$$\psi(x, y) = \int_{(x_0,y_0)}^{(x,y)} -\phi_t(s,t)\, ds + \phi_s(s,t)\, dt,$$

where the integration is independent of path. With the aid of equations (6), Sec. 90, we can, therefore, write

(4) $$\psi(x, y) = \int_C -q(s,t)\, ds + p(s,t)\, dt,$$

where C is any contour in D from (x_0, y_0) to (x, y).

Now it is shown in advanced calculus that the right-hand side of equation (4) represents the integral with respect to arc length σ along C of the normal component $V_N(x, y)$ of the vector whose x and y components are $p(x, y)$ and $q(x, y)$, respectively. So expression (4) can be written

(5) $$\psi(x, y) = \int_C V_N(s,t)\, d\sigma.$$

Physically, then, $\psi(x, y)$ represents the time rate of flow of the fluid across C. More precisely, $\psi(x, y)$ denotes the rate of flow, by volume, across a surface of unit height standing perpendicular to the xy plane on the curve C.

EXAMPLE.   When the complex potential is the function

(6) $$F(z) = Az,$$

where A is a positive real constant,

(7) $$\phi(x, y) = Ax \qquad \text{and} \qquad \psi(x, y) = Ay.$$

The streamlines $\psi(x, y) = c_2$ are the horizonal lines $y = c_2/A$, and the velocity at any point is

$$V = \overline{F'(z)} = A.$$

Here a point (x_0, y_0) at which $\psi(x, y) = 0$ is any point on the x axis. If the point (x_0, y_0) is taken as the origin, then $\psi(x, y)$ is the rate of flow across any contour drawn from the origin to the point (x, y) (Fig. 123). The flow is uniform and to the right. It can be interpreted as the uniform flow in the upper half plane bounded by the x axis, which is a streamline, or as the uniform flow between two parallel lines $y = y_1$ and $y = y_2$.

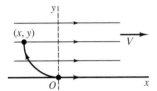

FIGURE 123

The stream function ψ characterizes a definite flow in a region. The question of whether just one such function exists corresponding to a given region, except possibly for a constant factor or an additive constant, is not examined here. In some of the examples to follow, where the velocity is uniform far from the obstruction, or in Chap. 11, where sources and sinks are involved, the physical situation indicates that the flow is uniquely determined by the conditions given in the problem.

A harmonic function is not always uniquely determined, even up to a constant factor, by simply prescribing its values on the boundary of a region. In the above example, the function $\psi(x, y) = Ay$ is harmonic in the half plane $y > 0$ and has zero values on the boundary. The function $\psi_1(x, y) = Be^x \sin y$ also satisfies those conditions. However, the streamline $\psi_1(x, y) = 0$ consists not only of the line $y = 0$ but also of the lines $y = n\pi$ $(n = 1, 2, \ldots)$. Here the function $F_1(z) = Be^z$ is the complex potential for the flow in the strip between the lines $y = 0$ and $y = \pi$, both lines making up the streamline $\psi(x, y) = 0$; if $B > 0$, the fluid flows to the right along the lower line and to the left along the upper one.

92. FLOWS AROUND A CORNER AND AROUND A CYLINDER

In analyzing a flow in the xy, or z, plane, it is often simpler to consider a corresponding flow in the uv, or w, plane. Then, if ϕ is a velocity potential and ψ a stream function for the flow in the uv plane, results in Secs. 82 and 83 can be applied to these harmonic functions. That is, when the domain of flow D_w in the uv plane is the image of a domain D_z under a transformation

$$w = f(z) = u(x, y) + i v(x, y),$$

where f is analytic, the functions

$$\phi[u(x, y), v(x, y)], \qquad \psi[u(x, y), v(x, y)]$$

are harmonic in D_z. These new functions may be interpreted as velocity potential and stream function in the xy plane. A streamline or natural boundary $\psi(u, v) = c_2$ in the

uv plane corresponds to a streamline or natural boundary $\psi[u(x, y), v(x, y)] = c_2$ in the xy plane.

In using this technique, it is usually most efficient to first write the complex potential function for the region in the w plane and then obtain from that the velocity potential and stream function for the corresponding region in the xy plane. More precisely, if the potential function in the uv plane is

$$F(w) = \phi(u, v) + i\psi(u, v),$$

then the composite function

$$F[f(z)] = \phi[u(x, y), v(x, y)] + i\psi[u(x, y), v(x, y)]$$

is the desired complex potential in the xy plane.

To avoid an excess of notation, we use the same symbols F, ϕ, and ψ for the complex potential, etc., in both the xy and the uv planes.

EXAMPLE 1. Consider a flow in the first quadrant $x > 0, y > 0$ that comes in downward parallel to the y axis but is forced to turn a corner near the origin, as shown in Fig. 124. To determine the flow, we recall (Example 3, Sec. 10) that the transformation

$$w = z^2 = x^2 - y^2 + i2xy$$

maps the first quadrant onto the upper half of the uv plane and the boundary of the quadrant onto the entire u axis.

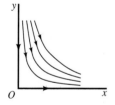

O x **FIGURE 124**

From the example in Sec. 91, we know that the complex potential for a uniform flow to the right in the upper half of the w plane is $F = Aw$, where A is a positive real constant. The potential in the quadrant is, therefore,

(1) $$F = Az^2 = A(x^2 - y^2) + i2Axy;$$

and it follows that the stream function for the flow there is

(2) $$\psi = 2Axy.$$

This stream function is, of course, harmonic in the first quadrant, and it vanishes on the boundary.

The streamlines are branches of the rectangular hyperbolas

$$2Axy = c_2.$$

According to equation (3), Sec. 91, the velocity of the fluid is

$$V = \overline{2Az} = 2A(x - iy).$$

Observe that the speed

$$|V| = 2A \sqrt{x^2 + y^2}$$

of a particle is directly proportional to its distance from the origin. The value of the stream function (2) at a point (x, y) can be interpreted as the rate of flow across a line segment extending from the origin to that point.

EXAMPLE 2.   Let a long circular cylinder of unit radius be placed in a large body of fluid flowing with a uniform velocity, the axis of the cylinder being perpendicular to the direction of flow. To determine the steady flow around the cylinder, we represent the cylinder by the circle $x^2 + y^2 = 1$ and let the flow distant from it be parallel to the x axis and to the right (Fig. 125). Symmetry shows that points on the x axis exterior to the circle may be treated as boundary points, and so we need to consider only the upper part of the figure as the region of flow.

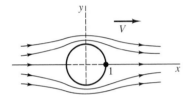

FIGURE 125

The boundary of this region of flow, consisting of the upper semicircle and the parts of the x axis exterior to the circle, is mapped onto the entire u axis by the transformation

$$w = z + \frac{1}{z}.$$

The region itself is mapped onto the upper half plane $v \geq 0$, as indicated in Fig. 17, Appendix 2. The complex potential for the corresponding uniform flow in that half plane is $F = Aw$, where A is a positive real constant. Hence the complex potential for the region exterior to the circle and above the x axis is

(3)
$$F = A\left(z + \frac{1}{z}\right).$$

The velocity

(4)
$$V = A\left(1 - \frac{1}{\bar{z}^2}\right)$$

approaches A as $|z|$ increases. Thus the flow is nearly uniform and parallel to the x axis at points distant from the circle, as one would expect. From expression (4), we see that $V(\bar{z}) = \overline{V(z)}$; hence that expression also represents velocities of flow in the lower region, the lower semicircle being a streamline.

According to equation (3), the stream function for the given problem is, in polar coordinates,

(5)
$$\psi = A\left(r - \frac{1}{r}\right)\sin\theta.$$

The streamlines

$$A\left(r - \frac{1}{r}\right)\sin\theta = c_2$$

are symmetric to the y axis and have asymptotes parallel to the x axis. Note that when $c_2 = 0$, the streamline consists of the circle $r = 1$ and the parts of the x axis exterior to the circle.

EXERCISES

1. State why the components of velocity can be obtained from the stream function by means of the equations

$$p(x, y) = \psi_y(x, y), \qquad q(x, y) = -\psi_x(x, y).$$

2. At an interior point of a region of flow and under the conditions that we have assumed, the fluid pressure cannot be less than the pressure at all other points in a neighborhood of that point. Justify this statement with the aid of statements in Secs. 90, 91, and 42.

3. For the flow around a corner described in Example 1, Sec. 92, at what point of the region $x \geq 0, y \geq 0$ is the fluid pressure greatest?

4. Show that the speed of the fluid at points on the cylindrical surface in Example 2, Sec. 92, is $2A|\sin\theta|$ and that the fluid pressure on the cylinder is greatest at the points $z = \pm 1$ and least at the points $z = \pm i$.

5. Write the complex potential for the flow around a cylinder $r = r_0$ when the velocity V at a point z approaches a real constant A as the point recedes from the cylinder.

6. Obtain the stream function $\psi = Ar^4 \sin 4\theta$ for a flow in the angular region $r \geq 0, 0 \leq \theta \leq \pi/4$ (Fig. 126), and sketch a few of the streamlines in the interior of that region.

x **FIGURE 126**

7. Obtain the complex potential $F = A \sin z$ for a flow inside the semi-infinite region $-\pi/2 \leq x \leq \pi/2, y \geq 0$ (Fig. 127). Write the equations of the streamlines.

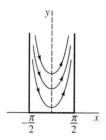

FIGURE 127

8. Show that if the velocity potential is $\phi = A \ln r \ (A > 0)$ for flow in the region $r \geq r_0$, then the streamlines are the half lines $\theta = c \ (r \geq r_0)$ and the rate of flow outward through each complete circle about the origin is $2\pi A$, corresponding to a source of that strength at the origin.

9. Obtain the complex potential $F = A(z^2 + z^{-2})$ for a flow in the region $r \geq 1, 0 \leq \theta \leq \pi/2$. Write expressions for V and ψ. Note how the speed $|V|$ varies along the boundary of the region, and verify that $\psi(x, y) = 0$ on the boundary.

10. Suppose that the flow at an infinite distance from the cylinder of unit radius in Example 2, Sec. 92, is uniform in a direction making an angle α with the x axis; that is,

$$\lim_{|z| \to \infty} V = A \exp(i\alpha) \qquad (A > 0).$$

Find the complex potential.

      Ans. $F = A[z \exp(-i\alpha) + z^{-1} \exp(i\alpha)]$.

11. The transformation $z = w + (1/w)$ maps the circle $|w| = 1$ onto the line segment joining the points $z = -2$ and $z = 2$, and it maps the domain outside that circle onto the rest of the z plane. [See Exercise 19, Sec. 69.] Write

$$z - 2 = r_1 \exp(i\theta_1), \qquad z + 2 = r_2 \exp(i\theta_2),$$

and

$$(z^2 - 4)^{1/2} = \sqrt{r_1 r_2} \exp \frac{i(\theta_1 + \theta_2)}{2} \qquad (0 \leq \theta_1 < 2\pi, 0 \leq \theta_2 < 2\pi);$$

the function $(z^2 - 4)^{1/2}$ is then single-valued and analytic everywhere except on the branch cut consisting of the segment of the x axis joining the points $z = \pm 2$. Show that the inverse of the transformation $z = w + (1/w)$, such that $|w| > 1$ for every point z not on the branch cut, can be written

$$w = \frac{1}{2}[z + (z^2 - 4)^{1/2}] = \frac{1}{4}\left(\sqrt{r_1} \exp \frac{i\theta_1}{2} + \sqrt{r_2} \exp \frac{i\theta_2}{2}\right)^2.$$

The transformation and this inverse establish a one to one correspondence between points in the two domains.

12. With the aid of the results found in Exercises 10 and 11, derive the expression

$$F = A[z \cos \alpha - i(z^2 - 4)^{1/2} \sin \alpha]$$

for the complex potential of the steady flow around a long plate whose width is 4 and whose cross section is the line segment joining the two points $z = \pm 2$ in Fig. 128, assuming that the velocity of the fluid at an infinite distance from the plate is $A \exp(i\alpha)$. The branch of $(z^2 - 4)^{1/2}$ that is used is the one described in Exercise 11, and $A > 0$.

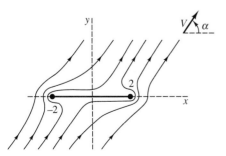

FIGURE 128

324   APPLICATIONS OF CONFORMAL MAPPING

13. Show that if $\sin \alpha \neq 0$ in Exercise 12, then the speed of the fluid along the line segment joining the points $z = \pm 2$ is infinite at the ends and is equal to $A|\cos \alpha|$ at the midpoint.

14. For the sake of simplicity, suppose that $0 < \alpha \leq \pi/2$ in Exercise 12. Then show that the velocity of the fluid along the upper side of the line segment representing the plate in Fig. 128 is zero at the point $x = 2 \cos \alpha$ and that the velocity along the lower side of the segment is zero at the point $x = -2 \cos \alpha$.

15. A circle with its center at a point x_0 ($0 < x_0 < 1$) on the x axis and passing through the point $z = -1$ is subjected to the transformation $w = z + (1/z)$. Individual nonzero points $z = \exp(i\theta)$ can be mapped geometrically by adding the vector $1/z = (1/r)\exp(-i\theta)$ to the vector z. Indicate by mapping some points that the image of the circle is a profile of the type shown in Fig. 129 and that points exterior to the circle map onto points exterior to the profile. This is a special case of the profile of a *Joukowski airfoil*. (See also Exercises 16 and 17 below.)

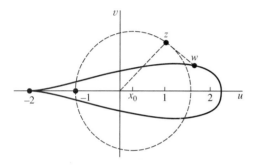

FIGURE 129

16. (*a*) Show that the mapping of the circle in Exercise 15 is conformal except at the point $z = -1$.

(*b*) Let the complex numbers

$$t = \lim_{\Delta z \to 0} \frac{\Delta z}{|\Delta z|}, \qquad \tau = \lim_{\Delta w \to 0} \frac{\Delta w}{|\Delta w|}$$

represent unit vectors tangent to a smooth directed arc at $z = -1$ and that arc's image, respectively, under the transformation $w = z + (1/z)$. Show that $\tau = -t^2$ and hence that the Joukowski profile in Fig. 129 has a cusp at the point $w = -2$, the angle between the tangents at the cusp being zero.

17. The inverse of the transformation $w = z + (1/z)$ used in Exercise 15 is given, with z and w interchanged, in Exercise 11. Find the complex potential for the flow around the airfoil in Exercise 15 when the velocity V of the fluid at an infinite distance from the origin is a real constant A.

18. Note that under the transformation

$$w = e^z + z,$$

both halves, where $x \geq 0$ and $x \leq 0$, of the line $y = \pi$ are mapped onto the half line $v = \pi$ ($u \leq -1$). Similarly, the line $y = -\pi$ is mapped onto the half line $v = -\pi$ ($u \leq -1$); and the strip $-\pi \leq y \leq \pi$ is mapped onto the w plane. Also, note that the change of directions, $\arg(dw/dz)$, under this transformation approaches zero as x tends to $-\infty$. Show that the streamlines of a fluid flowing through the open channel formed by the half lines in the w plane (Fig. 130) are the images of the lines $y = c_2$ in the strip. These streamlines also represent the equipotential curves of the electrostatic field near the edge of a parallel-plate capacitor.

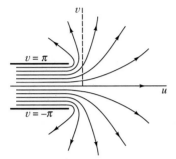

FIGURE 130

CHAPTER
11

THE SCHWARZ-CHRISTOFFEL
TRANSFORMATION

In this chapter, we construct a transformation, known as the Schwarz-Christoffel transformation, which maps the x axis and the upper half of the z plane onto a given simple closed polygon and its interior in the w plane. Applications are made to the solution of problems in fluid flow and electrostatic potential theory.

93. MAPPING THE REAL AXIS
ONTO A POLYGON

We represent the unit vector which is tangent to a smooth arc C at a point z_0 by the complex number t. Let the number τ denote the unit vector tangent to the image Γ of C at the corresponding point w_0 under a transformation $w = f(z)$. We assume that f is analytic at z_0 and that $f'(z_0) \neq 0$. According to Sec. 79,

$$(1) \qquad \qquad \arg \tau = \arg f'(z_0) + \arg t.$$

In particular, if C is a segment of the x axis with positive sense to the right, then $t = 1$ and $\arg t = 0$ at each point $z_0 = x$ on C. In that case, equation (1) becomes

$$(2) \qquad \qquad \arg \tau = \arg f'(x).$$

If $f'(z)$ has a constant argument along that segment, it follows that $\arg \tau$ is constant. Hence the image Γ of C is also a segment of a straight line.

Let us now construct a transformation $w = f(z)$ that maps the whole x axis onto a polygon of n sides, where $x_1, x_2, \ldots, x_{n-1}$, and ∞ are the points on that axis whose images are to be the vertices of the polygon and where

$$x_1 < x_2 < \cdots < x_{n-1}.$$

The vertices are the points $w_j = f(x_j)$ $(j = 1, 2, \ldots, n - 1)$ and $w_n = f(\infty)$. The function f should be such that $\arg f'(z)$ jumps from one constant value to another at the points $z = x_j$ as the point z traces out the x axis (Fig. 131).

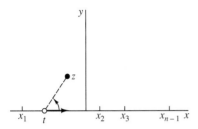 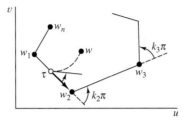

FIGURE 131

If the function f is chosen such that

$$(3) \qquad f'(z) = A(z - x_1)^{-k_1}(z - x_2)^{-k_2} \cdots (z - x_{n-1})^{-k_{n-1}},$$

where A is a complex constant and each k_j is a real constant, then the argument of $f'(z)$ changes in the prescribed manner as z describes the real axis; for the argument of the derivative (3) can be written

$$(4) \qquad \arg f'(z) = \arg A - k_1 \arg(z - x_1)$$
$$- k_2 \arg(z - x_2) - \cdots - k_{n-1} \arg(z - x_{n-1}).$$

When $z = x$ and $x < x_1$,

$$\arg(z - x_1) = \arg(z - x_2) = \cdots = \arg(z - x_{n-1}) = \pi.$$

When $x_1 < x < x_2$, $\arg(z - x_1) = 0$ and each of the other arguments is π. According to equation (4), then, $\arg f'(z)$ increases abruptly by the angle $k_1\pi$ as z moves to the right through the point $z = x_1$. It again jumps in value, by the amount $k_2\pi$, as z passes through the point x_2, etc.

In view of equation (2), the unit vector τ is constant in direction as z moves from x_{j-1} to x_j; w thus moves in that fixed direction along a straight line. The direction of τ changes abruptly, by the angle $k_j\pi$, at the image point w_j of x_j, as shown in Fig. 131. Those angles $k_j\pi$ are the exterior angles of the polygon described by the point w.

The exterior angles can be limited to angles between $-\pi$ and π; that is, $-1 < k_j < 1$. We assume that the sides of the polygon never cross one another and that the polygon is given a positive, or counterclockwise, orientation. The sum of the exterior angles of a *closed* polygon is, then, 2π; and the exterior angle at the vertex w_n, which is the image of the point $z = \infty$, can be written

$$k_n\pi = 2\pi - (k_1 + k_2 + \cdots + k_{n-1})\pi.$$

Thus the numbers k_j must necessarily satisfy the conditions

$$(5) \qquad k_1 + k_2 + \cdots + k_{n-1} + k_n = 2, \qquad -1 < k_j < 1 \quad (j = 1, 2, \ldots, n).$$

Note that $k_n = 0$ if

$$(6) \qquad k_1 + k_2 + \cdots + k_{n-1} = 2.$$

In that case, the direction of τ does not change at the point w_n. So w_n is not a vertex, and the polygon has $n - 1$ sides.

The existence of a mapping function f whose derivative is given by equation (3) will be established in the next section.

94. SCHWARZ-CHRISTOFFEL TRANSFORMATION

In our expression (Sec. 93)

$$(1) \qquad f'(z) = A(z - x_1)^{-k_1}(z - x_2)^{-k_2} \cdots (z - x_{n-1})^{-k_{n-1}}$$

for the derivative of a function that is to map the x axis onto a polygon, let the factors $(z - x_j)^{-k_j}$ represent branches of power functions with branch cuts extending below that axis. To be specific, write

$$(2) \qquad (z - x_j)^{-k_j} = |z - x_j|^{-k_j} \exp(-ik_j\theta_j) \qquad \left(-\frac{\pi}{2} < \theta_j < \frac{3\pi}{2}\right),$$

where $\theta_j = \arg(z - x_j)$ and $j = 1, 2, \ldots, n - 1$. Then $f'(z)$ is analytic everywhere in the half plane $y \geq 0$ except at the $n - 1$ branch points x_j.

If z_0 is a point in that region of analyticity, denoted here by R, then the function

$$(3) \qquad F(z) = \int_{z_0}^{z} f'(s)\,ds$$

is single-valued and analytic throughout the same region, where the path of integration from z_0 to z is any contour lying within R. Moreover, $F'(z) = f'(z)$ (see Sec. 34).

To define the function F at the point $z = x_1$ so that it is continuous there, we note that $(z - x_1)^{-k_1}$ is the only factor in expression (1) that is not analytic at x_1. Hence, if $\phi(z)$ denotes the product of the rest of the factors in that expression, $\phi(z)$ is analytic at x_1 and is represented throughout an open disk $|z - x_1| < R_1$ by its Taylor series about x_1. So we can write

$$f'(z) = (z - x_1)^{-k_1}\phi(z)$$

$$= (z - x_1)^{-k_1}\left[\phi(x_1) + \frac{\phi'(x_1)}{1!}(z - x_1) + \frac{\phi''(x_1)}{2!}(z - x_1)^2 + \cdots\right],$$

or

$$(4) \qquad f'(z) = \phi(x_1)(z - x_1)^{-k_1} + (z - x_1)^{1-k_1}\psi(z)$$

where ψ is analytic and, therefore, continuous throughout the entire open disk. Since $1 - k_1 > 0$, the last term on the right in equation (4) thus represents a continuous function of z throughout the upper half of the disk, where $\operatorname{Im} z \geq 0$, if we assign it the value zero at $z = x_1$. It follows that the integral

$$\int_{Z_1}^{z} (s - x_1)^{1-k_1}\psi(s)\,ds$$

of that last term along a contour from Z_1 to z, where Z_1 and the contour lie in the half disk, is a continuous function of z at $z = x_1$. The integral

$$\int_{Z_1}^{z} (s - x_1)^{-k_1} ds = \frac{1}{1 - k_1}[(z - x_1)^{1-k_1} - (Z_1 - x_1)^{1-k_1}]$$

along the same path also represents a continuous function of z at x_1 if we define the value of the integral there as its limit as z approaches x_1 in the half disk. The integral of the function (4) along the stated path from Z_1 to z is, then, continuous at $z = x_1$; and the same is true of integral (3) since it can be written as an integral along a contour in R from z_0 to Z_1 plus the integral from Z_1 to z.

The above argument applies at each of the $n-1$ points x_j to make F continuous throughout the region $y \geq 0$.

From equation (1), we can show that, for a sufficiently large positive number R, a positive constant M exists such that if $\operatorname{Im} z \geq 0$, then

$$(5) \qquad |f'(z)| < \frac{M}{|z|^{2-k_n}} \qquad \text{whenever} \qquad |z| > R.$$

Since $2 - k_n > 1$, this order property of the integrand in equation (3) ensures the existence of the limit of the integral there as z tends to infinity; that is, a number W_n exists such that

$$(6) \qquad \lim_{z \to \infty} F(z) = W_n \qquad (\operatorname{Im} z \geq 0).$$

Details of the argument are left to Exercises 10 and 11, Sec. 96.

Our mapping function, whose derivative is given by equation (1), can be written $f(z) = F(z) + B$, where B is a complex constant. The resulting transformation,

$$(7) \qquad w = A \int_{z_0}^{z} (s - x_1)^{-k_1}(s - x_2)^{-k_2} \cdots (s - x_{n-1})^{-k_{n-1}} ds + B,$$

is the *Schwarz-Christoffel transformation*, named in honor of the two German mathematicians H. A. Schwarz (1843–1921) and E. B. Christoffel (1829–1900), who discovered it independently.

Transformation (7) is continuous throughout the half plane $y \geq 0$ and is conformal there except for the points x_j. We have assumed that the numbers k_j satisfy conditions (5), Sec. 93. In addition, we suppose that the constants x_j and k_j are such that the sides of the polygon do not cross, so that the polygon is a simple closed contour. Then, according to Sec. 93, as the point z describes the x axis in the positive direction, its image w describes the polygon P in the positive sense; and there is a one to one correspondence between points on that axis and points on P. According to condition (6), the image w_n of the point $z = \infty$ exists and $w_n = W_n + B$.

If z is an interior point of the upper half plane $y \geq 0$ and x_0 is any point on the x axis other than one of the x_j, then the angle from the vector t at x_0 up to the line segment joining x_0 and z is positive and less than π (Fig. 131). At the image w_0 of x_0, the corresponding angle from the vector τ to the image of the line segment joining x_0 and z has that same value. Thus the images of interior points in the half plane lie to the left of the sides of the polygon, taken counterclockwise. A proof that the transformation establishes a one to one correspondence between the interior points of the half plane and the points within the polygon is left to the reader (Exercise 12, Sec. 96).

Given a specific polygon P, let us examine the number of constants in the Schwarz-Christoffel transformation that must be determined in order to map the x axis onto P. For this purpose, we may write $z_0 = 0, A = 1$, and $B = 0$ and simply require that the x axis be mapped onto some polygon P' similar to P. The size and position of P' can then be adjusted to match those of P by introducing the appropriate constants A and B.

The numbers k_j are all determined from the exterior angles at the vertices of P. The $n - 1$ constants x_j remain to be chosen. The image of the x axis is some polygon P' that has the same angles as P. But if P' is to be similar to P, then $n - 2$ connected sides must have a common ratio to the corresponding sides of P; this condition is expressed by means of $n - 3$ equations in the $n - 1$ real unknowns x_j. Thus *two of the numbers x_j, or two relations between them, can be chosen arbitrarily*, provided those $n - 3$ equations in the remaining $n - 3$ unknowns have real-valued solutions.

When a finite point $z = x_n$ on the x axis, instead of the point at infinity, represents the point whose image is the vertex w_n, it follows from Sec. 93 that the Schwarz-Christoffel transformation takes the form

(8)
$$w = A \int_{z_0}^{z} (s - x_1)^{-k_1} (s - x_2)^{-k_2} \cdots (s - x_n)^{-k_n} ds + B,$$

where $k_1 + k_2 + \cdots + k_n = 2$. The exponents k_j are determined from the exterior angles of the polygon. But, in this case, there are n real constants x_j that must satisfy the $n - 3$ equations noted above. Thus *three of the numbers x_j, or three conditions on those n numbers, can be chosen arbitrarily* in transformation (8) of the x axis onto a given polygon.

95.   TRIANGLES AND RECTANGLES

The Schwarz-Christoffel transformation is written in terms of the points x_j and not in terms of their images, the vertices of the polygon. No more than three of those points can be chosen arbitrarily; so, when the given polygon has more than three sides, some of the points x_j must be determined in order to make the given polygon, or any polygon similar to it, be the image of the x axis. The selection of conditions for the determination of those constants, conditions that are convenient to use, often requires ingenuity.

Another limitation in using the transformation is due to the integration that is involved. Often the integral cannot be evaluated in terms of a finite number of elementary functions. In such cases, the solution of problems by means of the transformation can become quite involved.

If the polygon is a triangle with vertices at the points w_1, w_2, and w_3 (Fig. 132), the transformation can be written

(1)
$$w = A \int_{z_0}^{z} (s - x_1)^{-k_1} (s - x_2)^{-k_2} (s - x_3)^{-k_3} ds + B,$$

where $k_1 + k_2 + k_3 = 2$. In terms of the interior angles θ_j,

$$k_j = 1 - \frac{1}{\pi}\theta_j \qquad (j = 1, 2, 3).$$

Here we have taken all three points x_j as finite points on the x axis. Arbitrary values can be assigned to each of them. The complex constants A and B, which are associated with the size and position of the triangle, can be determined so that the upper half plane is mapped onto the given triangular region.

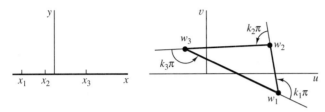

FIGURE 132

If we take the vertex w_3 as the image of the point at infinity, the transformation becomes

$$(2) \qquad w = A \int_{z_0}^{z} (s - x_1)^{-k_1}(s - x_2)^{-k_2} ds + B,$$

where arbitrary real values can be assigned to x_1 and x_2.

The integrals in equations (1) and (2) do not represent elementary functions unless the triangle is degenerate with one or two of its vertices at infinity. The integral in equation (2) becomes an *elliptic integral* when the triangle is equilateral or when it is a right triangle with one of its angles equal to either $\pi/3$ or $\pi/4$.

EXAMPLE 1. For an equilateral triangle, $k_1 = k_2 = k_3 = 2/3$. It is convenient to write $x_1 = -1, x_2 = 1$, and $x_3 = \infty$ and to use equation (2), where $z_0 = 1, A = 1$, and $B = 0$. The transformation then becomes

$$(3) \qquad w = \int_{1}^{z} (s + 1)^{-2/3}(s - 1)^{-2/3} ds.$$

The image of the point $z = 1$ is clearly $w = 0$; that is, $w_2 = 0$. When $z = -1$ in the integral, we can write $s = x$, where $-1 < x < 1$. Then $x + 1 > 0$ and $\arg(x + 1) = 0$, while $|x - 1| = 1 - x$ and $\arg(x - 1) = \pi$. Hence

$$(4) \qquad w = \int_{1}^{-1} (x + 1)^{-2/3}(1 - x)^{-2/3} \exp\left(-\frac{2\pi i}{3}\right) dx$$

$$= \exp\left(\frac{\pi i}{3}\right) \int_{0}^{1} \frac{2\, dx}{(1 - x^2)^{2/3}}.$$

With the substitution $x = \sqrt{t}$, the last integral here reduces to a special case of the one used in defining the beta function (Exercise 7, Sec. 64). Let b denote its value, which is positive:

$$(5) \qquad b = \int_0^1 \frac{2\,dx}{(1-x^2)^{2/3}} = \int_0^1 t^{-1/2}(1-t)^{-2/3}\,dt = B\left(\frac{1}{2}, \frac{1}{3}\right).$$

The vertex w_1 is, therefore, the point (Fig. 133)

$$(6) \qquad w_1 = b\exp\frac{\pi i}{3}.$$

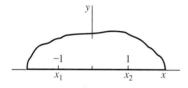
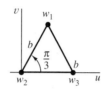

FIGURE 133

The vertex w_3 is on the positive u axis because

$$w_3 = \int_1^\infty (x+1)^{-2/3}(x-1)^{-2/3}\,dx = \int_1^\infty \frac{dx}{(x^2-1)^{2/3}}.$$

But the value of w_3 is also represented by integral (3) when z tends to infinity along the negative x axis; that is,

$$w_3 = \int_1^{-1} (|x+1||x-1|)^{-2/3}\exp\left(-\frac{2\pi i}{3}\right)dx$$

$$+ \int_{-1}^{-\infty} (|x+1||x-1|)^{-2/3}\exp\left(-\frac{4\pi i}{3}\right)dx.$$

In view of the first of expressions (4) for w_1, then,

$$w_3 = w_1 + \exp\left(-\frac{4\pi i}{3}\right)\int_{-1}^{-\infty} (|x+1||x-1|)^{-2/3}\,dx$$

$$= b\exp\frac{\pi i}{3} + \exp\left(-\frac{\pi i}{3}\right)\int_1^\infty \frac{dx}{(x^2-1)^{2/3}},$$

or

$$w_3 = b\exp\frac{\pi i}{3} + w_3\exp\left(-\frac{\pi i}{3}\right).$$

Solving for w_3, we find that

$$(7) \qquad w_3 = b.$$

We have thus verified that the image of the x axis is the equilateral triangle of side b shown in Fig. 133. We can see also that $w = (b/2)\exp(\pi i/3)$ when $z = 0$.

When the polygon is a rectangle, each $k_j = 1/2$. If we choose ± 1 and $\pm a$ as the points x_j whose images are the vertices and write

(8) $$g(z) = (z + a)^{-1/2}(z + 1)^{-1/2}(z - 1)^{-1/2}(z - a)^{-1/2},$$

where $0 \le \arg(z - x_j) \le \pi$, the Schwarz-Christoffel transformation becomes

(9) $$w = -\int_0^z g(s)\, ds,$$

except for a transformation $W = Aw + B$ to adjust the size and position of the rectangle. Integral (9) is a constant times the elliptic integral

$$\int_0^z (1 - s^2)^{-1/2}(1 - k^2 s^2)^{-1/2}ds \qquad \left(k = \frac{1}{a}\right);$$

but the form (8) of the integrand indicates more clearly the appropriate branches of the power functions involved.

EXAMPLE 2. Let us locate the vertices of the rectangle when $a > 1$. As shown in Fig. 134, $x_1 = -a, x_2 = -1, x_3 = 1,$ and $x_4 = a$. All four vertices can be described in terms of two positive numbers b and c that depend on the value of a in the following manner:

(10) $$b = \int_0^1 |g(x)|dx = \int_0^1 \frac{dx}{\sqrt{(1 - x^2)(a^2 - x^2)}},$$

(11) $$c = \int_1^a |g(x)|dx = \int_1^a \frac{dx}{\sqrt{(x^2 - 1)(a^2 - x^2)}}.$$

If $-1 < x < 0$, then

$$\arg(x + a) = \arg(x + 1) = 0 \qquad \text{and} \qquad \arg(x - 1) = \arg(x - a) = \pi;$$

hence

$$g(x) = \left[\exp\left(-\frac{\pi i}{2}\right)\right]^2 |g(x)| = -|g(x)|.$$

If $-a < x < -1$, then

$$g(x) = \left[\exp\left(-\frac{\pi i}{2}\right)\right]^3 |g(x)| = i|g(x)|.$$

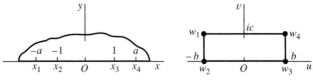

FIGURE 134

Thus

$$w_1 = -\int_0^{-a} g(x)\,dx = -\int_0^{-1} g(x)\,dx - \int_{-1}^{-a} g(x)\,dx$$

$$= \int_0^{-1} |g(x)|\,dx - i\int_{-1}^{-a} |g(x)|\,dx = -b + ic.$$

It is left to the exercises to show that

(12) $w_2 = -b, \qquad w_3 = b, \qquad w_4 = b + ic.$

The position and dimensions of the rectangle are shown in Fig. 134.

96. DEGENERATE POLYGONS

We now apply the Schwarz-Christoffel transformation to some degenerate polygons for which the integrals represent elementary functions. For purposes of illustration, the examples here result in transformations that we have already seen in Chap. 8.

  EXAMPLE 1.   Let us map the half plane $y \geq 0$ onto the semi-infinite strip

$$-\frac{\pi}{2} \leq u \leq \frac{\pi}{2}, \qquad v \geq 0.$$

We consider the strip as the limiting form of a triangle with vertices $w_1, w_2,$ and w_3 (Fig. 135) as the imaginary part of w_3 tends to infinity.

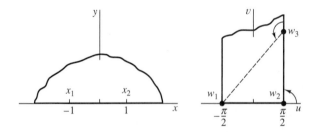

FIGURE 135

  The limiting values of the exterior angles are

$$k_1\pi = k_2\pi = \frac{\pi}{2} \qquad \text{and} \qquad k_3\pi = \pi.$$

We choose the points $x_1 = -1, x_2 = 1,$ and $x_3 = \infty$ as the points whose images are the vertices. Then the derivative of the mapping function can be written

$$\frac{dw}{dz} = A(z+1)^{-1/2}(z-1)^{-1/2} = A'(1-z^2)^{-1/2}.$$

Hence $w = A' \sin^{-1} z + B$. If we write $A' = 1/a$ and $B = b/a$, it follows that

$$z = \sin(aw - b).$$

This transformation from the w to the z plane satisfies the conditions $z = -1$ when $w = -\pi/2$ and $z = 1$ when $w = \pi/2$ if $a = 1$ and $b = 0$. The resulting transformation is

$$z = \sin w,$$

which we verified in Sec. 74 as one that maps the strip onto the half plane.

EXAMPLE 2. Consider the strip $0 < v < \pi$ as the limiting form of a rhombus with vertices at the points $w_1 = \pi i, w_2, w_3 = 0$, and w_4 as the points w_2 and w_4 are moved infinitely far to the left and right, respectively (Fig. 136). In the limit, the exterior angles become

$$k_1 \pi = 0, \qquad k_2 \pi = \pi, \qquad k_3 \pi = 0, \qquad k_4 \pi = \pi.$$

We leave x_1 to be determined and choose the values $x_2 = 0, x_3 = 1$, and $x_4 = \infty$. The derivative of the Schwarz-Christoffel mapping function then becomes

$$\frac{dw}{dz} = A(z - x_1)^0 z^{-1}(z - 1)^0 = \frac{A}{z};$$

thus

$$w = A \operatorname{Log} z + B.$$

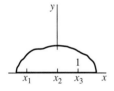

FIGURE 136

Now $B = 0$ because $w = 0$ when $z = 1$. The constant A must be real because the point w lies on the real axis when $z = x$ and $x > 0$. The point $w = \pi i$ is the image of the point $z = x_1$, where x_1 is a negative number; therefore

$$\pi i = A \operatorname{Log} x_1 = A \ln |x_1| + A \pi i.$$

By identifying real and imaginary parts here, we see that $|x_1| = 1$ and $A = 1$. Hence the transformation becomes

$$w = \operatorname{Log} z;$$

also, $x_1 = -1$. We already know from the discussion following Example 2 in Sec. 73 that this transformation maps the half plane onto the strip.

The procedure used in these two examples is not rigorous because limiting values of angles and coordinates were not introduced in an orderly way. Limiting values were used whenever it seemed expedient to do so. But, if we verify the mapping obtained, it is not essential that we justify the steps in our derivation of the mapping function. The formal method used here is shorter and less tedious than rigorous methods.

EXERCISES

1. In transformation (1), Sec. 95, write $B = z_0 = 0$ and

$$A = \exp \frac{3\pi i}{4}, \qquad x_1 = -1, \qquad x_2 = 0, \qquad x_3 = 1,$$

$$k_1 = \frac{3}{4}, \qquad k_2 = \frac{1}{2}, \qquad k_3 = \frac{3}{4}$$

to map the x axis onto an *isosceles right triangle*. Show that the vertices of that triangle are the points

$$w_1 = bi, \qquad w_2 = 0, \qquad \text{and} \qquad w_3 = b,$$

where b is the positive constant

$$b = \int_0^1 (1 - x^2)^{-3/4} x^{-1/2} \, dx.$$

Also show that

$$2b = B\left(\frac{1}{4}, \frac{1}{4}\right),$$

where B is the beta function.

2. Obtain expressions (12) in Sec. 95 for the rest of the vertices of the rectangle shown in Fig. 134.

3. Show that when $0 < a < 1$ in equations (8) and (9), Sec. 95, the vertices of the rectangle are those shown in Fig. 134, where b and c now have the values

$$b = \int_0^a |g(x)| \, dx, \qquad c = \int_a^1 |g(x)| \, dx.$$

4. Show that the special case

$$w = i \int_0^z (s + 1)^{-1/2} (s - 1)^{-1/2} s^{-1/2} \, ds$$

of the Schwarz-Christoffel transformation (7), Sec. 94, maps the x axis onto the *square* with vertices

$$w_1 = bi, \qquad w_2 = 0, \qquad w_3 = b, \qquad w_4 = b + ib,$$

where the positive number b is given in terms of the beta function:

$$b = \frac{1}{2} B\left(\frac{1}{4}, \frac{1}{2}\right).$$

5. Use the Schwarz-Christoffel transformation to arrive at the transformation $w = z^m$ ($0 < m < 1$), which maps the half plane $y \geq 0$ onto the angular region $|w| \geq 0, 0 \leq \arg w \leq$

$m\pi$ and transforms the point $z = 1$ into the point $w = 1$. Consider the angular region as the limiting case of the triangular one shown in Fig. 137 as the angle α tends to 0.

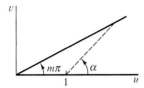

FIGURE 137

6. Refer to Fig. 26, Appendix 2. As the point z moves to the right along the negative real axis, its image point w is to move to the right along the entire u axis. As z describes the segment $0 \le x \le 1$ of the real axis, its image point w is to move to the left along the half line $v = \pi i$ $(u \ge 1)$; and, as z moves to the right along that part of the positive real axis where $x \ge 1$, its image point w is to move to the right along the same half line $v = \pi i$ $(u \ge 1)$. Note the changes in direction of the motion of w at the images of the points $z = 0$ and $z = 1$. These changes suggest that the derivative of a mapping function should be

$$f'(z) = A(z - 0)^{-1}(z - 1),$$

where A is some constant; thus obtain formally the mapping function

$$w = \pi i + z - \text{Log}\, z,$$

which can be verified as one that maps the half plane Re $z > 0$ as indicated in the figure.

7. As the point z moves to the right along that part of the negative real axis where $x \le -1$, its image point is to move to the right along the negative real axis in the w plane. As z moves on the real axis to the right along the segment $-1 \le x \le 0$ and then along the segment $0 \le x \le 1$, its image point w is to move in the direction of increasing v along the segment $0 \le v \le 1$ of the v axis and then in the direction of decreasing v along the same segment. Finally, as z moves to the right along that part of the positive real axis where $x \ge 1$, its image point is to move to the right along the positive real axis in the w plane. Note the changes in direction of the motion of w at the images of the points $z = -1, z = 0$, and $z = 1$. A mapping function whose derivative is

$$f'(z) = A(z + 1)^{-1/2}(z - 0)^{1}(z - 1)^{-1/2},$$

where A is some constant, is thus indicated. Obtain formally the mapping function

$$w = \sqrt{z^2 - 1},$$

where $0 < \arg \sqrt{z^2 - 1} < \pi$. By considering the successive mappings $Z = z^2, W = Z - 1$, and $w = \sqrt{W}$, verify that the resulting transformation maps the half plane Re $z > 0$ onto the half plane Im $w > 0$, with a cut along the segment $0 < v \le 1$ of the v axis.

8. The inverse of the linear fractional transformation

$$Z = \frac{i - z}{i + z}$$

maps the unit disk $|Z| \le 1$ conformally, except at the point $Z = -1$, onto the half plane Im $z \ge 0$. (See Fig. 13, Appendix 2.) Let Z_j be points on the circle $|Z| = 1$ whose images are the points $z = x_j$ $(j = 1, 2, \ldots, n)$ that are used in the Schwarz-Christoffel transformation (8), Sec. 94. Show formally, without determining the branches of the power functions, that

$$\frac{dw}{dZ} = A'(Z - Z_1)^{-k_1}(Z - Z_2)^{-k_2}\cdots(Z - Z_n)^{-k_n},$$

where A' is a constant. Thus show that *the transformation*

$$w = A'\int_0^Z (S - Z_1)^{-k_1}(S - Z_2)^{-k_2}\cdots(S - Z_n)^{-k_n}dS + B$$

maps the interior of the circle $|Z| = 1$ *onto the interior of a polygon*, the vertices of the polygon being the images of the points Z_j on the circle.

9. In the integral of Exercise 8, let the numbers Z_j $(j = 1, 2, \ldots, n)$ be the nth roots of unity. Write $\omega = \exp(2\pi i/n)$ and $Z_1 = 1, Z_2 = \omega, \ldots, Z_n = \omega^{n-1}$. Let each of the numbers k_j $(j = 1, 2, \ldots, n)$ have the value $2/n$. The integral in Exercise 8 then becomes

$$w = A'\int_0^Z \frac{dS}{(S^n - 1)^{2/n}} + B.$$

Show that when $A' = 1$ and $B = 0$, this transformation maps the interior of the unit circle $|Z| = 1$ onto the interior of a regular polygon of n sides and that the center of the polygon is the point $w = 0$.

   Suggestion: The image of each of the points Z_j $(j = 1, 2, \ldots, n)$ is a vertex of some polygon with an exterior angle of $2\pi/n$ at that vertex. Write

$$w_1 = \int_0^1 \frac{dS}{(S^n - 1)^{2/n}},$$

where the path of the integration is along the positive real axis from $Z = 0$ to $Z = 1$ and the principal value of the nth root of $(S^n - 1)^2$ is to be taken. Then show that the images of the points $Z_2 = \omega, \ldots, Z_n = \omega^{n-1}$ are the points $\omega w_1, \ldots, \omega^{n-1}w_1$, respectively. Thus verify that the polygon is regular and is centered at $w = 0$.

10. Obtain inequality (5), Sec. 94.

   Suggestion: Let R be larger than any of the numbers $|x_j|$ $(j = 1, 2, \ldots, n - 1)$. Note that if R is sufficiently large, the inequalities $|z|/2 < |z - x_j| < 2|z|$ hold for each x_j when $|z| > R$. Then use equation (1), Sec. 94, along with conditions (5), Sec. 93.

11. Use condition (5), Sec. 94, and sufficient conditions for the existence of improper integrals of real-valued functions to show that $F(x)$ has some limit W_n as x tends to infinity, where $F(z)$ is defined by equation (3) in that section. Also, show that the integral of $f'(z)$ over each arc of a semicircle $|z| = R$, $\operatorname{Im} z \geq 0$ approaches 0 as R tends to ∞. Then deduce that

$$\lim_{z \to \infty} F(z) = W_n \qquad (\operatorname{Im} z \geq 0),$$

as stated in equation (6) of Sec. 94.

12. According to Sec. 65, the expression

$$N = \frac{1}{2\pi i}\int_C \frac{g'(z)}{g(z)}\,dz$$

can be used to determine the number (N) of zeros of a function g interior to a positively oriented simple closed contour C when $g(z) \neq 0$ on C and when C lies in a simply connected domain D throughout which g is analytic and $g'(z)$ is never zero. In that expression, write $g(z) = f(z) - w_0$, where $f(z)$ is the Schwarz-Christoffel mapping function (7), Sec. 94, and the point w_0 is either interior to or exterior to the polygon P that is the image of the x axis; thus $f(z) \neq w_0$. Let the contour C consist of the upper half of a circle $|z| = R$ and a segment $-R < x < R$ of the x axis that contains all $n - 1$ points x_j, except that a small segment about each point x_j is replaced by the upper half of a circle $|z - x_j| = \rho_j$ with that segment as its diameter. Then the number of points z interior to C such that $f(z) = w_0$ is

$$N_C = \frac{1}{2\pi i} \int_C \frac{f'(z)}{f(z) - w_0} \, dz.$$

Note that $f(z) - w_0$ approaches the nonzero point $W_n - w_0$ when $|z| = R$ and R tends to ∞, and recall the order property (5), Sec. 94, for $|f'(z)|$. Let the ρ_j tend to zero, and prove that the number of points in the upper half of the z plane at which $f(z) = w_0$ is

$$N = \frac{1}{2\pi i} \lim_{R \to \infty} \int_{-R}^{R} \frac{f'(x)}{f(x) - w_0} \, dx.$$

Deduce that since

$$\int_P \frac{dw}{w - w_0} = \lim_{R \to \infty} \int_{-R}^{R} \frac{f'(x)}{f(x) - w_0} \, dx,$$

$N = 1$ if w_0 is interior to P and that $N = 0$ if w_0 is exterior to P. Thus show that the mapping of the half plane $\operatorname{Im} z > 0$ onto the interior of P is one to one.

97. FLUID FLOW IN A CHANNEL THROUGH A SLIT

We now present a further example of the idealized steady flow treated in Chap. 10, an example that will help show how sources and sinks can be accounted for in problems of fluid flow. In this and the following two sections, the problems are posed in the uv plane, rather than the xy plane. That allows us to refer directly to earlier results in this chapter without interchanging the planes.

Consider the two-dimensional steady flow of fluid between two parallel planes $v = 0$ and $v = \pi$ when the fluid is entering through a narrow slit along the line in the first plane that is perpendicular to the uv plane at the origin (Fig. 138). Let the rate of flow of fluid into the channel through the slit be Q units of volume per unit time for each unit of depth of the channel, where the depth is measured perpendicular to the uv plane. The rate of flow out at either end is, then, $Q/2$.

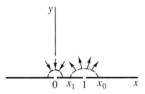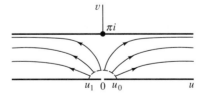

FIGURE 138

The transformation $w = \operatorname{Log} z$, derived in Example 2, Sec. 96, is a one to one mapping of the upper half of the z plane onto the strip in the w plane. The inverse transformation

(1) $$z = e^w = e^u e^{iv}$$

maps the strip onto the half plane (see Example 2, Sec. 73). Under transformation (1), the image of the u axis is the positive half of the x axis, and the image of the line $v = \pi$ is the negative half of the x axis. Hence the boundary of the strip is transformed into the boundary of the half plane.

The image of the point $w = 0$ is the point $z = 1$. The image of a point $w = u_0$, where $u_0 > 0$, is a point $z = x_0$, where $x_0 > 1$. The rate of flow of fluid across a curve joining the point $w = u_0$ to a point (u, v) within the strip is a stream function $\psi(u, v)$ for the flow (Sec. 91). If u_1 is a negative real number, then the rate of flow into the channel through the slit can be written

$$\psi(u_1, 0) = Q.$$

Now, under a conformal transformation, the function ψ is transformed into a function of x and y that represents the stream function for the flow in the corresponding region of the z plane; that is, the rate of flow is the same across corresponding curves in the two planes. As in Chap. 10, the same symbol ψ is used to represent the different stream functions in the two planes. Since the image of the point $w = u_1$ is a point $z = x_1$, where $0 < x_1 < 1$, the rate of flow across any curve connecting the points $z = x_0$ and $z = x_1$ and lying in the upper half of the z plane is also equal to Q. Hence there is a source at the point $z = 1$ equal to the source at $w = 0$.

The above argument applies in general to show that *under a conformal transformation, a source or sink at a given point corresponds to an equal source or sink at the image of that point.*

As Re w tends to $-\infty$, the image of w approaches the point $z = 0$. A sink of strength $Q/2$ at the latter point corresponds to the sink infinitely far to the left in the strip. To apply the above argument in this case, we consider the rate of flow across a curve connecting the boundary lines $v = 0$ and $v = \pi$ of the left-hand part of the strip and the rate of flow across the image of that curve in the z plane.

The sink at the right-hand end of the strip is transformed into a sink at infinity in the z plane.

The stream function ψ for the flow in the upper half of the z plane in this case must be a function whose values are constant along each of the three parts of the x axis. Moreover, its value must increase by Q as the point z moves around the point $z = 1$ from the position $z = x_0$ to the position $z = x_1$, and its value must decrease by $Q/2$ as z moves about the origin in the corresponding manner. We see that the function

$$\psi = \frac{Q}{\pi}\left[\operatorname{Arg}(z - 1) - \frac{1}{2}\operatorname{Arg} z \right]$$

satisfies those requirements. Furthermore, this function is harmonic in the half plane Im $z > 0$ because it is the imaginary component of the function

$$F = \frac{Q}{\pi}\left[\operatorname{Log}(z - 1) - \frac{1}{2}\operatorname{Log} z \right] = \frac{Q}{\pi}\operatorname{Log}(z^{1/2} - z^{-1/2}).$$

The function F is a complex potential function for the flow in the upper half of the z plane. Since $z = e^w$, a complex potential function $F(w)$ for the flow in the channel is

$$F(w) = \frac{Q}{\pi} \text{Log}(e^{w/2} - e^{-w/2}).$$

By dropping an additive constant, we can write

(2) $$F(w) = \frac{Q}{\pi} \text{Log}\left(\sinh \frac{w}{2}\right).$$

We have used the same symbol F to denote three distinct functions, once in the z plane and twice in the w plane.

The velocity vector $\overline{F'(w)}$ is given by the equation

(3) $$V = \frac{Q}{2\pi} \coth \frac{\overline{w}}{2}.$$

From this, it can be seen that

$$\lim_{|u| \to \infty} V = \frac{Q}{2\pi}.$$

Also, the point $w = \pi i$ is a *stagnation point*; that is, the velocity is zero there. Hence the fluid pressure along the wall $v = \pi$ of the channel is greatest at points opposite the slit.

The stream function $\psi(u, v)$ for the channel is the imaginary component of the function $F(w)$ given by equation (2). The streamlines $\psi(u, v) = c_2$ are, therefore, the curves

$$\frac{Q}{\pi} \text{Arg}\left(\sinh \frac{w}{2}\right) = c_2.$$

This equation reduces to

(4) $$\tan \frac{v}{2} = c \tanh \frac{u}{2},$$

where c is any real constant. Some of these streamlines are indicated in Fig. 138.

98. FLOW IN A CHANNEL WITH AN OFFSET

To further illustrate the use of the Schwarz-Christoffel transformation, let us find the complex potential for the flow of a fluid in a channel with an abrupt change in its breadth (Fig. 139). We take our unit of length such that the breadth of the wide part of the channel is π units; then $h\pi$, where $0 < h < 1$, represents the breadth of the narrow part. Let the real constant V_0 denote the velocity of the fluid far from the offset in the wide part; that is,

$$\lim_{u \to -\infty} V = V_0,$$

where the complex variable V represents the velocity vector. The rate of flow per unit depth through the channel, or the strength of the source on the left and of the sink on the right, is then

$$(1) \qquad\qquad\qquad Q = \pi V_0.$$

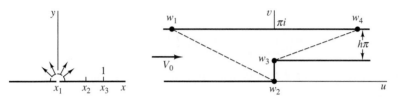

FIGURE 139

The cross section of the channel can be considered as the limiting case of the quadrilateral with the vertices w_1, w_2, w_3, and w_4 shown in Fig. 139 as the first and last of these vertices are moved infinitely far to the left and to the right, respectively. In the limit, the exterior angles become

$$k_1\pi = \pi, \qquad k_2\pi = \frac{\pi}{2}, \qquad k_3\pi = -\frac{\pi}{2}, \qquad k_4\pi = \pi.$$

As before, we proceed formally, using limiting values whenever it is convenient to do so. If we write $x_1 = 0, x_3 = 1, x_4 = \infty$ and leave x_2 to be determined, where $0 < x_2 < 1$, the derivative of the mapping function becomes

$$(2) \qquad\qquad \frac{dw}{dz} = Az^{-1}(z - x_2)^{-1/2}(z - 1)^{1/2}.$$

To simplify the determination of the constants A and x_2 here, we proceed at once to the complex potential of the flow. The source of the flow in the channel infinitely far to the left corresponds to an equal source at $z = 0$ (Sec. 97). The entire boundary of the cross section of the channel is the image of the x axis. In view of equation (1), then, the function

$$(3) \qquad\qquad F = V_0 \operatorname{Log} z = V_0 \ln r + iV_0\theta$$

is the potential for the flow in the upper half of the z plane, with the required source at the origin. Here the stream function is $\psi = V_0\theta$. It increases in value from 0 to $V_0\pi$ over each semicircle $z = Re^{i\theta}(0 \le \theta \le \pi)$, where $R > 0$, as θ varies from 0 to π. [Compare equation (5), Sec. 91, and Exercise 8, Sec. 92.]

The complex conjugate of the velocity V in the w plane can be written

$$\overline{V(w)} = \frac{dF}{dw} = \frac{dF}{dz}\frac{dz}{dw}.$$

Thus, by referring to equations (2) and (3), we can write

$$(4) \qquad\qquad \overline{V(w)} = \frac{V_0}{A}\left(\frac{z - x_2}{z - 1}\right)^{1/2}.$$

At the limiting position of the point w_1, which corresponds to $z = 0$, the velocity is the real constant V_0. It therefore follows from equation (4) that

$$V_0 = \frac{V_0}{A} \sqrt{x_2}.$$

At the limiting position of w_4, which corresponds to $z = \infty$, let the real number V_4 denote the velocity. Now it seems plausible that as a vertical line segment spanning the narrow part of the channel is moved infinitely far to the right, V approaches V_4 at each point on that segment. We could establish this conjecture as a fact by first finding w as a function of z from equation (2); but, to shorten our discussion, we assume that this is true. Then, since the flow is steady,

$$\pi h V_4 = \pi V_0 = Q,$$

or $V_4 = V_0/h$. Letting z tend to infinity in equation (4), we find that

$$\frac{V_0}{h} = \frac{V_0}{A}.$$

Thus

(5) $$A = h, \qquad x_2 = h^2$$

and

(6) $$\overline{V(w)} = \frac{V_0}{h} \left(\frac{z - h^2}{z - 1} \right)^{1/2}.$$

From equation (6), we can see that the magnitude $|V|$ of the velocity becomes infinite at the corner w_3 of the offset since it is the image of the point $z = 1$. Also, the corner w_2 is a stagnation point, a point where $V = 0$. Along the boundary of the channel, the fluid pressure is, therefore, greatest at w_2 and least at w_3.

To write the relation between the potential and the variable w, we must integrate equation (2), which can now be written

(7) $$\frac{dw}{dz} = \frac{h}{z} \left(\frac{z - 1}{z - h^2} \right)^{1/2}.$$

By substituting a new variable s, where

$$\frac{z - h^2}{z - 1} = s^2,$$

one can show that equation (7) reduces to

$$\frac{dw}{ds} = 2h \left(\frac{1}{1 - s^2} - \frac{1}{h^2 - s^2} \right).$$

Hence

(8) $$w = h \operatorname{Log} \frac{1 + s}{1 - s} - \operatorname{Log} \frac{h + s}{h - s}.$$

The constant of integration here is zero because when $z = h^2$, s is zero and so, therefore, is w.

In terms of s, the potential F of equation (3) becomes

$$F = V_0 \operatorname{Log} \frac{h^2 - s^2}{1 - s^2};$$

consequently,

$$(9) \qquad\qquad s^2 = \frac{\exp(F/V_0) - h^2}{\exp(F/V_0) - 1}.$$

By substituting s from this equation into equation (8), we obtain an implicit relation that defines the potential F as a function of w.

99. ELECTROSTATIC POTENTIAL ABOUT AN EDGE OF A CONDUCTING PLATE

Two parallel conducting plates of infinite extent are kept at the electrostatic potential $V = 0$, and a parallel semi-infinite plate, placed midway between them, is kept at the potential $V = 1$. The coordinate system and the unit of length are chosen so that the plates lie in the planes $v = 0$, $v = \pi$, and $v = \pi/2$ (Fig. 140). Let us determine the potential function $V(u, v)$ in the region between those plates.

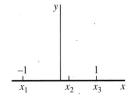

FIGURE 140

The cross section of that region in the uv plane has the limiting form of the quadrilateral bounded by the dashed lines in the figure as the points w_1 and w_3 move out to the right and w_4 to the left. In applying the Schwarz-Christoffel transformation here, we let the point x_4, corresponding to the vertex w_4, be the point at infinity. We choose the points $x_1 = -1$, $x_3 = 1$ and leave x_2 to be determined. The limiting values of the exterior angles of the quadrilateral are

$$k_1 \pi = \pi, \qquad k_2 \pi = -\pi, \qquad k_3 \pi = k_4 \pi = \pi.$$

Thus

$$\frac{dw}{dz} = A(z + 1)^{-1}(z - x_2)(z - 1)^{-1} = A\left(\frac{z - x_2}{z^2 - 1}\right) = \frac{A}{2}\left(\frac{1 + x_2}{z + 1} + \frac{1 - x_2}{z - 1}\right),$$

and so the transformation of the upper half of the z plane into the divided strip in the w plane has the form

(1)
$$w = \frac{A}{2}[(1 + x_2)\operatorname{Log}(z + 1) + (1 - x_2)\operatorname{Log}(z - 1)] + B.$$

Let A_1, A_2 and B_1, B_2 denote the real and imaginary parts of the constants A and B. When $z = x$, the point w lies on the boundary of the divided strip; and, according to equation (1),

(2)
$$u + iv = \frac{A_1 + iA_2}{2}\{(1 + x_2)[\ln|x + 1| + i\arg(x + 1)]$$

$$+ (1 - x_2)[\ln|x - 1| + i\arg(x - 1)]\} + B_1 + iB_2.$$

To determine the constants here, we first note that the limiting position of the line segment joining the points w_1 and w_4 is the u axis. That segment is the image of the part of the x axis to the left of the point $x_1 = -1$; this is because the line segment joining w_3 and w_4 is the image of the part of the x axis to the right of $x_3 = 1$, and the other two sides of the quadrilateral are the images of the remaining two segments of the x axis. Hence when $v = 0$ and u tends to infinity through positive values, the corresponding point x approaches the point $z = -1$ from the left. Thus

$$\arg(x + 1) = \pi, \qquad \arg(x - 1) = \pi,$$

and $\ln|x + 1|$ tends to $-\infty$. Also, since $-1 < x_2 < 1$, the real part of the quantity inside the braces in equation (2) tends to $-\infty$. Since $v = 0$, it follows that $A_2 = 0$; otherwise, the imaginary part on the right would become infinite. By equating imaginary parts on the two sides, we now see that

$$0 = \frac{A_1}{2}[(1 + x_2)\pi + (1 - x_2)\pi] + B_2.$$

Hence

(3)
$$-\pi A_1 = B_2, \qquad A_2 = 0.$$

The limiting position of the line segment joining the points w_1 and w_2 is the half line $v = \pi/2$ ($u \geq 0$). Points on that half line are images of the points $z = x$, where $-1 < x \leq x_2$; consequently,

$$\arg(x + 1) = 0, \qquad \arg(x - 1) = \pi.$$

Identifying the imaginary parts on the two sides of equation (2), we thus arrive at the relation

(4)
$$\frac{\pi}{2} = \frac{A_1}{2}(1 - x_2)\pi + B_2.$$

Finally, the limiting positions of the points on the line segment joining w_3 to w_4 are the points $u + \pi i$, which are the images of the points x when $x > 1$. By identifying, for those points, the imaginary parts in equation (2), we find that

$$\pi = B_2.$$

Then, in view of equations (3) and (4),

$$A_1 = -1, \qquad x_2 = 0.$$

Thus $x = 0$ is the point whose image is the vertex $w = \pi i/2$; and, upon substituting these values into equation (2) and identifying real parts, we see that $B_1 = 0$.

Transformation (1) now becomes

(5)
$$w = -\frac{1}{2}[\text{Log}(z + 1) + \text{Log}(z - 1)] + \pi i,$$

or

(6)
$$z^2 = 1 + e^{-2w}.$$

Under this transformation, the required harmonic function $V(u, v)$ becomes a harmonic function of x and y in the half plane $y > 0$; and the boundary conditions indicated in Fig. 141 are satisfied. Note that $x_2 = 0$ now. The harmonic function in that half plane which assumes those values on the boundary is the imaginary component of the analytic function

$$\frac{1}{\pi} \text{Log} \frac{z-1}{z+1} = \frac{1}{\pi} \ln \frac{r_1}{r_2} + \frac{i}{\pi}(\theta_1 - \theta_2),$$

where θ_1 and θ_2 range from 0 to π. Writing the tangents of these angles as functions of x and y and simplifying, we find that

(7)
$$\tan \pi V = \tan(\theta_1 - \theta_2) = \frac{2y}{x^2 + y^2 - 1}.$$

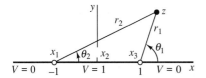

FIGURE 141

Equation (6) furnishes expressions for $x^2 + y^2$ and $x^2 - y^2$ in terms of u and v. Then, from equation (7), we find that the relation between the potential V and the coordinates u and v can be written

(8)
$$\tan \pi V = \frac{1}{s} \sqrt{e^{-4u} - s^2},$$

where

$$s = -1 + \sqrt{1 + 2e^{-2u} \cos 2v + e^{-4u}}.$$

EXERCISES

1. Use the Schwarz-Christoffel transformation to obtain formally the mapping function given with Fig. 22, Appendix 2.

2. Explain why the solution of the problem of flow in a channel with a semi-infinite rectangular obstruction (Fig. 142) is included in the solution of the problem treated in Sec. 98.

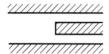

FIGURE 142

3. Refer to Fig. 29, Appendix 2. As the point z moves to the right along the negative part of the real axis where $x \leq -1$, its image point w is to move to the right along the half line $v = h$ ($u \leq 0$). As the point z moves to the right along the segment $-1 \leq x \leq 1$ of the x axis, its image point w is to move in the direction of decreasing v along the segment $0 \leq v \leq h$ of the v axis. Finally, as z moves to the right along the positive part of the real axis where $x \geq 1$, its image point w is to move to the right along the positive real axis. Note the changes in the direction of motion of w at the images of the points $z = -1$ and $z = 1$. These changes indicate that the derivative of a mapping function might be

$$\frac{dw}{dz} = A\left(\frac{z+1}{z-1}\right)^{1/2},$$

where A is some constant. Thus obtain formally the transformation given with the figure. Verify that the transformation, written in the form

$$w = \frac{h}{\pi}\{(z+1)^{1/2}(z-1)^{1/2} + \text{Log}[z + (z+1)^{1/2}(z-1)^{1/2}]\}$$

where $0 \leq \arg(z \pm 1) \leq \pi$, maps the boundary in the manner indicated in the figure.

4. Let $T(u, v)$ denote the bounded steady-state temperatures in the shaded region of the w plane in Fig. 29, Appendix 2, with the boundary conditions $T(u, h) = 1$ when $u < 0$ and $T = 0$ on the rest ($B'C'D'$) of the boundary. In terms of the real parameter α ($0 < \alpha < \pi/2$), show that the image of each point $z = i \tan \alpha$ on the positive y axis is the point

$$w = \frac{h}{\pi}\left[\ln(\tan \alpha + \sec \alpha) + i\left(\frac{\pi}{2} + \sec \alpha\right)\right]$$

(see Exercise 3) and that the temperature at that point w is

$$T(u, v) = \frac{\alpha}{\pi} \qquad \left(0 < \alpha < \frac{\pi}{2}\right).$$

5. Let $F(w)$ denote the complex potential function for the flow of a fluid over a step in the bed of a deep stream represented by the shaded region of the w plane in Fig. 29, Appendix 2, where the fluid velocity V approaches a real constant V_0 as $|w|$ tends to infinity in that region. The transformation that maps the upper half of the z plane onto that region is noted in Exercise 3. Using the identity $dF/dw = (dF/dz)(dz/dw)$, show that

$$\overline{V(w)} = V_0(z-1)^{1/2}(z+1)^{-1/2};$$

and, in terms of the points $z = x$ whose images are the points along the bed of the stream, show that

$$|V| = |V_0|\sqrt{\left|\frac{x-1}{x+1}\right|}.$$

Note that the speed increases from $|V_0|$ along $A'B'$ until $|V| = \infty$ at B', then diminishes to zero at C', and increases toward $|V_0|$ from C' to D'; note, too, that the speed is $|V_0|$ at the point

$$w = i\left(\frac{1}{2} + \frac{1}{\pi}\right)h,$$

between B' and C'.

INTEGRAL
FORMULAS
OF THE
POISSON
TYPE

In this chapter, we develop a theory that enables us to obtain solutions to a variety of boundary value problems where those solutions are expressed in terms of definite or improper integrals. Many of the integrals occurring are then readily evaluated.

100. POISSON INTEGRAL FORMULA

Let C_0 denote a positively oriented circle centered at the origin, and suppose that a function f is analytic within and on C_0. The Cauchy integral formula (Sec. 39)

$$(1) \qquad f(z) = \frac{1}{2\pi i} \int_{C_0} \frac{f(s)\, ds}{s - z}$$

expresses the value of f at any point z interior to C_0 in terms of the values of f at points s on C_0. In this section, we shall obtain from formula (1) a corresponding formula for the real part of the function f; and, in Sec. 101, we shall use that result to solve the Dirichlet problem (Sec. 82) for the disk bounded by C_0.

We let r_0 denote the radius of C_0 and write $z = r \exp(i\theta)$, where $0 < r < r_0$ (Fig. 143). The inverse of the nonzero point z with respect to the circle is the point z_1 lying on the same ray from the origin as z and satisfying the condition $|z_1||z| = r_0^2$; thus, if s is a point on C_0,

$$(2) \qquad z_1 = \frac{r_0^2}{r} \exp(i\theta) = \frac{r_0^2}{\bar{z}} = \frac{s\bar{s}}{\bar{z}}.$$

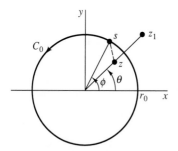

FIGURE 143

Since z_1 is exterior to the circle C_0, it follows from the Cauchy-Goursat theorem that the value of the integral in equation (1) is zero when z is replaced by z_1 in the integrand. Hence

$$f(z) = \frac{1}{2\pi i} \int_{C_0} \left(\frac{1}{s - z} - \frac{1}{s - z_1} \right) f(s)\, ds;$$

and, using the parametric representation $s = r_0 \exp(i\phi)$ $(0 \le \phi \le 2\pi)$ for C_0, we can write

$$f(z) = \frac{1}{2\pi} \int_0^{2\pi} \left(\frac{s}{s - z} - \frac{s}{s - z_1} \right) f(s)\, d\phi,$$

where, for convenience, we retain the s instead of writing $r_0 \exp(i\phi)$.

Note that, in view of the last of expressions (2) for z_1, the factor inside the parentheses here can be written

(3)
$$\frac{s}{s - z} - \frac{1}{1 - (\bar{s}/\bar{z})} = \frac{s}{s - z} + \frac{\bar{z}}{\bar{s} - \bar{z}} = \frac{r_0^2 - r^2}{|s - z|^2}.$$

An alternative form of the Cauchy integral formula (1) is, therefore,

(4)
$$f(re^{i\theta}) = \frac{r_0^2 - r^2}{2\pi} \int_0^{2\pi} \frac{f(r_0 e^{i\phi})}{|s - z|^2}\, d\phi$$

when $0 < r < r_0$. This form is also valid when $r = 0$; in that case, it reduces directly to

$$f(0) = \frac{1}{2\pi} \int_0^{2\pi} f(r_0 e^{i\phi})\, d\phi,$$

which is just the parametric form of equation (1) with $z = 0$.

The quantity $|s - z|$ is the distance between the points s and z, and the law of cosines can be used to write (see Fig. 143)

(5)
$$|s - z|^2 = r_0^2 - 2r_0 r \cos(\phi - \theta) + r^2.$$

Hence, if u is the real part of the analytic function f, it follows from formula (4) that

(6)
$$u(r, \theta) = \frac{1}{2\pi} \int_0^{2\pi} \frac{(r_0^2 - r^2) u(r_0, \phi)}{r_0^2 - 2r_0 r \cos(\phi - \theta) + r^2}\, d\phi \qquad (r < r_0).$$

This is the *Poisson integral formula* for the harmonic function u in the open disk bounded by the circle $r = r_0$.

Formula (6) defines a linear integral transformation of $u(r_0, \phi)$ into $u(r, \theta)$. The kernel of the transformation is, except for the factor $1/(2\pi)$, the real-valued function

$$(7) \qquad P(r_0, r, \phi - \theta) = \frac{r_0^2 - r^2}{r_0^2 - 2r_0 r \cos(\phi - \theta) + r^2},$$

which is known as the *Poisson kernel*. In view of equation (5), we can also write

$$(8) \qquad P(r_0, r, \phi - \theta) = \frac{r_0^2 - r^2}{|s - z|^2};$$

and, since $r < r_0$, it is clear that P is a positive function. Moreover, since $\bar{z}/(\bar{s} - \bar{z})$ and its complex conjugate $z/(s - z)$ have the same real parts, we find from the second of equations (3) that

$$(9) \qquad P(r_0, r, \phi - \theta) = \mathrm{Re}\left(\frac{s}{s - z} + \frac{z}{s - z}\right) = \mathrm{Re}\left(\frac{s + z}{s - z}\right).$$

Thus $P(r_0, r, \phi - \theta)$ is a harmonic function of r and θ interior to C_0 for each fixed s on C_0. From equation (7), we see that $P(r_0, r, \phi - \theta)$ is an even periodic function of $\phi - \theta$, with period 2π; and its value is 1 when $r = 0$.

The Poisson integral formula (6) can now be written

$$(10) \qquad u(r, \theta) = \frac{1}{2\pi} \int_0^{2\pi} P(r_0, r, \phi - \theta) u(r_0, \phi)\, d\phi \qquad (r < r_0).$$

When $f(z) = u(r, \theta) = 1$, equation (10) shows that P has the property

$$(11) \qquad \frac{1}{2\pi} \int_0^{2\pi} P(r_0, r, \phi - \theta)\, d\phi = 1 \qquad (r < r_0).$$

We have assumed that f is analytic not only interior to C_0 but also on C_0 itself and that u is, therefore, harmonic in a domain which includes all points on that circle. In particular, u is continuous on C_0. The conditions will now be relaxed.

101. DIRICHLET PROBLEM FOR A DISK

Let F be a piecewise continuous function of θ on the interval $0 \leq \theta \leq 2\pi$. The *Poisson integral transform* of F is defined in terms of the Poisson kernel $P(r_0, r, \phi - \theta)$, introduced in Sec. 100, by means of the equation

$$(1) \qquad U(r, \theta) = \frac{1}{2\pi} \int_0^{2\pi} P(r_0, r, \phi - \theta) F(\phi)\, d\phi \qquad (r < r_0).$$

In this section, we shall prove that *the function $U(r, \theta)$ is harmonic inside the circle* $r = r_0$ *and*

$$(2) \qquad \lim_{\substack{r \to r_0 \\ r < r_0}} U(r, \theta) = F(\theta)$$

for each fixed θ at which F is continuous. Thus U is a solution of the Dirichlet problem for the disk $r < r_0$ in the sense that $U(r,\theta)$ approaches the boundary value $F(\theta)$ as the point (r,θ) approaches (r_0,θ) along a radius, except at the finite number of points (r_0,θ) where discontinuities of F may occur.

EXAMPLE. Before proving the above statement, let us apply it to find the potential $V(r,\theta)$ inside a long hollow circular cylinder of unit radius, split lengthwise into two equal parts, when $V = 1$ on one of the parts and $V = 0$ on the other. This problem was solved by conformal mapping in Sec. 89; and we recall how it was interpreted as a Dirichlet problem for the disk $r < 1$, where $V = 0$ on the upper half of the boundary $r = 1$ and $V = 1$ on the lower half.

In equation (1), write V for U, $r_0 = 1$, and $F(\phi) = 0$ when $0 < \phi < \pi$ and $F(\phi) = 1$ when $\pi < \phi < 2\pi$ to obtain

$$(3) \qquad V(r,\theta) = \frac{1}{2\pi} \int_{\pi}^{2\pi} P(1,r,\phi - \theta)\, d\phi,$$

where

$$P(1,r,\phi - \theta) = \frac{1 - r^2}{1 + r^2 - 2r\cos(\phi - \theta)}.$$

An antiderivative of $P(1,r,\psi)$ is

$$(4) \qquad \int P(1,r,\psi)\, d\psi = 2\arctan\left(\frac{1+r}{1-r}\tan\frac{\psi}{2}\right),$$

the integrand here being the derivative with respect to ψ of the function on the right. So it follows from expression (3) that

$$\pi V(r,\theta) = \arctan\left(\frac{1+r}{1-r}\tan\frac{2\pi - \theta}{2}\right) - \arctan\left(\frac{1+r}{1-r}\tan\frac{\pi - \theta}{2}\right).$$

After simplifying the expression for $\tan[\pi V(r,\theta)]$ obtained from this last equation (see Exercise 3, Sec. 102), we find that

$$(5) \qquad V(r,\theta) = \frac{1}{\pi}\arctan\left(\frac{1-r^2}{2r\sin\theta}\right) \qquad (0 \le \arctan t \le \pi),$$

where the stated restriction on the values of the arctangent function is physically evident. When expressed in rectangular coordinates, the solution here is the same as solution (5) in Sec. 89.

We turn now to the proof that the function U defined in equation (1) satisfies the Dirichlet problem for the disk $r < r_0$, as asserted just prior to this example. First of all, U is harmonic inside the circle $r = r_0$ because P is a harmonic function of r and θ there. More precisely, since F is piecewise continuous, integral (1) can be written as the sum of a finite number of definite integrals each of which has an integrand that is continuous in r, θ, and ϕ. The partial derivatives of those integrands with respect to r and θ are also continuous. Since the order of integration and differentiation with respect to r and θ can, then, be interchanged and since P satisfies Laplace's equation

$$r^2 P_{rr} + r P_r + P_{\theta\theta} = 0$$

in the polar coordinates r and θ (Exercise 12, Sec. 22), it follows that U satisfies that equation too.

   In order to verify limit (2), we need to show that if F is continuous at θ, there corresponds to each positive number ε a positive number δ such that

(6) $|U(r,\theta) - F(\theta)| < \varepsilon$ whenever $0 < r_0 - r < \delta$.

We start by referring to property (11), Sec. 100, of the Poisson kernel and writing

$$U(r,\theta) - F(\theta) = \frac{1}{2\pi} \int_0^{2\pi} P(r_0, r, \phi - \theta)[F(\phi) - F(\theta)]\, d\phi.$$

For convenience, we let F be extended periodically, with period 2π, so that the integrand here is periodic in ϕ with that same period. Also, we may assume that $0 < r < r_0$ because of the nature of the limit to be established.

   Next, we observe that, since F is continuous at θ, there is a small positive number α such that

(7) $|F(\phi) - F(\theta)| < \dfrac{\varepsilon}{2}$ whenever $|\phi - \theta| \le \alpha$.

Evidently,

(8) $U(r,\theta) - F(\theta) = I_1(r) + I_2(r),$

where

$$I_1(r) = \frac{1}{2\pi} \int_{\theta-\alpha}^{\theta+\alpha} P(r_0, r, \phi - \theta)[F(\phi) - F(\theta)]\, d\phi,$$

$$I_2(r) = \frac{1}{2\pi} \int_{\theta+\alpha}^{\theta-\alpha+2\pi} P(r_0, r, \phi - \theta)[F(\phi) - F(\theta)]\, d\phi.$$

   The fact that P is a positive function (Sec. 100), together with the first of inequalities (7) just above and property (11), Sec. 100, of that function, enables us to write

$$|I_1(r)| \le \frac{1}{2\pi} \int_{\theta-\alpha}^{\theta+\alpha} P(r_0, r, \phi - \theta)|F(\phi) - F(\theta)|\, d\phi$$

$$< \frac{\varepsilon}{4\pi} \int_0^{2\pi} P(r_0, r, \phi - \theta)\, d\phi = \frac{\varepsilon}{2}.$$

As for the integral $I_2(r)$, we can see from Fig. 143 in Sec. 100 that the denominator $|s - z|^2$ in expression (8) for $P(r_0, r, \phi - \theta)$ in that section has a (positive) minimum value m as the argument ϕ of s varies over the closed interval

$$\theta + \alpha \le \phi \le \theta - \alpha + 2\pi.$$

So, if M denotes an upper bound of the piecewise continuous function $|F(\phi) - F(\theta)|$ on the interval $0 \leq \phi \leq 2\pi$, it follows that

$$|I_2(r)| \leq \frac{(r_0^2 - r^2)M}{2\pi m} 2\pi < \frac{2Mr_0}{m}(r_0 - r) < \frac{2Mr_0}{m}\delta = \frac{\varepsilon}{2}$$

whenever $r_0 - r < \delta$, where

(9)
$$\delta = \frac{m\varepsilon}{4Mr_0}.$$

Finally, the results in the two preceding paragraphs tell us that

$$|U(r, \theta) - F(\theta)| \leq |I_1(r)| + |I_2(r)| < \frac{\varepsilon}{2} + \frac{\varepsilon}{2} = \varepsilon$$

whenever $r_0 - r < \delta$, where δ is the positive number defined by equation (9). That is, statement (6) holds when that choice of δ is made.

According to expression (1), the value of U at $r = 0$ is

$$\frac{1}{2\pi} \int_0^{2\pi} F(\phi)\, d\phi.$$

Thus *the value of a harmonic function at the center of the circle $r = r_0$ is the average of the boundary values on the circle.*

It is left to the exercises to prove that P and U can be represented by series involving the elementary harmonic functions $r^n \cos n\theta$ and $r^n \sin n\theta$ as follows:

(10)
$$P(r_0, r, \phi - \theta) = 1 + 2 \sum_{n=1}^{\infty} \left(\frac{r}{r_0}\right)^n \cos n(\phi - \theta) \qquad (r < r_0)$$

and

(11)
$$U(r, \theta) = \frac{1}{2}a_0 + \sum_{n=1}^{\infty} \left(\frac{r}{r_0}\right)^n (a_n \cos n\theta + b_n \sin n\theta) \qquad (r < r_0),$$

where

(12)
$$a_n = \frac{1}{\pi} \int_0^{2\pi} F(\phi) \cos n\phi\, d\phi, \qquad b_n = \frac{1}{\pi} \int_0^{2\pi} F(\phi) \sin n\phi\, d\phi.$$

102. RELATED BOUNDARY VALUE PROBLEMS

Details of proofs of results given below are left to the exercises. The function F representing boundary values on the circle $r = r_0$ is assumed to be piecewise continuous.

Suppose that $F(2\pi - \theta) = -F(\theta)$. The Poisson integral formula (1) of Sec. 101 then becomes

(1)
$$U(r, \theta) = \frac{1}{2\pi} \int_0^{\pi} [P(r_0, r, \phi - \theta) - P(r_0, r, \phi + \theta)]F(\phi)\, d\phi.$$

This function U has zero values on the horizontal radii $\theta = 0$ and $\theta = \pi$ of the circle, as one would expect when U is interpreted as a steady temperature. Formula (1) thus solves *the Dirichlet problem for the semicircular region $r < r_0, 0 < \theta < \pi$, where $U = 0$ on the diameter AB, shown in Fig. 144, and*

$$(2) \qquad \lim_{\substack{r \to r_0 \\ r < r_0}} U(r, \theta) = F(\theta) \qquad (0 < \theta < \pi)$$

for each fixed θ at which F is continuous.

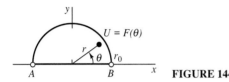

FIGURE 144

If $F(2\pi - \theta) = F(\theta)$, then

$$(3) \qquad U(r, \theta) = \frac{1}{2\pi} \int_0^\pi [P(r_0, r, \phi - \theta) + P(r_0, r, \phi + \theta)] F(\phi)\, d\phi;$$

and $U_\theta(r, \theta) = 0$ when $\theta = 0$ or $\theta = \pi$. Hence formula (3) furnishes a function U that is *harmonic in the semicircular region $r < r_0, 0 < \theta < \pi$ and satisfies condition (2) as well as the condition that its normal derivative be zero on the diameter AB, shown in Fig. 144.*

The analytic function $z = r_0^2/Z$ maps the circle $|Z| = r_0$ in the Z plane onto the circle $|z| = r_0$ in the z plane, and it maps the exterior of the first circle onto the interior of the second. Writing $z = r \exp(i\theta)$ and $Z = R \exp(i\psi)$, we note that $r = r_0^2/R$ and $\theta = 2\pi - \psi$. The harmonic function $U(r, \theta)$ represented by formula (1), Sec. 101, is, then, transformed into the function

$$U\left(\frac{r_0^2}{R}, 2\pi - \psi\right) = -\frac{1}{2\pi} \int_0^{2\pi} \frac{r_0^2 - R^2}{r_0^2 - 2r_0 R \cos(\phi + \psi) + R^2} F(\phi)\, d\phi,$$

which is harmonic in the domain $R > r_0$. Now, in general, if $u(r, \theta)$ is harmonic, then so is $u(r, -\theta)$ (see Exercise 11). Hence the function $H(R, \psi) = U(r_0^2/R, \psi - 2\pi)$, or

$$(4) \qquad H(R, \psi) = -\frac{1}{2\pi} \int_0^{2\pi} P(r_0, R, \phi - \psi) F(\phi)\, d\phi \qquad (R > r_0),$$

is also harmonic. For each fixed ψ at which $F(\psi)$ is continuous, we find from condition (2), Sec. 101, that

$$(5) \qquad \lim_{\substack{R \to r_0 \\ R > r_0}} H(R, \psi) = F(\psi).$$

Thus formula (4) solves *the Dirichlet problem for the region exterior to the circle $R = r_0$ in the Z plane* (Fig. 145). We note from expression (8), Sec. 100, that

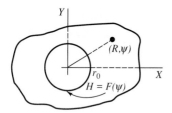

FIGURE 145

the Poisson kernel $P(r_0, R, \phi - \psi)$ is negative when $R > r_0$. Also,

(6)
$$\frac{1}{2\pi} \int_0^{2\pi} P(r_0, R, \phi - \psi) \, d\phi = -1 \qquad (R > r_0)$$

and

(7)
$$\lim_{R \to \infty} H(R, \psi) = \frac{1}{2\pi} \int_0^{2\pi} F(\phi) \, d\phi.$$

EXERCISES

1. Use the Poisson integral formula (1), Sec. 101, to derive the expression
$$V(x, y) = \frac{1}{\pi} \arctan \left[\frac{1 - x^2 - y^2}{(x - 1)^2 + (y - 1)^2 - 1} \right] \qquad (0 \le \arctan t \le \pi)$$

   for the electrostatic potential interior to a cylinder $x^2 + y^2 = 1$ when $V = 1$ on the first quadrant ($x > 0, y > 0$) of the cylindrical surface and $V = 0$ on the rest of that surface. Also, point out why $1 - V$ is the solution to Exercise 8, Sec. 89.

2. Let T denote the steady temperatures in a disk $r \le 1$, with insulated faces, when $T = 1$ on the arc $0 < \theta < 2\theta_0$ ($0 < \theta_0 < \pi/2$) of the edge $r = 1$ and $T = 0$ on the rest of the edge. Use the Poisson integral formula to show that
$$T(x, y) = \frac{1}{\pi} \arctan \left[\frac{(1 - x^2 - y^2)y_0}{(x - 1)^2 + (y - y_0)^2 - y_0^2} \right] \qquad (0 \le \arctan t \le \pi),$$

   where $y_0 = \tan \theta_0$. Verify that this function T satisfies the boundary conditions.

3. With the aid of the trigonometric identities
$$\tan(\alpha - \beta) = \frac{\tan \alpha - \tan \beta}{1 + \tan \alpha \tan \beta}, \qquad \tan \alpha + \cot \alpha = \frac{2}{\sin 2\alpha},$$

   show how solution (5) in the example in Sec. 101 is obtained from the expression for $\pi V(r, \theta)$ just prior to that solution.

4. Let I denote this *finite unit impulse function* (Fig. 146):
$$I(h, \theta - \theta_0) = \begin{cases} 1/h & \text{when } \theta_0 \le \theta \le \theta_0 + h, \\ 0 & \text{when } 0 \le \theta < \theta_0 \text{ or } \theta_0 + h < \theta \le 2\pi, \end{cases}$$

   where h is a positive number and $0 \le \theta_0 < \theta_0 + h < 2\pi$. Note that
$$\int_{\theta_0}^{\theta_0 + h} I(h, \theta - \theta_0) \, d\theta = 1.$$

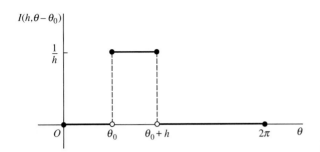

FIGURE 146

With the aid of a mean value theorem for definite integrals, show that

$$\int_0^{2\pi} P(r_0, r, \phi - \theta)I(h, \phi - \theta_0)\, d\phi = P(r_0, r, c - \theta)\int_{\theta_0}^{\theta_0 + h} I(h, \phi - \theta_0)\, d\phi,$$

where $\theta_0 \leq c \leq \theta_0 + h$, and hence that

$$\lim_{\substack{h \to 0 \\ h > 0}} \int_0^{2\pi} P(r_0, r, \phi - \theta)I(h, \phi - \theta_0)\, d\phi = P(r_0, r, \theta - \theta_0) \qquad (r < r_0).$$

Thus the Poisson kernel $P(r_0, r, \theta - \theta_0)$ is the limit, as h approaches 0 through positive values, of the harmonic function inside the circle $r = r_0$ whose boundary values are represented by the impulse function $2\pi I(h, \theta - \theta_0)$.

5. Show that the expression in Exercise 8(b), Sec. 47, for the sum of a certain cosine series can be written

$$1 + 2\sum_{n=1}^{\infty} a^n \cos n\theta = \frac{1 - a^2}{1 - 2a\cos\theta + a^2} \qquad (-1 < a < 1).$$

Then show that the Poisson kernel has the series representation (10), Sec. 101.

6. Show that the series in representation (10), Sec. 101, for the Poisson kernel converges uniformly with respect to ϕ. Then obtain from formula (1) of that section the series representation (11) for $U(r, \theta)$ there.*

7. Use expressions (11) and (12) in Sec. 101 to find the steady temperatures $T(r, \theta)$ in a solid cylinder $r \leq r_0$ of infinite length if $T(r_0, \theta) = A\cos\theta$. Show that no heat flows across the plane $y = 0$.
   Ans. $T = A(r/r_0)\cos\theta = Ax/r_0$.

8. Obtain the special case

(a) $H(R, \psi) = \dfrac{1}{2\pi}\displaystyle\int_0^{\pi} [P(r_0, R, \phi + \psi) - P(r_0, R, \phi - \psi)]F(\phi)\, d\phi;$

(b) $H(R, \psi) = -\dfrac{1}{2\pi}\displaystyle\int_0^{\pi} [P(r_0, R, \phi + \psi) + P(r_0, R, \phi - \psi)]F(\phi)\, d\phi$

---

*This result is obtained when $r_0 = 1$ by the method of separation of variables in the authors' "Fourier Series and Boundary Value Problems," 5th ed., Sec. 40, 1993.

of formula (4), Sec. 102, for the harmonic function H in the unbounded region $R > r_0$, $0 < \psi < \pi$, shown in Fig. 147, if that function satisfies the boundary condition

$$\lim_{\substack{R \to r_0 \\ R > r_0}} H(R, \psi) = F(\psi) \qquad (0 < \psi < \pi)$$

on the semicircle and (a) it is zero on the rays BA and DE; (b) its normal derivative is zero on the rays BA and DE.

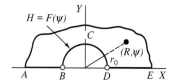

FIGURE 147

9. Give the details needed in establishing formula (1) in Sec. 102 as a solution of the Dirichlet problem stated there for the region shown in Fig. 144.
10. Give the details needed in establishing formula (3) in Sec. 102 as a solution of the boundary value problem stated there.
11. Obtain formula (4), Sec. 102, as a solution of the Dirichlet problem for the region exterior to a circle (Fig. 145). To show that $u(r, -\theta)$ is harmonic when $u(r, \theta)$ is harmonic, use the polar form

$$r^2 u_{rr}(r, \theta) + r u_r(r, \theta) + u_{\theta\theta}(r, \theta) = 0$$

of Laplace's equation.
12. State why equation (6), Sec. 102, is valid.
13. Establish the limit (7), Sec. 102.

103. SCHWARZ INTEGRAL FORMULA

Let f be an analytic function of z throughout the half plane $\operatorname{Im} z \geq 0$ such that, for some positive constants a and M, f satisfies the order property

(1) $$|z^a f(z)| < M \qquad (\operatorname{Im} z \geq 0).$$

For a fixed point z above the real axis, let C_R denote the upper half of a positively oriented circle of radius R centered at the origin, where $R > |z|$ (Fig. 148). Then, according to the Cauchy integral formula,

(2) $$f(z) = \frac{1}{2\pi i} \int_{C_R} \frac{f(s)\,ds}{s - z} + \frac{1}{2\pi i} \int_{-R}^{R} \frac{f(t)\,dt}{t - z}.$$

We find that the first of these integrals approaches 0 as R tends to ∞ since, in view of condition (1),

$$\left| \int_{C_R} \frac{f(s)\,ds}{s - z} \right| < \frac{M}{R^a(R - |z|)} \pi R = \frac{\pi M}{R^a(1 - |z|/R)}.$$

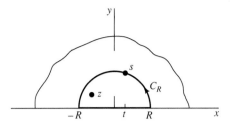

FIGURE 148

Thus

$$(3) \qquad f(z) = \frac{1}{2\pi i} \int_{-\infty}^{\infty} \frac{f(t)\, dt}{t - z} \qquad (\operatorname{Im} z > 0).$$

Condition (1) also ensures that the improper integral here converges.* The number to which it converges is the same as its Cauchy principal value (see Sec. 60), and representation (3) is a *Cauchy integral formula for the half plane* $\operatorname{Im} z > 0$.

When the point z lies below the real axis, the right-hand side of equation (2) is zero; hence integral (3) is zero for such a point. Thus, when z is above the real axis, we have the following formula, where c is an arbitrary complex constant:

$$(4) \qquad f(z) = \frac{1}{2\pi i} \int_{-\infty}^{\infty} \left(\frac{1}{t - z} + \frac{c}{t - \overline{z}} \right) f(t)\, dt \qquad (\operatorname{Im} z > 0).$$

In the two cases $c = -1$ and $c = 1$, this reduces, respectively, to

$$(5) \qquad f(z) = \frac{1}{\pi} \int_{-\infty}^{\infty} \frac{y f(t)}{|t - z|^2}\, dt \qquad (y > 0)$$

and

$$(6) \qquad f(z) = \frac{1}{\pi i} \int_{-\infty}^{\infty} \frac{(t - x) f(t)}{|t - z|^2}\, dt \qquad (y > 0).$$

If $f(z) = u(x, y) + i v(x, y)$, it follows from formulas (5) and (6) that the harmonic functions u and v are represented in the half plane $y > 0$ in terms of the boundary values of u by the formulas

$$(7) \qquad u(x, y) = \frac{1}{\pi} \int_{-\infty}^{\infty} \frac{y u(t, 0)}{|t - z|^2}\, dt = \frac{1}{\pi} \int_{-\infty}^{\infty} \frac{y u(t, 0)}{(t - x)^2 + y^2}\, dt \qquad (y > 0)$$

and

$$(8) \qquad v(x, y) = \frac{1}{\pi} \int_{-\infty}^{\infty} \frac{(x - t) u(t, 0)}{(t - x)^2 + y^2}\, dt \qquad (y > 0).$$

---

*See, for instance, A. E. Taylor and W. R. Mann, "Advanced Calculus," 3d ed., Chap. 22, 1983.

Formula (7) is known as the *Schwarz integral formula*, or the Poisson integral formula for the half plane. In the next section, we shall relax the conditions for the validity of formulas (7) and (8).

104. DIRICHLET PROBLEM FOR A HALF PLANE

Let F denote a real-valued function of x that is bounded for all x and continuous except for at most a finite number of finite jumps. When $y \geq \varepsilon$ and $|x| \leq 1/\varepsilon$, where ε is any positive constant, the integral

$$I(x, y) = \int_{-\infty}^{\infty} \frac{F(t)\, dt}{(t - x)^2 + y^2}$$

converges uniformly with respect to x and y, as do the integrals of the partial derivatives of the integrand with respect to x and y. Each of these integrals is the sum of a finite number of improper or definite integrals over intervals where F is continuous; hence the integrand of each component integral is a continuous function of t, x, and y when $y \geq \varepsilon$. Consequently, each partial derivative of $I(x, y)$ is represented by the integral of the corresponding derivative of the integrand whenever $y > 0$.

We write $U(x, y) = yI(x, y)/\pi$. Thus U is the *Schwarz integral transform* of F, suggested by the second of expressions (7), Sec. 103:

$$(1) \qquad U(x, y) = \frac{1}{\pi} \int_{-\infty}^{\infty} \frac{yF(t)}{(t - x)^2 + y^2} \, dt \qquad (y > 0).$$

Except for the factor $1/\pi$, the kernel here is $y/|t - z|^2$. It is the imaginary component of the function $1/(t - z)$, which is analytic in z when $y > 0$. It follows that the kernel is harmonic, and so it satisfies Laplace's equation in x and y. Because the order of differentiation and integration can be interchanged, the function (1) then satisfies that equation. Consequently, *U is harmonic when $y > 0$.*

To prove that

$$(2) \qquad \lim_{\substack{y \to 0 \\ y > 0}} U(x, y) = F(x)$$

for each fixed x at which F is continuous, we substitute $t = x + y \tan \tau$ in formula (1) and write

$$(3) \qquad U(x, y) = \frac{1}{\pi} \int_{-\pi/2}^{\pi/2} F(x + y \tan \tau)\, d\tau \qquad (y > 0).$$

Then, if

$$G(x, y, \tau) = F(x + y \tan \tau) - F(x)$$

and α is some small positive constant,

$$(4) \qquad \pi[U(x, y) - F(x)] = \int_{-\pi/2}^{\pi/2} G(x, y, \tau)\, d\tau = I_1(y) + I_2(y) + I_3(y),$$

where

$$I_1(y) = \int_{-\pi/2}^{(-\pi/2)+\alpha} G(x, y, \tau)\, d\tau, \qquad I_2(y) = \int_{(-\pi/2)+\alpha}^{(\pi/2)-\alpha} G(x, y, \tau)\, d\tau,$$

$$I_3(y) = \int_{(\pi/2)-\alpha}^{\pi/2} G(x, y, \tau)\, d\tau.$$

If M denotes an upper bound for $|F(x)|$, then $|G(x, y, \tau)| \leq 2M$. For a given positive number ε, we select α so that $6M\alpha < \varepsilon$; and this means that

$$|I_1(y)| \leq 2M\alpha < \frac{\varepsilon}{3} \qquad \text{and} \qquad |I_3(y)| \leq 2M\alpha < \frac{\varepsilon}{3}.$$

We next show that, corresponding to ε, there is a positive number δ such that

$$|I_2(y)| < \frac{\varepsilon}{3} \qquad \text{whenever} \qquad 0 < y < \delta.$$

To do this, we observe that, since F is continuous at x, there is a positive number γ such that

$$|G(x, y, \tau)| < \frac{\varepsilon}{3\pi} \qquad \text{whenever} \qquad 0 < y|\tan \tau| < \gamma.$$

Now the maximum value of $|\tan \tau|$ as τ ranges from $(-\pi/2) + \alpha$ to $(\pi/2) - \alpha$ is $\tan[(\pi/2) - \alpha] = \cot \alpha$. Hence, if we write $\delta = \gamma \tan \alpha$, it follows that

$$|I_2(y)| < \frac{\varepsilon}{3\pi}(\pi - 2\alpha) < \frac{\varepsilon}{3} \qquad \text{whenever} \qquad 0 < y < \delta.$$

We have thus shown that

$$|I_1(y)| + |I_2(y)| + |I_3(y)| < \varepsilon \qquad \text{whenever} \qquad 0 < y < \delta.$$

Condition (2) now follows from this result and equation (4).

Formula (1) therefore solves *the Dirichlet problem for the half plane $y > 0$,* with the boundary condition (2). It is evident from the form (3) of expression (1) that $|U(x, y)| \leq M$ in the half plane, where M is an upper bound of $|F(x)|$; that is, U is bounded. We note that $U(x, y) = F_0$ when $F(x) = F_0$, where F_0 is a constant.

According to formula (8) of Sec. 103, under certain conditions on F the function

(5) $$V(x, y) = \frac{1}{\pi} \int_{-\infty}^{\infty} \frac{(x - t)F(t)}{(t - x)^2 + y^2}\, dt \qquad (y > 0)$$

is a harmonic conjugate of the function U given by formula (1). Actually, *formula (5) furnishes a harmonic conjugate of U if F is everywhere continuous, except for at most a finite number of finite jumps, and if F satisfies an order property $|x^a F(x)| < M$, where $a > 0$.* For, under those conditions, we find that U and V satisfy the Cauchy-Riemann equations when $y > 0$.

Special cases of formula (1) when F is an odd or an even function are left to the exercises.

105. NEUMANN PROBLEM FOR A DISK

As in Sec. 100 and Fig. 143, we write $s = r_0 \exp(i\phi)$ and $z = r\exp(i\theta)$, where $r < r_0$. When s is fixed, the function

$$(1) \quad Q(r_0, r, \phi - \theta) = -2r_0 \ln|s - z| = -r_0 \ln[r_0^2 - 2r_0 r \cos(\phi - \theta) + r^2]$$

is harmonic interior to the circle $|z| = r_0$ because it is the real component of $-2r_0 \log(z - s)$, where the branch cut of $\log(z - s)$ is an outward ray from the point s. If, moreover, $r \neq 0$,

$$(2) \qquad Q_r(r_0, r, \phi - \theta) = -\frac{r_0}{r}\left[\frac{2r^2 - 2r_0 r \cos(\phi - \theta)}{r_0^2 - 2r_0 r \cos(\phi - \theta) + r^2}\right]$$

$$= \frac{r_0}{r}[P(r_0, r, \phi - \theta) - 1],$$

where P is the Poisson kernel (7) of Sec. 100.

These observations suggest that the function Q may be used to write an integral representation for a harmonic function U whose normal derivative U_r on the circle $r = r_0$ assumes prescribed values $G(\theta)$.

If G is piecewise continuous and U_0 is an arbitrary constant, the function

$$(3) \qquad U(r, \theta) = \frac{1}{2\pi}\int_0^{2\pi} Q(r_0, r, \phi - \theta)G(\phi)\, d\phi + U_0 \qquad (r < r_0)$$

is harmonic because the integrand is a harmonic function of r and θ. If the mean value of G over the circle $|z| = r_0$ is zero, or

$$(4) \qquad\qquad\qquad \int_0^{2\pi} G(\phi)\, d\phi = 0,$$

then, in view of equation (2),

$$U_r(r, \theta) = \frac{1}{2\pi}\int_0^{2\pi}\frac{r_0}{r}[P(r_0, r, \phi - \theta) - 1]G(\phi)\, d\phi$$

$$= \frac{r_0}{r}\cdot\frac{1}{2\pi}\int_0^{2\pi} P(r_0, r, \phi - \theta)G(\phi)\, d\phi.$$

Now, according to equations (1) and (2) of Sec. 101,

$$\lim_{\substack{r\to r_0 \\ r<r_0}} \frac{1}{2\pi}\int_0^{2\pi} P(r_0, r, \phi - \theta)G(\phi)\, d\phi = G(\theta).$$

Hence

$$(5) \qquad\qquad\qquad \lim_{\substack{r\to r_0 \\ r<r_0}} U_r(r, \theta) = G(\theta)$$

for each value of θ at which G is continuous.

When G is piecewise continuous and satisfies condition (4), the formula

$$(6) \quad U(r, \theta) = -\frac{r_0}{2\pi}\int_0^{2\pi} \ln[r_0^2 - 2r_0 r \cos(\phi - \theta) + r^2]G(\phi)\, d\phi + U_0 \quad (r < r_0),$$

therefore, solves *the Neumann problem for the region interior to the circle* $r = r_0$, where $G(\theta)$ is the normal derivative of the harmonic function $U(r, \theta)$ at the boundary in the sense of condition (5). Note how it follows from equations (4) and (6) that, since $\ln r_0^2$ is constant, U_0 is the value of U at the center $r = 0$ of the circle $r = r_0$.

The values $U(r, \theta)$ may represent steady temperatures in a disk $r < r_0$ with insulated faces. In that case, condition (5) states that the flux of heat into the disk through its edge is proportional to $G(\theta)$. Condition (4) is the natural physical require-ment that the total rate of flow of heat into the disk be zero, since temperatures do not vary with time.

A corresponding formula for a harmonic function H in the region *exterior* to the circle $r = r_0$ can be written in terms of Q as

$$(7) \qquad H(R, \psi) = -\frac{1}{2\pi} \int_0^{2\pi} Q(r_0, R, \phi - \psi) G(\phi) \, d\phi + H_0 \qquad (R > r_0),$$

where H_0 is a constant. As before, we assume that G is piecewise continuous and that condition (4) holds. Then

$$H_0 = \lim_{R \to \infty} H(R, \psi)$$

and

$$(8) \qquad \lim_{\substack{R \to r_0 \\ R > r_0}} H_R(R, \psi) = G(\psi)$$

for each ψ at which G is continuous.

The verification of formula (7), as well as special cases of formula (3) that apply to semicircular regions, is left to the exercises.

106. NEUMANN PROBLEM FOR A HALF PLANE

Let $G(x)$ be continuous for all real x, except for at most a finite number of finite jumps, and let it satisfy an order property

$$(1) \qquad |x^a G(x)| < M \qquad (-\infty < x < \infty),$$

where $a > 1$. For each fixed real number t, the function $\mathrm{Log}\,|z - t|$ is harmonic in the half plane $\mathrm{Im}\, z > 0$. Consequently, the function

$$(2) \qquad U(x, y) = \frac{1}{\pi} \int_{-\infty}^{\infty} \ln|z - t| G(t) \, dt + U_0$$

$$= \frac{1}{2\pi} \int_{-\infty}^{\infty} \ln[(t - x)^2 + y^2] G(t) \, dt + U_0 \qquad (y > 0),$$

where U_0 is a real constant, is harmonic in that half plane.

Formula (2) was written with the Schwarz integral transform (1), Sec. 104, in mind; for it follows from formula (2) that

$$(3) \qquad U_y(x, y) = \frac{1}{\pi} \int_{-\infty}^{\infty} \frac{y G(t)}{(t - x)^2 + y^2} \, dt \qquad (y > 0).$$

In view of equations (1) and (2) of Sec. 104, then,

(4) $$\lim_{\substack{y \to 0 \\ y > 0}} U_y(x, y) = G(x)$$

at each point x where G is continuous.

Integral formula (2) evidently solves *the Neumann problem for the half plane* $y > 0$, with boundary condition (4). But we have not presented conditions on G that are sufficient to ensure that the harmonic function U is bounded as $|z|$ increases.

When G is an odd function, formula (2) can be written

(5) $$U(x, y) = \frac{1}{2\pi} \int_0^\infty \ln \left[\frac{(t - x)^2 + y^2}{(t + x)^2 + y^2} \right] G(t)\, dt \qquad (x > 0, y > 0).$$

This represents a function that is harmonic in the *first quadrant* $x > 0, y > 0$ and satisfies the boundary conditions

(6) $$U(0, y) = 0 \qquad (y > 0),$$
(7) $$\lim_{\substack{y \to 0 \\ y > 0}} U_y(x, y) = G(x) \qquad (x > 0).$$

The kernels of the integral formulas for harmonic functions presented in this chapter can be described in terms of a single real-valued function of the complex variables $z = x + iy$ and $w = u + iv$:

(8) $$K(z, w) = \ln |z - w| \qquad (z \neq w).$$

This is *Green's function* for the *logarithmic potential* in the z plane. The function is symmetric; that is, $K(w, z) = K(z, w)$. Expressions for kernels used earlier, in terms of K and its derivatives, are given in the exercises.

EXERCISES

1. Obtain as a special case of formula (1), Sec. 104, the expression

$$U(x, y) = \frac{y}{\pi} \int_0^\infty \left[\frac{1}{(t - x)^2 + y^2} - \frac{1}{(t + x)^2 + y^2} \right] F(t)\, dt \qquad (x > 0, y > 0)$$

for a bounded function U that is harmonic in the *first quadrant* and satisfies the boundary conditions

$$U(0, y) = 0 \qquad (y > 0),$$
$$\lim_{\substack{y \to 0 \\ y > 0}} U(x, y) = F(x) \qquad (x > 0, x \neq x_j),$$

where F is bounded for all positive x and continuous except for at most a finite number of finite jumps at the points x_j $(j = 1, 2, \ldots, n)$.

2. Obtain as a special case of formula (1), Sec. 104, the expression

$$U(x, y) = \frac{y}{\pi} \int_0^\infty \left[\frac{1}{(t - x)^2 + y^2} + \frac{1}{(t + x)^2 + y^2} \right] F(t)\, dt \qquad (x > 0, y > 0)$$

for a bounded function U that is harmonic in the *first quadrant* and satisfies the boundary conditions

$$U_x(0, y) = 0 \qquad (y > 0),$$
$$\lim_{\substack{y \to 0 \\ y > 0}} U(x, y) = F(x) \qquad (x > 0, x \neq x_j),$$

where F is bounded for all positive x and continuous except possibly for finite jumps at a finite number of points $x = x_j$ ($j = 1, 2, \ldots, n$).

3. Interchange the x and y axes in Sec. 104 to write the solution

$$U(x, y) = \frac{1}{\pi} \int_{-\infty}^{\infty} \frac{xF(t)}{(t - y)^2 + x^2} \, dt \qquad (x > 0)$$

of the Dirichlet problem for the half plane $x > 0$. Then write

$$F(y) = \begin{cases} 1 & \text{when } -1 < y < 1, \\ 0 & \text{when } |y| > 1, \end{cases}$$

and obtain these expressions for U and its harmonic conjugate $-V$:

$$U(x, y) = \frac{1}{\pi}\left(\arctan \frac{y + 1}{x} - \arctan \frac{y - 1}{x}\right), \qquad V(x, y) = \frac{1}{2\pi} \ln \frac{x^2 + (y + 1)^2}{x^2 + (y - 1)^2},$$

where $-\pi/2 \leq \arctan t \leq \pi/2$. Also, show that

$$V(x, y) + iU(x, y) = \frac{1}{\pi}[\text{Log}(z + i) - \text{Log}(z - i)],$$

where $z = x + iy$.

4. Let $T(x, y)$ denote the bounded steady temperatures in a plate $x > 0, y > 0$, with insulated faces, when

$$\lim_{\substack{y \to 0 \\ y > 0}} T(x, y) = F_1(x) \qquad (x > 0),$$

$$\lim_{\substack{x \to 0 \\ x > 0}} T(x, y) = F_2(y) \qquad (y > 0)$$

(Fig. 149). Here F_1 and F_2 are bounded and continuous except for at most a finite number of finite jumps. Write $x + iy = z$ and show with the aid of the expression obtained in Exercise 1 that

$$T(x, y) = T_1(x, y) + T_2(x, y) \qquad (x > 0, y > 0),$$

where

$$T_1(x, y) = \frac{y}{\pi} \int_0^{\infty} \left(\frac{1}{|t - z|^2} - \frac{1}{|t + z|^2}\right) F_1(t) \, dt,$$

$$T_2(x, y) = \frac{y}{\pi} \int_0^{\infty} \left(\frac{1}{|it - z|^2} - \frac{1}{|it + z|^2}\right) F_2(t) \, dt.$$

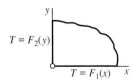

$T = F_2(y)$

$T = F_1(x)$ x **FIGURE 149**

5. Establish formula (7), Sec. 105, as a solution of the Neumann problem for the region exterior to a circle $r = r_0$, using earlier results found in that section.

6. Obtain as a special case of formula (3), Sec. 105, the expression

$$U(r,\theta) = \frac{1}{2\pi} \int_0^\pi [Q(r_0,r,\phi - \theta) - Q(r_0,r,\phi + \theta)]G(\phi)\,d\phi$$

for a function U that is harmonic in the *semicircular region* $r < r_0, 0 < \theta < \pi$ and satisfies the boundary conditions

$$U(r,0) = U(r,\pi) = 0 \qquad (r < r_0),$$
$$\lim_{\substack{r \to r_0 \\ r < r_0}} U_r(r,\theta) = G(\theta) \qquad (0 < \theta < \pi)$$

for each θ at which G is continuous.

7. Obtain as a special case of formula (3), Sec. 105, the expression

$$U(r,\theta) = \frac{1}{2\pi} \int_0^\pi [Q(r_0,r,\phi - \theta) + Q(r_0,r,\phi + \theta)]G(\phi)\,d\phi + U_0$$

for a function U that is harmonic in the *semicircular region* $r < r_0, 0 < \theta < \pi$ and satisfies the boundary conditions

$$U_\theta(r,0) = U_\theta(r,\pi) = 0 \qquad (r < r_0),$$
$$\lim_{\substack{r \to r_0 \\ r < r_0}} U_r(r,\theta) = G(\theta) \qquad (0 < \theta < \pi)$$

for each θ at which G is continuous, provided that

$$\int_0^\pi G(\phi)\,d\phi = 0.$$

8. Let $T(x,y)$ denote the steady temperatures in a plate $x \geq 0, y \geq 0$. The faces of the plate are insulated, and $T = 0$ on the edge $x = 0$. The flux of heat (Sec. 84) into the plate along the segment $0 < x < 1$ of the edge $y = 0$ is a constant A, and the rest of that edge is insulated. Use formula (5), Sec. 106, to show that the flux out of the plate along the edge $x = 0$ is

$$\frac{A}{\pi} \ln\left(1 + \frac{1}{y^2}\right).$$

9. Show that the Poisson kernel (Sec. 100) is given in terms of Green's function

$$K(z,w) = \ln|z - w| = \frac{1}{2}\ln[\rho^2 - 2\rho r \cos(\phi - \theta) + r^2],$$

where $z = r\exp(i\theta)$ and $w = \rho\exp(i\phi)$, by the equation

$$P(\rho,r,\phi - \theta) = 2\rho\frac{\partial K}{\partial \rho} - 1.$$

10. Show that the kernel used in the Schwarz integral transform (Sec. 104) can be written in terms of Green's function

$$K(z,w) = \ln|z - w| = \frac{1}{2}\ln[(x - u)^2 + (y - v)^2],$$

where $z = x + iy$ and $w = u + iv$, as

$$\frac{y}{|u - z|^2} = \frac{\partial K}{\partial y}\bigg]_{v=0} = -\frac{\partial K}{\partial v}\bigg]_{v=0}.$$

Here K is to be interpreted as a function of the four real variables x, y, u, and v.

APPENDIX

1

BIBLIOGRAPHY

The following list of supplementary books is far from exhaustive. Further references can be found in many of the books listed here.

Theory

Ahlfors, L. V.: "Complex Analysis," 3d ed., McGraw-Hill, Inc., New York, 1979.

Bieberbach, L.: "Conformal Mapping," Chelsea Publishing Co., New York, 1986.

Boas, R. P.: "Invitation to Complex Analysis," McGraw-Hill, Inc., New York, 1987.

———: "Yet Another Proof of the Fundamental Theorem of Algebra," *Amer. Math. Monthly,* Vol. 71, No. 2, p. 180, 1964.

Carathéodory, C.: "Conformal Representation," Cambridge University Press, London, 1952.

———: "Theory of Functions of a Complex Variable," Vols. 1 and 2, Chelsea Publishing Co., New York, 1954.

Conway, J. B.: "Functions of One Complex Variable," 2d ed., 6th Printing, Springer-Verlag, New York, 1993.

Copson, E. T.: "Theory of Functions of a Complex Variable," Oxford University Press, London, 1962.

Evans, G. C.: "The Logarithmic Potential," 3 vols. in one, 2d ed., Chelsea Publishing Co., New York, 1927 (reprinted).

Fisher, S. D.: "Complex Variables," Wadsworth, Inc., Belmont, CA, 1986.

Flanigan, F. J.: "Complex Variables: Harmonic and Analytic Functions," Dover Publications, Inc., New York, 1983.

Hille, E.: "Analytic Function Theory," Vols. 1 and 2, 2d ed., Chelsea Publishing Co., New York, 1973.

Kaplan, W.: "Advanced Calculus," 4th ed., Addison-Wesley Publishing Company, Inc., Reading, MA, 1991.

———: "Advanced Mathematics for Engineers," Addison-Wesley Publishing Company, Inc., Reading, MA, 1981.

Kellogg, O. D.: "Foundations of Potential Theory," Dover Publications, Inc., New York, 1953.

Knopp, K.: "Elements of the Theory of Functions," Dover Publications, Inc., New York, 1952.

Krantz, S. G.: "Complex Analysis: The Geometric Viewpoint," Carus Mathematical Monograph Series, The Mathematical Association of America, Washington, DC, 1990.

Krzyż, J. G.: "Problems in Complex Variable Theory," American Elsevier Publishing Company, Inc., New York, 1971.

Lang, S.: "Complex Analysis," 3d ed., Springer-Verlag, New York, 1993.

Levinson, N., and R. M. Redheffer: "Complex Variables," McGraw-Hill, Inc., New York, 1988.

Markushevich, A. I.: "Theory of Functions of a Complex Variable," 3 vols. in one, 2d ed., Chelsea Publishing Co., New York, 1977.

Marsden, J. E., and M. J. Hoffman: "Basic Complex Analysis," 2d ed., W. H. Freeman and Company, New York, 1987.

Mathews, J. H.: "Complex Variables for Mathematics and Engineering," 2d ed., Wm. C. Brown Publishers, Dubuque, IA, 1988.

Mitrinović, D. S.: "Calculus of Residues," P. Noordhoff, Ltd., Groningen, 1966.

Nehari, Z.: "Conformal Mapping," Dover Publications, Inc., New York, 1975.

Newman, M. H. A.: "Elements of the Topology of Plane Sets of Points," Dover Publications, Inc., New York, 1992.

Palka, B. P.: "An Introduction to Complex Function Theory," Springer-Verlag, New York, 1990.

Pennisi, L. L.: "Elements of Complex Variables," Holt, Rinehart and Winston, Inc., New York, 1963.

Rubenfeld, L. A.: "A First Course in Applied Complex Variables," John Wiley & Sons, Inc., New York, 1985.

Saff, E. B., and A. D. Snider: "Fundamentals of Complex Analysis for Mathematics, Science, and Engineering," 2d ed., Prentice-Hall, Inc., Englewood Cliffs, NJ, 1993.

Silverman, R. A.: "Complex Analysis with Applications," Dover Publications, Inc., New York, 1984.

Springer, G.: "Introduction to Riemann Surfaces," 2d ed., Chelsea Publishing Co., New York, 1981.

Taylor, A. E., and W. R. Mann: "Advanced Calculus," 3d ed., John Wiley & Sons, Inc., New York, 1983.

Thron, W. J.: "Introduction to the Theory of Functions of a Complex Variable," John Wiley & Sons, Inc., New York, 1953.

Titchmarsh, E. C.: "Theory of Functions," 2d ed., Oxford University Press, London, 1939.

Volkovyskii, L. I., G. L. Lunts, and I. G. Aramanovich: "A Collection of Problems on Complex Analysis," Dover Publications, Inc., New York, 1991.

Whittaker, E. T., and G. N. Watson: "A Course of Modern Analysis," 4th ed., Cambridge University Press, London, 1963.

Applications

Bowman, F.: "Introduction to Elliptic Functions, with Applications," English Universities Press, London, 1953.

Brown, G. H., C. N. Hoyler, and R. A. Bierwirth: "Theory and Application of Radio-Frequency Heating," D. Van Nostrand Company, Inc., New York, 1947.

Brown, J. W., and R. V. Churchill: "Fourier Series and Boundary Value Problems," 5th ed., McGraw-Hill, Inc., New York, 1993.

Churchill, R. V.: "Operational Mathematics," 3d ed., McGraw-Hill, Inc., New York, 1972.

Dettman, J. W.: "Applied Complex Variables," Dover Publications, Inc., New York, 1984.

Hayt, W. H., Jr.: "Engineering Electromagnetics," 5th ed., McGraw-Hill, Inc., New York, 1989.

Henrici, P.: "Applied and Computational Complex Analysis," Vols. 1, 2, and 3, John Wiley & Sons, Inc., 1988, 1991, and 1993.

Kober, H.: "Dictionary of Conformal Representations," Dover Publications, Inc., New York, 1952.

Lamb, H.: "Hydrodynamics," 6th ed., Dover Publications, Inc., New York, 1945.

Lebedev, N. N.: "Special Functions and Their Applications," rev. ed., Dover Publications, Inc., New York, 1972.

Love, A. E.: "Treatise on the Mathematical Theory of Elasticity," 4th ed., Dover Publications, Inc., New York, 1944.

Milne-Thomson, L. M.: "Theoretical Hydrodynamics," Macmillan & Co., Ltd., London, 1955.

Oberhettinger, F., and W. Magnus: "Anwendung der elliptischen Funktionen in Physik und Technik," Springer-Verlag OHG, Berlin, 1949.

Oppenheim, A. V., and R. W. Schafer: "Discrete-Time Signal Processing," Prentice-Hall, Inc., Englewood Cliffs, NJ, 1989.

Sokolnikoff, I. S.: "Mathematical Theory of Elasticity," 2d ed., Krieger Publishing Co., Inc., Melbourne, FL, 1983.

Streeter, V. L., and E. B. Wylie: "Fluid Mechanics," 8th ed., McGraw-Hill, Inc., New York, 1985.

Timoshenko, S. P., and J. N. Goodier: "Theory of Elasticity," 3d ed., McGraw-Hill, Inc., New York, 1970.

Wen, G.-C.: "Conformal Mappings and Boundary Value Problems," Translations of Mathematical Monographs, Vol. 106, American Mathematical Society, Providence, RI, 1992.

TABLE OF TRANSFORMATIONS
OF REGIONS
(See Chap. 8)

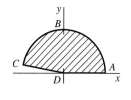
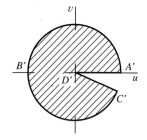

FIGURE 1
$w = z^2.$

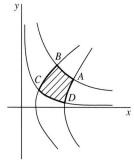
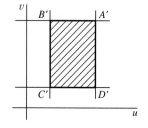

FIGURE 2
$w = z^2.$

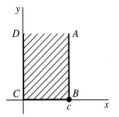

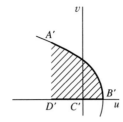

FIGURE 3
$w = z^2$;
$A'B'$ on parabola $v^2 = -4c^2(u - c^2)$.

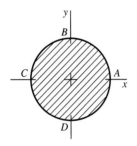

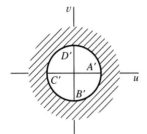

FIGURE 4
$w = 1/z$.

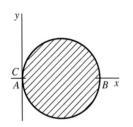

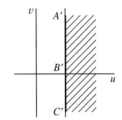

FIGURE 5
$w = 1/z$.

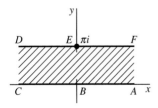

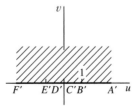

FIGURE 6
$w = \exp z$.

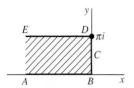
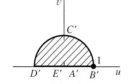

FIGURE 7
$w = \exp z$.

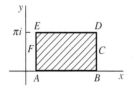
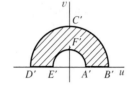

FIGURE 8
$w = \exp z$.

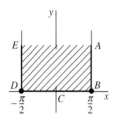
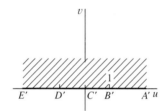

FIGURE 9
$w = \sin z$.

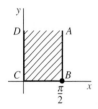
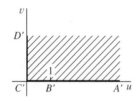

FIGURE 10
$w = \sin z$.

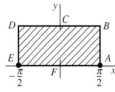
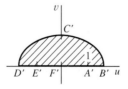

FIGURE 11
$w = \sin z$; BCD on line $y = b \, (b > 0)$,
$B'C'D'$ on ellipse $\dfrac{u^2}{\cosh^2 b} + \dfrac{v^2}{\sinh^2 b} = 1$.

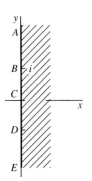
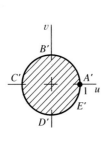

FIGURE 12

$$w = \frac{z - 1}{z + 1}.$$

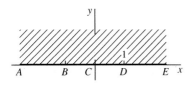
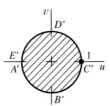

FIGURE 13

$$w = \frac{i - z}{i + z}.$$

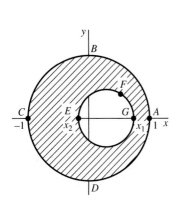
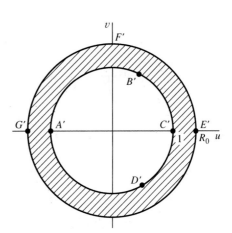

FIGURE 14

$$w = \frac{z - a}{az - 1}; \quad a = \frac{1 + x_1 x_2 + \sqrt{(1 - x_1^2)(1 - x_2^2)}}{x_1 + x_2},$$

$$R_0 = \frac{1 - x_1 x_2 + \sqrt{(1 - x_1^2)(1 - x_2^2)}}{x_1 - x_2} \quad (a > 1 \text{ and } R_0 > 1 \text{ when } -1 < x_2 < x_1 < 1).$$

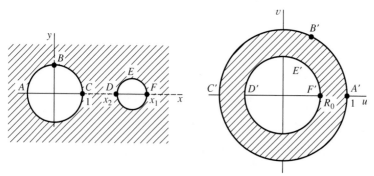

FIGURE 15

$$w = \frac{z - a}{az - 1}; \quad a = \frac{1 + x_1 x_2 + \sqrt{(x_1^2 - 1)(x_2^2 - 1)}}{x_1 + x_2},$$

$$R_0 = \frac{x_1 x_2 - 1 - \sqrt{(x_1^2 - 1)(x_2^2 - 1)}}{x_1 - x_2} \quad (x_2 < a < x_1 \text{ and } 0 < R_0 < 1 \text{ when } 1 < x_2 < x_1).$$

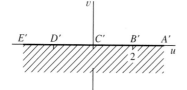

FIGURE 16

$$w = z + \frac{1}{z}.$$

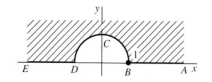

FIGURE 17

$$w = z + \frac{1}{z}.$$

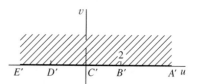

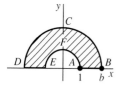

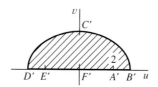

FIGURE 18

$$w = z + \frac{1}{z}; \quad B'C'D' \text{ on ellipse } \frac{u^2}{(b + 1/b)^2} + \frac{v^2}{(b - 1/b)^2} = 1.$$

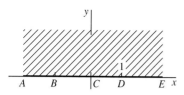 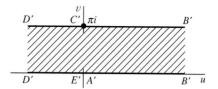

FIGURE 19

$$w = \operatorname{Log} \frac{z-1}{z+1}; \ z = -\coth \frac{w}{2}.$$

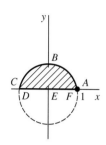 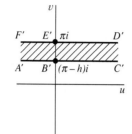

FIGURE 20

$$w = \operatorname{Log} \frac{z-1}{z+1};$$

ABC on circle $x^2 + (y + \cot h)^2 = \csc^2 h \ (0 < h < \pi)$.

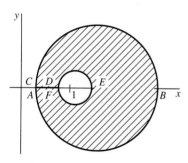 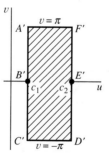

FIGURE 21

$$w = \operatorname{Log} \frac{z+1}{z-1}; \text{ centers of circles at } z = \coth c_n,$$

radii: $\operatorname{csch} c_n \quad (n = 1, 2)$.

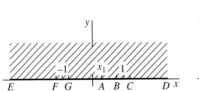
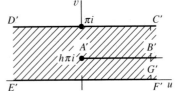

FIGURE 22

$$w = h \ln \frac{h}{1-h} + \ln 2(1-h) + i\pi - h \operatorname{Log}(z+1) - (1-h)\operatorname{Log}(z-1);$$
$$x_1 = 2h - 1.$$

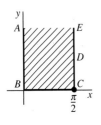
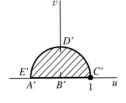

FIGURE 23

$$w = \left(\tan \frac{z}{2}\right)^2 = \frac{1-\cos z}{1+\cos z}.$$

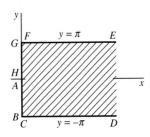
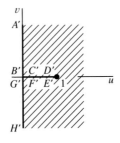

FIGURE 24

$$w = \coth \frac{z}{2} = \frac{e^z + 1}{e^z - 1}.$$

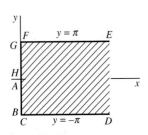
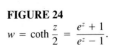
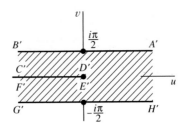

FIGURE 25

$$w = \operatorname{Log}\left(\coth \frac{z}{2}\right).$$

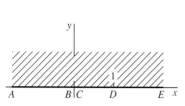

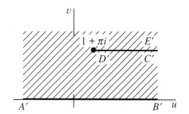

FIGURE 26

$w = \pi i + z - \text{Log}\, z.$

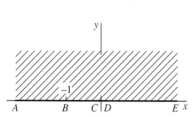

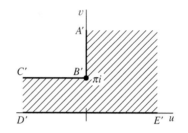

FIGURE 27

$w = 2(z + 1)^{1/2} + \text{Log}\, \dfrac{(z + 1)^{1/2} - 1}{(z + 1)^{1/2} + 1}.$

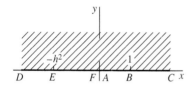

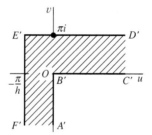

FIGURE 28

$w = \dfrac{i}{h} \text{Log}\, \dfrac{1 + iht}{1 - iht} + \text{Log}\, \dfrac{1 + t}{1 - t}; t = \left(\dfrac{z - 1}{z + h^2}\right)^{1/2}.$

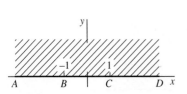

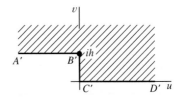

FIGURE 29

$$w = \frac{h}{\pi}[(z^2 - 1)^{1/2} + \cosh^{-1} z].^*$$

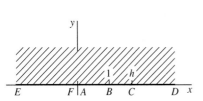

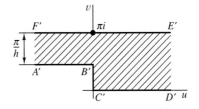

FIGURE 30

$$w = \cosh^{-1}\left(\frac{2z - h - 1}{h - 1}\right) - \frac{1}{h}\cosh^{-1}\left[\frac{(h + 1)z - 2h}{(h - 1)z}\right].$$

---

* See Exercise 3, Sec. 99.

INDEX